Developmental Cognitive Neuroscience

Developmental Cognitive Neuroscience

An Introduction

Fifth Edition

Michelle de Haan, Iroise Dumontheil, and Mark H. Johnson

WILEY Blackwell

Library of Congress Cataloging-in-Publication Data
Names: de Haan, Michelle, 1969– author. | Dumontheil, Iroise,
    author. | Johnson, Mark H. (Mark Henry), 1960– author.
Title: Developmental cognitive neuroscience : an introduction / Michelle de Haan,
    Iroise Dumontheil, and Mark H. Johnson.
Description: Fifth edition. | Hoboken, NJ, USA : John Wiley & Sons, Inc.,
    2023. | Revised edition of: Developmental cognitive neuroscience / Mark
    H. Johnson and Michelle de Haan. Fourth edition. 2015. | Includes
    bibliographical references and index.
Identifiers: LCCN 2022056812 (print) | LCCN 2022056813 (ebook) | ISBN
    9781119904694 (Paperback) | ISBN 9781119904700 (ePDF) | ISBN
    9781119904717 (epub)
Subjects: LCSH: Cognitive neuroscience. | Developmental psychology.
Classification: LCC QP360.5 .J64 2023 (print) | LCC QP360.5 (ebook) | DDC
    612.8/233–dc23/eng/20221222
LC record available at https://lccn.loc.gov/2022056812
LC ebook record available at https://lccn.loc.gov/2022056813

Cover Design: Wiley
Cover Image: Courtesy of Danyal Akarca, Alexa Mousley, Alicja Monaghan, Sofia Carozza and Duncan Astle

Set in 9.5/12.5pt STIXTwoText by Straive, Pondicherry, India
Printed and bound by CPI Group (UK) Ltd, Croydon, CR0 4YY

C9781119904694_170323

To our parents, who provided both our nature and our nuture.

Contents

List of Figures

Figures listed below without a page number appear in the color plate section.
The color plate section appears between pages 136 and 137.

List of Tables

Abbreviations

ASL	American Sign Language
ADHD	attention deficit/hyperactivity disorder
BOLD	blood oxygen level dependent
CA1	cornu ammonis 1 area of the hippocampus
CA3	cornu ammonis 3 area of the hippocampus
CANTAB	Cambridge Neuropsychological Testing Automated Battery
COMT	catechol-O-methyltransferase gene
DAT1	dopamine active transporter 1 gene
DCD	developmental coordination disorder
DLPFC	dorsolateral prefrontal cortex
DNA	deoxyribonucleic acid
DTI	diffusion tensor imaging
EEG	electroencephalography
ERO	event-related oscillations
ERP	event-related potential
FEF	frontal eye fields
FFA	fusiform face area
FG	fusiform gyrus
fMRI	functional magnetic resonance imaging
FMR1	fragile X mental retardation 1 gene
fNIRS	functional near-infrared spectroscopy
FOXP2	forkhead box protein P2 gene
GABA	gamma-aminobutyric acid
GWAS	genome-wide association studies
HD	high density
HD-ERP	high-density event-related potential
HM	initials of a patient with amnesia
IPS	intraparietal suclus
IS	interactive specialization
ISI	inter-stimulus interval
IQ	intelligence quotient
KBCC	knowledge-based cascade correlation

KE	Family with an inherited disordered involved the FOXP2 gene that affects motor speech
LGN	lateral geniculate nucleus
LTC	lateral temporal complex
MDRs	marginalization-related diminished returns
MEG	magnetoencephalography
MRI	magnetic resonance imaging
MGN	medial geniculate nucleus
MNS	mirror neuron system
MPFC	medial prefrontal cortex
ms	milliseconds
MT	middle temporal area
MTL	medial temporal lobes
NIRS	near-infrared spectroscopy
PET	positron emission tomography
PFC	prefrontal cortex
RNA	ribonucleic acid
SES	socioeconomic status
SIPN	social information processing network
SLI	specific language impairment
SNP	single-nucleotide polymorphism
SOA	stimulus onset asynchrony
SP	spike potential
STS	superior temporal sulcus
TPH2	tryptophan hydroxylase gene 2
TPJ	temporo-parietal junction
TV	television
V1	primary visual cortex
VWFA	visual word form area
WEIRD	western, rich, industrialized, rich and democratic
WM	working memory
WS	Williams syndrome

Preface to the First Edition

In the first chapter of this book I describe some of the factors responsible for the recent emergence of a subdiscipline at the interface between developmental psychology and cognitive neuroscience. I have chosen to refer to this new field as "developmental cognitive neuroscience," though it has been known under a number of other terms such as "developmental neurocognition" (de Boysson-Bardies et al., 1993). Though a series of edited volumes on the topic has recently appeared, like most newly emerging disciplines there is a time lag before the first books suitable for teaching appear. This book and the Reader which I edited in 1993 (Johnson, 1993) are initial attempts to fill the gap. While some may believe these efforts to be premature, my own view is that the lifeblood of any new discipline is in the students and postdocs recruited to the cause. And the sooner they are recruited, the better.

Is developmental cognitive neuroscience really significantly different from other fields that have a more extended history, such as developmental neuropsychology or cognitive development? Clearly, it would be unwise to rigidly demarcate developmental cognitive neuroscience from related, and mutually informative, fields. However, it is my belief that the emerging field has a number of characteristics that make it distinctive. First, while there is some disagreement about exact definitions, the fields of developmental neuropsychology and developmental psychopathology focus on atypical development, while commonly comparing them to normal developmental trajectories. In contrast, cognitive neuroscience (including the developmental variant outlined in this book) focuses on normal cognitive functioning, but uses information from deviant functioning and development as "nature's experiments" which can shed light on the neural basis of normal cognition. This book is therefore not intended as an introduction to the neuropsychology of developmental disorders. For such information the reader is referred to the excellent introductions by Cicchetti and Cohen (1995) and Spreen et al. (1995).

Second, unlike many in cognitive development, this book adopts the premise that information from brain development is more than just a useful additional source of evidence for supporting particular cognitive theories. Rather, information about brain development is viewed as both changing and originating theories at the cognitive level. Third, developmental cognitive neuroscience restricts itself to issues at the neural, cognitive, and immediate environmental levels. In my view, it is a hazard of some interdisciplinary fields that the focus of interest is diffused across many different levels of explanation. This is not to deny the importance of these other levels, but a mechanistic interdisciplinary science needs to restrict both the domains (in this case aspects of cognitive processing) and levels of explanation with which it is concerned. Finally, developmental cognitive neuroscience is specifically concerned with understanding the relation between neural and cognitive phenomena. For this reason, I have not discussed evidence from the related field of developmental behavior genetics. In general, developmental behavior genetics tends to be concerned with correlations between the molecular level (genetics) and gross behavioral measures such as IQ. With some notable exceptions, little effort is made to specifically relate these two levels of explanation via the intermediate neural and cognitive levels. Having pointed out the different focus of developmental cognitive neuroscience, my hope is that this book is written to be both accessible and informative to those in related and overlapping disciplines.

The above comments go some way to explaining the choice of material that I have presented in the book. However, I have no doubt that there is a substantive amount of excellent experimentation and theorizing that could have been included but was not. Since this is intended as a brief introduction to the field, I have chosen to focus on a few particular issues in some detail. Of course, the choice of material also reflects my own biases and knowledge since the book is intended as an introductory survey of the field as viewed from my own perspective. I apologize in advance for the inevitable omissions and errors.

The book is aimed at the advanced-level student and assumes some introductory knowledge of both neuroscience and cognitive development. Students without this background will probably need to refer to more introductory textbooks in the appropriate areas. I also hope that the book will attract developmentalists with an interest in learning more about the brain, and cognitive neuroscientists curious as to how developmental data can help constrain their theories about adult functioning. But most of all I hope that the book inspires readers to find out more about the field, and to consider a developmental cognitive neuroscience approach to their own topic.

Preface to Fifth Edition

It is now several decades since the first edition of this book was published, and the field continues to grow rapidly, inspiring us to prepare this fifth edition of *Developmental Cognitive Neuroscience: An Introduction*. The expansion of the field has been driven by a number of factors, including development of new technologies and analysis techniques and increasing linking of developmental cognitive neuroscience with other fields such as clinical sciences and social and educational policy making.

The continuing growth in the field is exciting, and we hope that this introduction to it will motivate further work in this area. In order to better cover these rapid developments, particularly in the areas of mid-childhood and adolescent development, social cognition, and neuroimaging, we have been delighted to recruit a third author—Iroise Dumontheil. The abundance of studies published means that we will not be exhaustively reviewing the entire area; this book does inevitably reflect to some extent our biases—but always with the aim of best illustrating developmental cognitive neuroscience approaches and theory.

One area that has grown considerably since the publication of the last edition is applying developmental cognitive neuroscience in global and cross-cultural settings. Thus, in this fifth edition we have included a new chapter addressing this area. Building from the fourth edition we have continued to include clinical and educational issues as well, reflecting the continued research and applied interest around these topics.

There will always be topics that we cannot completely cover within this volume—as in previous volumes, we give pointers to further reading which can guide the way on broader issues. We also continue to include topics for further thought and discussion at the end of each chapter. The website with teachers' resources is also still available in an updated form—here there are multiple choice, short answer, and essay questions available to facilitate formulation of assessments in courses on developmental cognitive neuroscience.

We would like to thank our colleagues and collaborators for educating and informing us on so many topics. Likewise, we owe thanks to our publishers for their continued support and commitment to this book throughout the years.

About the Companion Website

This book is accompanied by a companion website:

www.wiley.com/go/johnson/devneuro5e

The website includes:

- Multiple choice questions, short answer questions and an answer guide

1

The Biology of Change

In this introductory chapter we discuss a number of background issues for developmental cognitive neuroscience, beginning with historical approaches to the nature–nurture debate. Constructivism, in which biological forms are an emergent product of complex dynamic interactions between genes and environment, is presented as an approach to development that is superior to accounts that seek to identify pre-existing information exclusively in either genes or the external environment. However, if we are to abandon existing ways of analyzing development into "innate" and "acquired" components, this raises the question of how we should best understand developmental processes. One scheme is proposed for taking account of the various levels of interaction between genes and environment. Following this, a number of factors are discussed that demonstrate the importance of the cognitive neuroscience approach to development, including the increasing availability of brain imaging and molecular approaches around the globe. Conversely, the importance of taking a developmental approach to analyzing the relation between brain structure and cognition is reviewed. In examining the ways in which development and cognitive neuroscience can be combined, three different perspectives on human functional brain development are discussed: a maturational view, a skill learning view, and an "interactive specialization" framework. We expand on the latter framework, which will be used to structure evidence discussed in later chapters, and revisited in the closing chapter. Finally, the contents of the rest of the book are outlined.

Viewpoints on Development

As many people know, the changes we can observe during the growth of children from birth to adolescence are truly amazing. Perhaps the most remarkable aspects of this growth involve the brain and mind. Accompanying the fourfold increase in the volume of the brain during this time are numerous, and sometimes surprising, changes in behavior, thought, and emotion. An understanding of how the developments in brain and mind relate to each other could potentially revolutionize our thinking about education, social policy, and disorders of mental development. It is no surprise, therefore, that there has been increasing

Developmental Cognitive Neuroscience: An Introduction, Fifth Edition. Michelle de Haan,
Iroise Dumontheil, and Mark H. Johnson.
© 2023 John Wiley & Sons Ltd. Published 2023 by John Wiley & Sons Ltd.
Companion website: www.wiley.com/go/johnson/devneuro5e

interest in this new branch of science, including from grant funding agencies, medical charities, and international governmental summits. Since the publication of the first edition of this book in 1997, this field has become known as *developmental cognitive neuroscience.*

Developmental cognitive neuroscience has emerged at the interface between two of the most fundamental questions that challenge humankind. The first of these questions concerns the relation between mind and body, and specifically between the physical substance of the brain and the mental processes it supports. This issue is fundamental to the scientific discipline of *cognitive neuroscience.* The second question concerns the origin of organized biological structures, such as the highly complex structure of the adult human brain. This issue is fundamental to the study of *development.* In this book we will show that light can be shed on these two fundamental questions by tackling them both simultaneously, and specifically by focusing on the relation between the postnatal development of the human brain and the emerging cognitive processes it supports.

The second of the two questions above, that of the origins of organized biological structure, can be posed in terms of *phylogeny* or *ontogeny.* The phylogenetic (evolutionary) version of this question concerns the origin of species, and has been addressed by Charles Darwin and many others since. The ontogenetic version of this question concerns individual development within a life span. The ontogenetic question has been somewhat neglected relative to phylogeny, since some influential scientists have held the view that once a particular set of genes have been selected by evolution, ontogeny is simply a process of executing the "instructions" coded for by those genes. By this view, the ontogenetic question essentially reduces to phylogeny (e.g., so-called "evolutionary psychology"). In contrast to this view, in this book we argue that ontogenetic development is an active process through which biological structure is constructed afresh in each individual by means of complex and variable interactions between genes and their respective environments. The information is not in the genes, but emerges from the constructive interaction between genes and their environment. However, since both ontogeny and phylogeny concern the emergence of biological structures, some of the same mechanisms of change have been invoked in the two cases.

Further Reading   Oyama (2000).

The debate about the extent to which the ontogenetic question (individual development) is subsidiary to the phylogenetic question (evolution) is otherwise known as the nature–nurture issue, and has been central in developmental psychology, philosophy, and neuroscience. Broadly speaking, at one extreme the belief is that most of the information necessary to build a human brain, and the mind it supports, is latent within the genes of the individual. While most of this information is common to the species, each individual has some specific information that will make them differ from others. By this view, development is a process of unfolding or triggering the expression of information already contained within the genes.

At the opposing extreme, others believe that most of the information that shapes the human mind comes from the structure of the external world. Some facets of the environment, such as gravity, patterned light, and so on, will be common throughout the species, while other aspects of the environment will be specific to that individual. It will become clear in this book that both of these extreme views are ill conceived, since they assume that the information for the structure of an organism exists (either in the genes or in the external world) prior to its construction. In contrast to this, it appears that biological structure emerges anew within each individual's development from constrained dynamic interactions between genes and various levels of environment, and is not easily reducible to simple genetic and experiential components (Scarr, 1992).

It is more commonly accepted these days that the mental abilities of adults are the result of complex interactions between genes and environment. However, the nature of this interaction remains controversial and poorly understood, although, as we shall see, light may be shed on it by simultaneously considering brain and psychological development. Before going further, however, it is useful briefly to review some historical perspectives on the nature–nurture debate. This journey into history may help us avoid slipping back into ways of thinking that are deeply embedded in the Western intellectual tradition.

Throughout the 17th century there was an ongoing debate in biology between the so-called "vitalists," on the one hand, and the "preformationists," on the other. The vitalists believed that ontogenetic change was driven by "vital" life forces. Belief in this somewhat mystical and ill-defined force was widespread and actively encouraged by some members of the clergy. Following the invention of the microscope, however, some of those who viewed themselves as being of a more rigorous scientific mind championed the preformationist viewpoint. This view argued that a complete human being was contained in either the male sperm ("spermists") or the female egg ("ovists"). In order to support their claim, spermists produced drawings of a tiny, but perfect, human form enclosed within the head of sperm (see Figure 1.1). They argued that there was a simple and direct mapping between the seed of the organism and its end state: simultaneous growth of all the body parts. Indeed, preformationists of a religious conviction argued that God, on the sixth day of his work, placed about 200,000 million fully formed human miniatures into the ovaries of Eve or sperm of Adam (Gottlieb, 1992)!

Of course, we now know that such drawings were the result of overactive imagination, and that no such perfectly formed miniature human forms exist in the sperm or ovaries. However, as we shall see, the general idea behind preformationism, that there is a pre-existing blueprint or plan of the final state, has remained a pervasive one for many decades in biological and psychological development. In fact, Oyama (2000) suggests that the same notion of a "plan" or "blueprint" that exists prior to the development process has persisted to the present day, with genes replacing the little man inside the sperm. As it became

Figure 1.1 Drawings such as this influenced a 17th-century school of thought, the "spermists," who believed that there was a complete preformed person in each male sperm and that development merely consisted of increasing size.

clear that genes do not contain a simple "code" for body parts, in more recent years, "regulator" and "switching" genes have been invoked to orchestrate the expression of the other genes. Common to all of these versions of the nativist viewpoint is the belief that there is a fixed mapping between a pre-existing set of coded instructions and the final form. We will see in Chapter 3 that we are discovering that the relationship between the genotype and its resulting phenotype is much more dynamic and flexible than traditionally supposed.

On the other side of the nature–nurture dichotomy, those who believe in the structuring role of experience also view the information as existing prior to the end state, only the source of that information is different. This argument has been applied to psychological development, since it is obviously less plausible for physical growth. An example of this approach came from some of the more extreme members of the behaviorist school of psychology who believed that a child's psychological abilities could be entirely shaped by its early environment. Since that time, some developmental psychologists who work with computer models of the brain have suggested that the infant's mind is shaped largely by the statistical regularities latent in the external environment. Such efforts can reveal hitherto unrecognized contributions from the environment, and it will become evident in this book that these computer models can also be an excellent method for exploring types of interaction between intrinsic and extrinsic structure.

Further Reading   Mareschal (2010); Munakata et al. (2008).

The viewpoints discussed above share the common assumption that the information necessary for constructing the final state (in this case, the adult mind) is present prior to the developmental process itself. While vitalists' beliefs were sometimes more dynamic in character than preformationists', the forces that guided development were still assumed to originate with an external creator. Preformationism in historical or modern guises involves the execution of plans or codes (from genes) or the incorporation of information from the structure of the environment. Oyama (2000) argues that these views on ontogenetic development resemble pre-Darwinian theories of evolution in which a creator was deemed to have planned all the species in existence. In both the ontogenetic and phylogenetic theories of this kind a plan for the final form of the species or individual exists prior to its emergence.

Following on from this, there have been steps forward in thinking about ontogenetic development, called constructivism. Constructivism differs from preformationist views in that biological structures are viewed as an emergent property of complex interactions between genes and environment. Perhaps the most famous proponent of such a view with regard to cognitive development was the Swiss psychologist Jean Piaget. The essence of constructivism is that the relationship between the initial state and the final product can only be understood by considering the progressive construction of information. This construction is a dynamic and emergent process to which multiple factors contribute. There is no simple sense in which information either exclusively in the genes or in the environment can specify the end product. Rather, these two factors combine in a constructive manner such that each developmental step will be greater than the sum of the factors that contributed to it. The upshot of this viewpoint is not that we can never understand the mapping between genetic (or environmental) information and the final product, but rather that this mapping can only be understood once we have unraveled some of the key interactions that occur between genetic and environmental factors during ontogeny. Unfortunately, this means that there are unlikely to be quick breakthroughs in understanding the functions of regions of the human genome for psychological development.

> **Further Reading**   Piaget (2002); Mareschal (2010).

Until recently the constructivist view suffered from the same problem as vitalism, in that the mechanisms of change were poorly specified and the emergence of new structures from old resembled the conjuror's trick of making a rabbit appear from a hat. Even the "mechanisms" proposed by Piaget appeared somewhat elusive on closer inspection. Another problem with the constructivist approach was that, despite its emphasis on interaction, it was unclear how to analyze development in the absence of the traditional dichotomy between innate and environmental factors. By taking a cognitive neuroscience approach to psychological development, in conjunction with a number of new theoretical approaches, we will see that it is possible to flesh out the constructivist approach to development and to provide new ways to analyze cognitive and brain development.

Analyzing Development

Viewpoints on cognitive development that involve reducing behavior to information derived from genes, on the one hand, and/or information derived from the external environment, on the other, have commonly used the distinction between "innate" and "acquired" components. The term "innate" is actually rarely explicitly defined, and has a somewhat checkered history in developmental science. Indeed, the term has been dropped from use, or even actively banned, in some areas of developmental biology. The main reason for the term having been dropped from use in fields of biology such as ethology and genetics is because it is simply no longer useful, since it has become evident that genes interact with their environment at many different levels, including the molecular. One compelling example of this point, discussed by Gottlieb (1992), concerns the formation of the beak in the chick embryo.

The production of the (toothless) beak in the chick embryo results from the coaction of two types of tissue. However, if, in an experimental situation, one of these types of tissue (mesenchyme) is replaced with the same tissue from a mouse, then teeth will form instead of a beak! Thus, as Gottlieb (1992) points out, the genetic component that is necessary for the chick to produce teeth has been retained from the reptilian ancestry of birds. More generally, the phenotype that emerges from these genes in the chick can vary dramatically according to the molecular and cellular context in which they are located.

> **Further Reading**   Greenough et al. (2002); Gottlieb (2007).

Thus, there is no aspect of development that can be said to be strictly "genetic," that is, exclusively a product of information contained within a particular gene. If the term "innate" is taken to refer to structure that is specified exclusively by genetic information, it refers to

Table 1.1   Levels of Interaction between Genes and their Environment

Levels of interaction	Term
Molecular	Internal environment
Cellular	Internal environment (innate)
Organism–external environment	Species-typical environment (primal)
	Individual-specific environment (learning)

nothing that exists in the natural world, except for genes themselves. In cognitive science, however, use of the term "innate" has persisted despite repeated calls for it to be dropped from use (e.g., Gottlieb, 1992; Hinde, 1974; Johnston, 1988; Oyama, 2000). Presumably its persistent usage reflects the need for a term to describe the interaction between factors intrinsic to the developing child and features of the external environment. In considering this issue, Johnson and Morton (1991a) suggested that it is useful to distinguish between the various levels of interaction between genes and their environment. Some of these are shown in Table 1.1. Within this analysis, the term "innate" refers only to changes that arise as a result of interactions that occur within the organism, and therefore does not equate with "genetic." That is, it refers to the *level of the interaction between genes and environment, and not to the source of the information that generates change.* We will adopt this working definition of the term in this book. Interactions between the organism and aspects of the external environment that are common to all members of the species, the species-typical environment (such as patterned light, gravity, etc.), were referred to as "primal" by Johnson and Morton. Interactions between the organism and aspects of the environment unique to an individual, or subset of members of a species, were referred to as "learning."

Based on a series of experiments on the effects on brain structure of rearing rats in impoverished or comparatively enriched early environments, Greenough et al. (2002) proposed a similar distinction between two types of information storage induced by the environment. Changes induced by aspects of the environment that are common to all members of a species were classified as "experience-expectant" information storage (="species-typical"), and were associated with selective synaptic loss (see Chapter 4). The second type of information incorporated by the brain through interaction with the environment was referred to as "experience-dependent" (="individual-specific"). This referred to interactions with the environment that are, or can be, specific to an individual and are associated with the generation of new synaptic connections. Clearly, the boundary between these types of experience is often difficult to ascertain, and there have been many instances from ethological studies where behaviors thought to be innate turn out to be primal on closer study.

Why Take a Cognitive Neuroscience Approach to Development?

Until the turn of the century, the majority of theories of perceptual and cognitive development were generated without recourse to evidence from the brain. Indeed, some authors argued strongly for the independence of cognitive-level theorizing from considerations of the brain mechanisms (e.g., Morton et al., 1984). Evidence from the brain was thought to be either distracting, irrelevant, or hopelessly complicating for the construction of psychological theories. However, our understanding of brain function has improved enormously over the past 30 years or so. Accordingly, the time is now ripe for exploring the interface between cognitive development and brain development, and a spate of books on the topic have appeared since 2000 (e.g., de Haan & Johnson, 2003; Goswami, 2019; Johnson, Munakata, et al., 2002; Nelson & Luciana, 2008; Nelson et al., 2006; Stiles, 2008). Further, the integration of information from biology and cognitive development sets the stage for a more comprehensive psychology and biology of ontogeny than was previously thought possible: a *developmental cognitive neuroscience*. By the term "cognitive neuroscience" we include not only evidence about brain development, such as that from neuroanatomy, brain imaging, and the behavioral or cognitive effects of brain lesions, but also evidence from *ethology*. Ethology, a science pioneered by Tinbergen, Lorenz, and others in the 1940s and 1950s, concerns the study of a whole organism within its natural environment (see Hinde, 1974; Lorenz, 1965; Tinbergen, 1951). Ethology can be a powerful complement to neuroscience, and that the two fields combined can change the way we think about critical issues in perceptual and cognitive development.

Further Reading   Morton & Johnson (1991a).

In general, insights from biology have begun to play a more central role in informing thinking about perceptual and cognitive development, for a number of reasons. First, a range of powerful new methods and tools have become available to cognitive neuroscientists. These techniques permit questions to be asked more directly than before about the biological basis of cognitive and perceptual development. These methods are discussed in Chapter 2.

Importantly, theories which incorporate and reveal relationships between brain structures and cognitive functions will be useful in understanding the effects of early brain injury or genetic disorder on cognitive development. Some of the different clinical and at-risk groups that have been studied are discussed in Chapter 2. In addition, evidence derived from infants with congenital and acquired brain damage will be discussed throughout later chapters. Beyond its clinical utility, this approach can also contribute to the development of theories about functional specification, critical periods, and plasticity in the brain. Thus, there is a two-way interaction between clinical evidence and basic research in developmental cognitive neuroscience.

Why Take a Developmental Approach to Cognitive Neuroscience?

Ontogeny is the constructive process by which genes interact with their environment at various levels to yield complex organic structures such as the brain and the cognitive processes it supports. The study of development is necessarily multidisciplinary since new levels of structure that emerge as a result of this process (such as particular neural systems) often require different levels and methods of analysis from those that preceded them. The flip side of this is that development can be used as a tool for unraveling the interaction between seemingly disparate levels of organization, such as that between the molecular biology of gene expression (see Chapter 3) and the development of cognitive abilities such as object recognition. Further, the human adult brain, and the mind it supports, is composed of a complex series of hierarchical and parallel systems that has proven very difficult to analyze in an exclusively "top-down" manner. Brain damage induced by surgical lesions, or by accident or stroke, is unlikely to cleanly dissociate different levels of hierarchical organization. The developmental approach may allow different levels of hierarchical control to be observed independently. Specifically, it presents the opportunity to observe how various neurocognitive systems emerge and become integrated during development. For example, in Chapter 5 we will see how different brain pathways underlying eye movement control emerge and become integrated during development.

The Cause of Developmental Change

Those inclined to see development as the unfolding of pre-existing information in the genes tend to adopt a maturational view of developmental psychology in which infants have reduced versions of the adult mind which increase by steps as particular brain pathways or structures mature. In contrast, taking a constructivist view of development involves attempting to unravel the dynamic relations between intrinsic and extrinsic structure that progressively restrict the phenotypes that can emerge. The distinction between these two approaches has also been noted by Gottlieb (1992), who refers to them as "predetermined epigenesis" and "probabilistic epigenesis." Predetermined epigenesis assumes that there is a unidirectional causal path from genes to structural brain changes to brain function and experience. In contrast, probabilistic epigenesis views the interactions between genes, structural brain changes, and function as bidirectional:

Predetermined epigenesis:

(Unidirectional structure–functional development)

genes $\rightarrow$ brain structure $\rightarrow$ brain function $\rightarrow$ experience

Probabilistic epigenesis:

(Bidirectional structure–functional development)

genes $\leftrightarrow$ brain structure $\leftrightarrow$ brain function $\leftrightarrow$ experience

(Gottlieb, 1992)

Thus, by the predetermined epigenesis view the infant mind is viewed as being comparable to adults with focal brain injury. That is, specific cognitive mechanisms are either present or absent at a given age. For example, parallels have been drawn between infants and adult patients with frontal lobe injury (see Chapter 10). Circuits that support components of the adult system are assumed to come "on-line" at various ages. However, while this approach is likely to provide a reasonable first approximation for normal developmental events, it is unlikely to provide a full account in the long run.

An alternative approach to investigating the relation between the developing brain and cognition is associated with a probabilistic epigenesis approach to biological development. This view assumes that development involves the progressive restriction of fate. Early in development a system, such as the brain/mind, has a range of possible developmental paths and end states. The developmental path and end state that actually result are dependent on the particular sets of constraints that operate. This type of analysis of ontogenetic development derives from much earlier work on the development of body structure by D'Arcy Thompson (1917) and C.H. Waddington (1975), among others.

Waddington (whose work greatly influenced Piaget) proposed that there are developmental pathways, or necessary epigenetic routes, which he termed "chreods." Chreods can be conceptualized as valleys in an epigenetic landscape such as that shown in Figure 1.2. Self-regulatory processes (which Waddington called "homeorhesis") ensure that the organism (conceptualized as a ball rolling down the landscape) returns to its channel following small perturbations or disturbances. Large perturbations, such as being reared in darkness, can result in a quite different valley route being taken, especially when these occur near a critical decision point. These decision points are regions of the epigenetic landscape where a small perturbation can lead to a different route being taken. Thus, while for the typically developing child the same end point will be reached despite the small perturbations that arise from slightly different rearing environments, a deviation from the normal path early in development (high up the hill), at a decision point, or a major perturbation later in development, may cause the child to take a different developmental path and reach one of a discrete set of possible alternative end states (phenotypes).

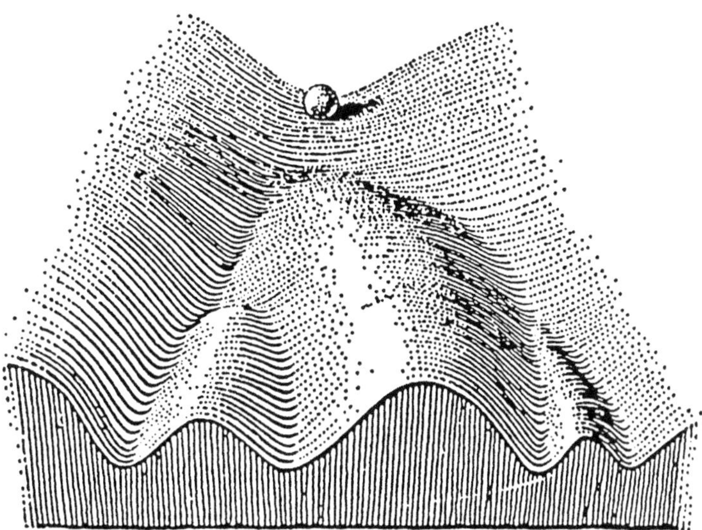

Figure 1.2 The epigenetic landscape of Waddington (1975).

Aside from Waddington's informal conceptualization, the constructivist (probabilistic epigenesis) approach to development is currently more difficult to work with since we have few theoretical tools for understanding emergent phenomena in complex dynamic systems. By this view, developmental disorders are possible developmental trajectories that are responses to different sets of constraints (see Chapter 2). This implies that from the moment when the developmental trajectory deviates from the typical one, a variety of new factors and adaptations will come into play, making it likely that some reorganization of brain functioning will take place. In contrast to the causal epigenesis approach, therefore, applying the mapping between brain regions and functions found in normal adults to such cases may be only partially informative. It should be stressed that the constructivist view just outlined does not seek to downplay the role of genetic factors. Rather, it seeks to understand the emergence of new structures and functions through the complex interactions between genes and their different environments.

Three Viewpoints on Human Functional Brain Development

Relating the neuroanatomical changes that occur during the development of the brain to the remarkable changes in motor, perceptual, and cognitive abilities during the first decade or so of human life presents a considerable challenge. Throughout this book, we will discuss evidence inspired by three distinct viewpoints on human functional brain development. These are: (a) a *maturational perspective*, (b) *interactive specialization*, and (c) a *skill learning viewpoint*.

As mentioned earlier, much of the research to date attempting to relate brain to behavioral development in humans has been from a maturational viewpoint in which the goal is to relate the "maturation" of particular regions of the brain, usually regions of the cerebral cortex, to newly emerging sensory, motor, and cognitive functions. Evidence concerning the differential neuroanatomical development of brain regions is used to determine an age when a particular region is likely to become functional. Success in a new behavioral task at this age may then be attributed to the maturation of this "new" brain region. By this view, functional brain development is the reverse of adult neuropsychology, with the difference that specific brain regions are added in instead of being damaged.

Despite the intuitive appeal and attractive simplicity of the maturational approach, we will see during the course of this book that it does not successfully explain some major aspects of human functional brain development. Further, associations between neural and cognitive changes based on age of maturation of brain areas are only weak predictions due to the great variety of neuroanatomical and neurochemical measures that change at different times in different regions of the brain.

In contrast to the above approach, a specific constructivist viewpoint, "interactive specialization" (IS), assumes that postnatal functional brain development, at least within the cerebral cortex, involves a process of organizing patterns of inter-regional interactions (Johnson, 2001, 2011). The process is constrained by initial processing biases in different cortical areas and pre-established major highways of structural connectivity. According to IS, how a region of the cortex responds to a sensory event or task is partly determined by its patterns of connectivity to other regions, and their respective patterns of activity. During postnatal development, changes in the response properties of cortical regions occur as they interact and compete with each other to acquire their role in new computational (thinking) abilities. From this perspective, some cortical regions may begin with relatively poorly defined functions, and consequently are partially activated in a wide range of different contexts and tasks. During development, activity-dependent interactions between regions hone the functions of regions such that their activity becomes restricted to a narrower set of circumstances (e.g., a region originally activated by a wide variety of visual objects may in time come to confine its response to human faces viewed in the typical right-side-up orientation). The onset of new behavioral competencies during childhood will therefore be associated with changes in activity over several regions, and not just with the onset of activity in one or more additional region(s). We will expand further on this theory in Chapter 13.

A third perspective on human functional brain development, skill learning, involves the proposal that the brain regions active in infants during the onset of new perceptual or motor abilities are similar, or identical, to those involved in complex skill acquisition in adults. For example, with regard to perceptual expertise, Isabel Gauthier and colleagues have shown that extensive training of adults with artificial objects (called "greebles") eventually results in activation of a cortical region previously associated with face processing, the "fusiform face area" (Gauthier et al., 1999). This suggests that the region is normally activated by faces in adults, not because it is prespecified for faces, but due to our extensive expertise with that class of stimulus. Further, it encourages parallels with the development of face-processing skills in infants (see Gauthier & Nelson, 2001). While it remains unclear how far parallels can be drawn between adult expertise and child development, to the extent that the skill learning hypothesis is correct, it presents a clear view of a continuity of mechanisms throughout the life span. Finally, skill learning is not always incompatible with IS, and sometimes the two viewpoints make similar predictions.

Interactive Specialization

In the following chapters we use the IS framework to help cohere observations and evidence from a variety of different brain imaging methods described in Chapter 2. According to IS, during development activity-dependent interactions between regions result in modifications of the intra-regional connectivity such that the activity of a given area becomes restricted to a narrower range of stimuli or circumstances—a process called "specialization." As a result of becoming more finely tuned (specialized), small-scale functional areas of cortex become increasingly distinct and differentiated from their surrounding cortical tissue, and this will be evident in functional brain imaging studies as increasing localization for specific functions. These same processes as described for specialization of regions can also apply to networks of densely interconnected regions, revealing the tuning up of specific networks that become engaged by particular tasks or stimuli. In the final chapter of this book, we will bring together evidence from the different domains of cognition, sensation, and emotion that we cover within the broad framework of IS. We will also discuss ways in which this broad approach can be translated into specific and testable hypotheses for the future.

Further Reading   Johnson (2011).

Looking Forward

The next chapter reviews some of the different methods currently used to study emerging brain structures and functions. Although atypical development is not the primary focus of this book, throughout the review of typical development, evidence from atypical development is discussed where relevant. We will see that apparently specific neurocognitive deficits can often result from diffuse damage to multiple brain systems. Brain damage in prenatal development can divert the child from one developmental path to another. However, it is possible that different types of brain damage can result in the same adult end state (equifinality of development, somewhat like the discrete number of valleys in Waddington's epigenetic landscape). Conversely, the same type of brain damage or developmental challenge (e.g., preterm birth) can result in different outcomes (multifinality of development). It is important to note that brain damage in early life (perinatal and early postnatal) may be compensated for by other parts of the brain. Thus, at this stage a focal brain lesion may have only mild diffuse cognitive consequences, resembling Waddington's self-organizing adaptation keeping the organism within a certain chreod and resulting in the same general phenotype.

In Chapter 3 we introduce some basic facts about genes and discuss what is known about their expression during human development, while in Chapter 4 we present the current state of knowledge about the pre- and postnatal development of the human brain. While the general sequence of developmental events is very similar for all mammals, the timing of human development, and especially human postnatal development, is protracted. This extended period of postnatal development is associated with a greater extent of area of the cerebral cortex, in particular the prefrontal regions. The more extended postnatal development observed in humans reveals differential rates of development in aspects of brain structure (e.g., different cortical areas and layers). The more differentiated picture of postnatal brain development in humans has also been used to make predictions about the emergence of function.

Focusing on the cerebral cortex, early in life large-scale regions of cortex have approximate biases that make them best suited to supporting particular types of computations. The fairly consistent structure–function relations observed in the cortex of normal human adults appear to be the consequence of multiple constraints both intrinsic and extrinsic to the organism, rather than of detailed intrinsic genetic specification. In the following chapters a number of domains of perceptual, cognitive, social, and motor development that have been associated with neural development are reviewed. In each of these domains we reveal some of the sources of constraint on the representations that emerge within cortical circuits of the brain. Examples of combinations of constraints from the correlational structure of the external environment, the basic architecture of the cortex, and the influence of subcortical circuits are discussed. We also consider the role of developmental cognitive neuroscience in informing approaches to education, and within a global, cross-cultural context.

In Chapter 13, mechanisms and types of changes in representations during human postnatal development are discussed, and the IS viewpoint on human functional brain development is expanded further. The emergence of specific functions in cortical areas is seen as a product of interactions within the brain (albeit constrained by initial regional biases and highways of structural connectivity), and between the brain and its external environment. Just as the child develops within a social and physical environment, and the brain within a body ("embodiment"), each cortical area develops its functionality within the context of the whole brain ("embrainment"). Lastly, a number of conclusions and recommendations for future research are made.

Key Issues for Discussion
• To what extent do researchers investigating issues in cognitive neuroscience in adults need to consider evidence from development? • What aspects of the typical developing child's environment are likely to be "experience-expectant" and "species-typical," across cultures? • To what extent can Waddington's "epigenetic landscape" satisfactorily account for the recovery of some cognitive functions following early brain damage?

2

Methods and Populations

This chapter provides background on the different methods used, and populations studied, in developmental cognitive neuroscience. Behavioral techniques for studying infants and children have been available for several decades, but recent advances in eye tracking and motion sensing have opened new possibilities for finer grained analyses. A relatively new set of tools relate to the generation of structural and functional maps of brain activity based on changes in either cerebral metabolism, blood oxygen level, or electrical activity. For some questions these neuroimaging methods have replaced research on animals, while for other important scientific questions, such as genetic manipulation or gene expression, there remains little substitute to animal research. In addition to typically developing infants and children from Western cultures, researchers in developmental cognitive neuroscience are now focusing on neural and cultural diversity, including studying children with developmental disorders such as autism and Williams Syndrome, inherited family risk, and cases of early deprivation due to sensory limitations or poor social context as well as different global and cultural contexts, including how all of the above might be expressed within these contexts. Only by employing a variety of methods and studying various different populations will we gain powerful leverage on key questions in the field.

Introduction

Progress in a field of science critically depends on three things: empirical discoveries, the development of theories to account for the available evidence and to make predictions for future work, and, finally, methods and analysis. The importance of the latter is often underestimated in the history of science. However, at least for developmental cognitive neuroscience, a strong case can be made that progress could have been much more rapid in the past had current technology been available. Over the past decades, new technology for data collection and analysis has arisen and it continues to be developed, and we have learned how to apply this technology to the special difficulties and constraints that come with studying infants and children. Behavioral methods for studying infants and children are

Developmental Cognitive Neuroscience: An Introduction, Fifth Edition. Michelle de Haan, Iroise Dumontheil, and Mark H. Johnson.
© 2023 John Wiley & Sons Ltd. Published 2023 by John Wiley & Sons Ltd.
Companion website: www.wiley.com/go/johnson/devneuro5e

incrementally being improved and extended. A relatively new set of tools relates to neuroimaging—the generation of "functional" maps of brain activity based on changes in either cerebral metabolism, blood oxygen level, or electrical activity. We will see that for some issues these neuroimaging methods have replaced research with animals, while for other important scientific questions there still remains no adequate substitute for animal research. In addition to typically developing infants and children from high-income Western societies, researchers in developmental cognitive neuroscience have also turned their attention to studying neural and cultural diversity (discussed further in Chapter 12). This gives powerful leverage to understanding some of the basic principles underlying neurocognitive development. In addition, inherited family risk and cases of early deprivation due to sensory limitations or poor social context are not just important to study for clinical or societal reasons, but they can also shed light on the importance of specific kinds of early experience for later development and inform theories of the relations between experience and brain and behavioral development.

Behavioral and Cognitive Tasks

Since the 1950s behavioral and cognitive analyses of child development have moved from an important early stage of "natural history" where Piaget and others described some of the striking phenomena associated with development, such as the apparent lack of object permanence (see Chapter 10), to a variety of ingenious experimental ways to gather information about psychological change in infants, toddlers and children. One of the major challenges to studying infants and toddlers that had to be overcome was the need for behavioral tasks that do not involve verbal instruction or require sophisticated motor responses like pressing specific keys or buttons. Further, infants and preschoolers have shorter attention spans in terms of the length of time they are prepared to cooperate with the experimenter compared to older children and adults. Thus, studies involving extensive training periods are not feasible either. Fortunately, a number of methods have been developed for testing infants and toddlers that build on their natural tendencies to look at conspicuous and novel visual stimuli. One of these procedures, called "preferential looking," involves presenting paired visual stimuli and recording the time that the infants choose to look at each. Another procedure, called "habituation," involves showing the same stimulus repeatedly until the infant shows a clear decrease in the time they spend looking at it. When a certain criterion for the decrement in looking is reached, a novel stimulus is presented and the increase or recovery in looking time is recorded. If there is significant recovery of looking time, we may infer that the infant can discriminate between the two stimuli. If there is little or no recovery, we may infer that the infant is unable to discriminate between them. Other techniques for eliciting discriminative responses from young infants include using rate of sucking to measure habituation, the use of an eye tracker to determine exact patterns of looking, and the use of heart rate measures. In particular, the advent of recent more robust eye-tracker technology is revolutionizing the study of visual attention in early development, as it allows data to be gathered both on general areas of interest within a visual presentation and on mechanisms of eye movement planning and control (see Chapter 5). Methods of gaze-contingent eye tracking now allow preverbal infants and toddlers to interface directly with a computer screen via eye movements, allowing for new designs of experiments involving training programs (e.g., Wass et al., 2011), as well as the use of head-mounted eye-tracking cameras to study more naturalistic interactions.

> **Further Reading**   Aslin (2007); Karatekin (2008); Franchuk & Yu (2022).

Another useful way of linking brain development to behavior across different age groups is the "marker task." This method involves devising specific behavioral tasks closely based on those that have been related to one or more brain regions or networks in adult humans and non-human primates by either neurophysiological or brain imaging studies. By studying the development of performance on the task at different ages and in different contexts, the researcher can gather evidence about how the observed behavioral change is accounted for by known patterns of brain development. In this book, we will see that the marker task approach has been taken in several different domains of cognition. There are also weaknesses to the marker task approach, such as that findings from one specific task sometimes do not generalize to others that on the surface seem closely related, and it can be difficult to directly compare results from groups of participants who differ significantly. Another challenge of the marker task approach stems from designing a task that is sufficiently limited in its demands of the participant to give interpretable results with infants or young children, and yet sufficiently demanding to call upon "higher" cognitive capacities. Finally, we will see in several subsequent chapters that different brain regions may be critical for the same task at different ages—this reflects a potential weakness in applying models of brain–behavior relations from models of adult humans and/or non-human primates. Thus, the interpretation of marker task results is made more complex. Nevertheless, the marker task approach is a useful methodology that can provide initial insights into the development of mental abilities.

Assessing Brain Function in Development

With the exception of one relatively new method, the techniques available for observing the functioning of the young human brain are those already commonly used in research on adults (see Figure 2.1 in the color plate section). High-density event-related potentials (HD-ERP) is a method of recording the electrical activity of the brain by means of sensitive electrodes that gently rest on the surface of the scalp (see Figure 2.2 in the color plate section). These sensors detect tiny changes in electrical voltage at the scalp surface caused by groups of neurons within the brain firing together. These recordings can be of either the spontaneous natural electrical rhythms of the brain (electroencephalography—EEG) or the electrical activity evoked by a stimulus presentation or action (event-related potentials—-ERPs). When studying ERPs the data from many trials is averaged so that the spontaneous EEG unrelated to the stimulus presentation averages out to zero. Another useful measure is the rapid bursts of high-frequency EEG (such as the gamma or 40 Hz frequency) that appear to be related to stages of information processing in the brain (event-related oscillations—EROs) (Csibra et al., 2000).

> **Further Reading**   Csibra & Johnson (2007); de Haan (2014); Rollins & Riggins (2021).

With a high density (HD) of sensors placed on the scalp, algorithms can be employed to infer the position and orientation of the brain sources of electrical activity (dipoles) for the particular pattern of scalp surface electrical activity. Some of the assumptions necessary for the successful use of these algorithms are actually more likely to be true of infants than adults. For example, lower levels of skull conductance and fewer cortical convolutions may improve the accuracy and interpretability of HD-ERP results in infants relative to adult subjects (see Johnson et al., 2001; Reynolds & Richards, 2009 for discussion of this methodology as applied to infants). Through measures of the extent to which different channels record electrical changes oscillating at the same frequency and in phase (coherence), HD-ERP can also provide pictures of functional connectivity between different regions (e.g., Bathelt et al., 2013). The relative low cost and portability of EEG/ERP allows it to be applied across cultural contexts, as we will see in more detail in Chapter 12 (Kihara, de Haan, et al., 2010; Kihara, Hogan, et al., 2010).

HD-ERP and related methods are an excellent way to study brain functions even in very young babies. However, while they offer excellent time resolution (of the order of milliseconds) it has been difficult to obtain anything other than coarse spatial resolution (e.g., frontal versus temporal lobes). Still, there have been developments looking at functional connectivity and localization that have improved this method over the past years (Bathelt et al., 2013; Reynolds & Richards, 2009). A method that has far greater spatial resolution, albeit at the expense of temporal resolution and economic cost of implementation, is magnetic resonance imaging (MRI) (see Figure 2.1 in the color plate section). In addition to excellent maps of brain structure, the method can also be used to detect functional activity (fMRI). As different regions of the brain are activated, the cells in that area require oxygen delivered through networks of tiny blood vessels. Oxygen is transported in the blood by a molecule called hemoglobin, and when a brain region is active it calls for more oxygen, resulting in a localized increase in oxygenated hemoglobin and a decrease in deoxygenated hemoglobin. The change in the blood-oxygen-level-dependent (BOLD) response is detected by MRI, thus allowing the non-invasive measurement of cerebral blood oxygen levels in the different parts of the brain with a spatial resolution on the order of millimeters and a time resolution at the scale of seconds. The low temporal resolution of fMRI is driven by the fact that changes in blood flow and oxygen level peak around 5 s after the neural activity; methods like EEG/ERP measure the neural activity at the scalp but suffer from poorer spatial resolution as this activity "blurs" when measured this way.

While this fMRI technique for studying brain function is now routinely being applied to children from 6 or 7 years old in several laboratories worldwide, for a variety of reasons, mainly that participants need to stay very still for good data to be collected, it remains technically challenging to use this method with children younger than this age. Studying younger children can still be feasible with thorough preparation, such as using a mock scanner and training children to stay still in a preliminary visit (de Bie et al., 2010). However, there remains a significant amount of data loss the younger children are, and there is also more data loss when studying clinical pediatric groups compared to typically developing children and adolescents (Yerys et al., 2009). Studying infants is even more challenging, yet possible, particularly if they are sleeping or drowsy while passively listening to auditory stimuli such as speech or music. A few visual studies have also been possible in some cases (e.g., Kosakowski et al., 2022). There are a variety of complex issues about the analysis of fMRI data from children and its comparison to that collected in adults (see Thomas & Tseng, 2008). In addition to conventional fMRI analysis focused on specific regions of interest, there are also analysis methods that allow the researcher to assess the degree of functional connectivity between different brain regions. As we will see in later chapters, this technique allows us to test hypotheses about the emergence of coordinated functional brain networks during development.

> **Further Reading**   de Haan (2014); Le et al. (2020); Thomas & Tseng (2008).

A more recent method that also measures brain activity through the levels of oxygenated hemoglobin in the blood, including the BOLD signal, is functional near-infrared spectroscopy (fNIRS). This is a form of optical imaging, meaning that it depends on measuring minute changes in the absorption and scatter or bending of weak light beams as they pass through the skull and brain (see Lloyd-Fox et al., 2010; Meek, 2002). Tiny light emitters and detectors are embedded within a cap and carefully placed on the child's head (see Figure 2.3 in the color plate section). Like fMRI, changes in blood oxygenation due to brain activity can be detected with this method, but fNIRS is less sensitive to motion artifacts and does not require confinement in an enclosed scanner. Therefore, fNIRS potentially provides an excellent alternative to fMRI for use with infants and toddlers. Indeed, the relatively thin skull of very young children means that light passes through more easily, and thus better optical signals are usually obtained. While the technique is still undergoing further development in several laboratories, there is now a rapidly increasing number of papers published that describe the use of this method to study brain functions in babies and toddlers. In addition, the relatively low cost and transportability of fNIRS equipment can allow the study of culturally varied populations which would typically not be involved in fMRI studies (Blasi et al., 2019 and see Chapter 12).

> **Further Reading**   Aslin et al. (2015); Lloyd-Fox et al. (2010); Mehler et al. (2008).

Observing Brain Structure in Development

Part of the goal of developmental cognitive neuroscience is to relate changes in brain function and cognition to changes in the underlying brain structure. For decades, the study of postnatal human brain structural development depended on traditional neuroanatomical methods applied to postmortem human or animal tissue. These methods involve staining neurons and their processes with substances that make them more clearly visible under the microscope. One of these stains (Golgi stain) was invented by a founding father of neuroscience and Nobel Prize winner, Camillo Golgi (1843–1926). For human postmortem tissue, such analyses are painstakingly slow and difficult, and tend to be based on relatively small numbers of children due to the difficulties associated in gaining such tissue. Furthermore, those children who unfortunately come to autopsy have often suffered from trauma or diseases that complicate generalizations to normal brain development. Perhaps the most notable series of studies conducted with this approach came from Conel, who, between 1939 and 1967, published several volumes of detailed drawings of the postnatal development of human cortex (Figure 4.5). More recently, the advent of the electron microscope has allowed scientists to study changes at even smaller scales, such as the formation or loss of synapses on dendrites (Chapter 4).

In addition to studies of brain activation, MRI allows the study of the development of brain structure in healthy living babies and children. While this is an enormous methodological advance on traditional postmortem neuroanatomy, until past decades such MRI methods only allowed a clear dissociation between the brain's gray matter (clusters of neurons and their local processes and connections) and white matter (bundles of connecting fibers) at an order of magnitude less detailed than microscope images. Nevertheless, as we will see in Chapter 4, there has been an explosion of knowledge about the trajectories of postnatal anatomical brain development using MRI. In recent years, new analytic methods have been emerging that allow us to go beyond the simple assessment of the shape and quantity of white and gray matter and begin to trace the pathways of major structural connections between regions in the brain. One such method, diffusion tensor imaging (DTI), uses measures of the motion of water molecules to give detailed pictures of fiber tracts and their development. Other methods of tracing fiber tracts during development are currently being investigated (see Figure 2.4 in the color plate section). Such methods will be important for testing predictions about the precise relation between structural connectivity and function in brain development.

Further Reading   O'Hare & Sowell (2008); Tamnes et al. (2018).

Animal Studies and Genetics

As we will see in Chapter 4, quite a lot of what we currently know about both pre- and postnatal brain development comes from research on other species, though fetal MRI imaging provides additional information. While there are obviously some differences between species, the overwhelming majority of early neurodevelopment is common to most or all species studied. Additionally, in behavioral development a number of animal models have reached a stage where some of the principles discovered are applicable to aspects of human development (see Blass, 1992). Studying simple animal models can both directly and indirectly inform our understanding of how brain development relates to the more complex cognitive and behavioral changes observed in humans. In one example of the application of animal models to human development, comparisons between the behavioral development of humans and other primates can illuminate the importance of language to changes in other domains of cognition. Consequently, while homologies between species must be made with great care, well-studied animal models increasingly provide useful theoretical and empirical insights into cognitive development in humans.

Further Reading   Bachevalier (2008); Matsuzawa (2007).

In the next chapter we will discuss genetic methods in more detail. However, at this point it is useful to note that techniques exist in molecular genetics which allow the deleting of particular genes from the genome of an animal. An example of this is the deletion of the alpha-calcium calmodulin kinase II gene which results in so-called "knockout" mice being unable to perform certain learning tasks when adults (Silva, Paylor, et al., 1992; Silva, Stevens, et al., 1992). This method opens further understanding in the analysis of genetic contributions to cognitive and perceptual change in animals, and may be particularly fruitful when applied to well-studied animal models of development such as visual imprinting in chicks and song learning in passerine birds. These animal models may also be useful when investigating the role of genetic deletions/atypicalities for neurocognitive development in disorders of known genetic origin.

Neurodiversity and Developmental Disorders

In addition to needing a variety of different methods to study typically developing infants and children, developmental cognitive neuroscience also involves the study of differences in development caused by genetics, early brain damage or trauma, familial risk factors, or atypical early experiences, to name some examples. While the study of atypical developmental pathways (neurodiversity, often called "developmental neuropsychology" or "developmental psychopathology") is clearly important for clinical and societal reasons, it can also inform us about the causal factors and basic processes important in neurotypical development.

While this book is mainly focused on the typical trajectory of development of the human brain, it is important to note that typical trajectories are not single unitary developmental pathways, but rather a range of similar trajectories. These can extend into the atypical range and this can be associated with the factors mentioned above. We will discuss several different developmental disorders or alternative trajectories of human brain development throughout this book. As we will discuss further in the next chapter, genetic deviations can involve mutations of single genes (such as Fragile-X and phenylketonuria), abnormalities in chromosomal structure (such as Down's syndrome), and microdeletions that involve several genes from one part of a chromosome (such as Prader-Willi and Williams syndrome). There are also several developmental disorders that are believed to have complex genetic bases involving multiple (hundreds) of genes of small effect, and that may sometimes represent the extreme ends of typical variation (e.g., autism and attention deficit/hyperactivity disorder).

Considering autism, this is a relatively common developmental disorder (with an incidence including the spectrum of related disorders of around 1% of the population). Genetic studies of autism have revealed a complex and evolving story with hundreds of different genes currently being implicated. Interestingly, although autism is among the most heritable of psychiatric conditions, reviews lead to the conclusion that the genetic contribution to the disorder involves very many genes, each of small effect size. However, even adding all of the known genetic factors together, these still only account for a minority of cases of autism (Abrahams & Geschwind, 2008). Attempts to resolve this issue have entailed arguments that there are probably multiple different combinations of genetic factors that contribute to autism (Happé et al., 2006), and the view that any disturbance in whole families of genes, such as those involved with the formation and maintenance of synapses, may be the relevant factor (Banerjee et al., 2014). However, rather than trying to rescue the idea of a genetic cause for autism it may be more useful to consider the condition as one that arises only from the interaction between genes and their environment, and that its causes do not reside in either factor alone (see Elsabbagh & Johnson, 2010). It is relevant to note that autism or autism-like behaviors are observed in a number of conditions that compromise brain or experiential development (e.g., preterm birth; children with congenital visual impairment). This highlights the multidimensional contributions to this complex disorder which are still not fully understood.

Many of the core difficulties of the developmental disorder of autism lie in the domain of social relations with others, though there are many non-social cognitive deficits as well. Major behavioral symptoms include avoidance of eye contact, difficulties in comprehending the thoughts of others, a reluctance to be held or touched by others, repetitive behaviors such as rocking or hand flapping, and commonly a fascination for certain objects or other inanimate aspects of their environment. Alongside these symptoms are often differences in other cognitive and language skills, such as echolalia (repeating words and sentences heard previously). Autism includes a whole spectrum of related disorders that share overlapping symptoms. To take the case of autism, Rumsey and Ernst (2000) summarize their review of functional imaging of autistic disorders as "... studies of brain metabolism and blood flow thus far have yet to yield consistent findings, but suggest considerable variability in regional patterns of cerebral synaptic activity" (p. 171). More recent reviews concur and conclude that neuroanatomical measures are unlikely to change our understanding of autism neuropathology as whole, although there may be subgroups with specific patterns (Fetita et al., 2021). We will consider aspects of autism in more depth in Chapter 7.

> **Further Reading**   Frith (2003); Akarca et al. (2021).

In contrast to the clear social difficulties seen in autism, in Williams syndrome (WS) participants appear at first sight to have good social skills. WS (also known as infantile hypercalcemia) is a relatively rare disorder of genetic origin effecting approximately 1 in 20,000 to 50,000 births (Greenberg, 1990). The disorder can now be diagnosed in early infancy through genetic or metabolic markers, and is typified by a number of physical and cognitive characteristics.

Evidence from structural neuroimaging indicates that WS brains are only about 80–85% of the overall volume of typical brains, but there are no obvious gross atypicalities or lesions (Jernigan & Bellugi, 1994). There is evidence for a specific focus of difference, which is that they show a relative increase in volume in particular lobules on the cerebellum. This cerebellar atypicality contrasts with autism, in which the same lobules are relatively smaller than normal (Jernigan & Bellugi, 1994). At the microanatomical level, one analysis by Galaburda et al. (1994) found disturbances within cortical layers and decreased myelination (Chapter 4). There is also evidence of disruption, particularly in the parieto-occipital regions, with a pattern of hyperconnectivity (Gagliardi et al., 2018). Further studies documenting structure–function relations, including cross-syndrome comparisons, can help understand these findings in a broader context. It is important to note that, while the contrast in skills in autism and WS is interesting to investigate, autism is diagnosed in a broader range of conditions, whereas WS is a more specifically genetically diagnosed syndrome.

> **Further Reading**   Karmiloff-Smith (2008); Gagliardi et al. (2018).

It is often suggested that the pattern of spared abilities in WS suggests approximately the opposite of deficits described for autism, and this raised the initial hypothesis that people with WS have intact a functional brain system corresponding to a "social module" (see Chapter 7). Specifically, one hypothesis is that the social brain network remains intact in WS, while being specifically damaged in autism. However, this hypothesis is now regarded as a gross over-simplification following more recent research (see Chapter 7). Nevertheless, the contrast between autism and WS illustrates the potential value of comparing and contrasting different developmental disorders in similar experimental paradigms.

Atypically Developing Brains

When considering the brain correlates of developmental disorders of genetic origin, there are at least four levels at which they can be described; (a) gross brain anatomy, (b) individual "deficit" areas, (c) functional neural systems and pathways, and (d) neurochemistry and microcircuitry (Table 2.1). Over the past decades there have been many specific hypotheses advanced about the neural deficits that cause, or correlate with, different disorders. Initially, these claims often involved a search for discrete localized cortical or subcortical "lesions" resulting from the atypical genetics. However, it has become clear that, at least when studied in later childhood or as adults, many developmental disorders involve widespread systemic subtle differences in brain structure and function.

> **Further Reading**   Thapar & Riglin (2020).

For example, attention deficit/hyperactivity disorder (ADHD) is associated with overall smaller global gray matter volumes, but also specifically smaller gray matter volumes in the right lentiform nucleus and caudate nucleus, and greater gray matter volumes in the left posterior cingulate cortex (Frodl & Skokauskas, 2012; Nakao et al., 2011). Interestingly, in a meta-analysis, structural differences in the right basal ganglia were found to be greater at younger ages and to be reduced when patients received stimulant medication (Nakao et al., 2011). It has been suggested that some of the structural differences observed in ADHD may be driven by a delayed cortical maturation, in particular in the prefrontal cortex (Shaw et al., 2007).

Schizophrenia is also associated with widespread differences in white matter microstructure, enlarged ventricles, and reduced temporal and frontal gray matter volumes (Karlsgodt et al., 2008). Like for ADHD, it is important to consider the developmental time course of brain structure changes in schizophrenia. There is evidence that although gray matter deficits have been established in prodromal and early stages of schizophrenia, the disorder is also associated with an accelerated ventricular expansion, and progressive frontal and temporal volume loss during adolescence and early adulthood (Fraguas et al., 2016; Karslgodt et al., 2008). Developmental cognitive neuroscience research is slowly trying to identify

Table 2.1   A Thumbnail Sketch of Some of the Major Developmental Disorders

Disorder	Genetic basis	Illustrative brain atypicality	Behavioral phenotype
Phenylketonuria	Single gene (mutated PAH)	Low dopamine levels in prefrontal cortex	Deficits in executive function, working memory
Fragile-X	Single gene (silenced FMRI gene)	Relative decrease in volume of cerebellum; atypical dendritic spine morphology throughout neocortex	Retardation, hyperactivity, attention problems, autistic symptoms, visuospatial and numeracy problems
Down's syndrome	Chromosomal (extra chromosome 21, translocation or mosaic)	Microcephaly	General retardation, but relative strengths in visuospatial cognition and weak verbal abilities
Turner syndrome	Chromosomal (absence of X in females)		Visuospatial and numeracy deficits, but verbal IQ in normal range
Prader-Willi syndrome	Microdeletion/ duplication (SNRPN, NDN)	Hypothalamus	Moderate retardation, but sociable; poor in most cognitive domains, and especially verbal skills
Williams syndrome	Microdeletion (chromosome 7, 24–30 genes)	Microcephaly (due to white matter reduction) but with increased proportional volume of cerebellum; cerebellar abnormalities	Relatively poor at spatial and number tasks, but behaviorally proficient at face processing and language
Autism	Polygenic	Macrocephaly due to increased white matter in many cortical regions	Impaired social function, impaired language, restricted repertoire of interests
Dyslexia/SLI	Polygenic	Increased cerebral white matter throughout cortex in SLI	Deficits in language and/or reading
ADHD	Polygenic	Atypical functioning of widespread cortical areas including frontostriatal and limbic circuitry	Deficits in attention; particularly response inhibition and delay avoidance

Note. For the sake of comparison, the table focuses on contrastive aspects of different disorders.

the risk factors for schizophrenia, which appear to converge on neurodevelopment (Birnbaum & Weinberger, 2017).

In sum, there is little support for the notion that discrete structural differences in functional cortical areas can be observed in common developmental disorders. Rather, we tend to see widespread effects in a brain that has actually developed atypically over a number of years.

> **Further Reading**   Penzes et al. (2011).

Another important point to note is that the majority of brain abnormalities associated with developmental disorders are not specific to one disorder. For example, an atypical cerebellum has been reported for autism, Fragile-X, WS, and dyslexia. In our view, these atypicalities that are shared across several developmental disorders are no less interesting, but they do raise the importance of studying general profiles of differences across several brain regions and disorders, rather than focusing heavily on specific phenomena within individual areas or regions in a single developmental disorder. In general, claims about specific neural deficits underlying a developmental disorder need to be treated with caution until other brain areas, and other developmental disorders, are examined in equal depth.

With regard to gross measures of brain structure, overall brain volume can be reduced (microcephaly) or increased (macrocephaly) relative to typical development, and may relate to overall differences in the rate of brain development (Chapter 4). Differences in overall volume can be due to changes in the gray (neurons and their local connectivity) or white (fiber bundles and long-range connectivity) matter. An interesting consequence of several major developmental disorders is that both microcephaly and macrocephaly are principally due to deviations in the extent of white matter. For example, a consistent finding from structural imaging of autism is larger cerebral volumes (see Filipek, 1999, for review), particularly in temporal, parietal, and occipital regions. Interestingly, however, this increased volume is due to white matter rather than gray matter (Filipek et al., 1992). In other words, the fiber bundles connecting regions and mediating inter-regional interaction were affected more than the cellular content of regions themselves. MRI studies of WS have shown that, unlike autism, overall brain and cerebral volume is smaller than aged-matched controls (Reiss et al., 2000). Assessment of tissue composition indicates that, compared to controls, individuals with WS have relative preservation of cerebral gray matter volume and disproportionate reduction in cerebral white matter volume. This pattern is restricted to the cerebral hemispheres, and is not found in the cerebellum (Galaburda & Bellugi, 2000; Reiss et al., 2000). Thus, in at least two major developmental disorders of genetic origin, differences in overall cerebral volume appear to be related to the extent of connectivity (white matter) between regions.

The most micro level of neuroanatomy considered for developmental disorders concerns microcircuitry, dendrites, and synapses. As it becomes evident that, at least for some syndromes, atypicalities are widespread across several regions or systems of the brain, some hypotheses have focused on putative differences in synaptic structure (Perisco & Bourgeron, 2006) or connectivity patterns (Just et al., 2007; Minshew & Williams, 2007). These differences in microstructure are assumed to have compounding effects during postnatal development, that then differentially affect some domains of cognition or types of computation more than others (Johnson, Jones, & Gliga, 2015). In later chapters we will review in more detail data from different disorders where these have been specifically associated with deficits in a particular domain (e.g., Chapter 5—ADHD; Chapter 7—Autism and Williams Syndrome).

Sensory and Environmental Variations

Several populations have been studied by developmental cognitive neuroscientists because they involve sensory or environmental deprivation for at least part of the individual's development. Thus, the scientist can potentially study the effects of experience resulting from sensory or environmental input on human brain and cognitive development. In later chapters of this book we will see examples of how studies on the congenitally deaf have illuminated the study of language acquisition and its brain basis (Chapter 9). Another example of this general approach comes from the work of Maurer and colleagues who have studied individuals who suffered visual deprivation for varying periods of time following birth due to dense cataracts in one or both eyes (further details in Chapter 7). These dense cataracts prevent structured visual input until they are reversed by surgery, which is usually within the first year. By studying aspects on face processing in this clinical population, this research group has shown that even after many years of normal experience of faces some subtle deficits remained (Le Grand et al., 2001). In other words, visual deprivation over the first months has detectable life-long effects on face processing. Further, by examining cases of unilateral deprivation it has been shown that these effects are more due to right hemisphere (left eye) deprivation. Data such as these present a severe challenge for the "skill learning" approach to human functional brain development, and suggest the right hemisphere may be biased toward face processing from the first months.

While the cataract and deaf populations can inform us about effects of sensory deprivation, other populations have been studied that suffered from social deprivation. For example, samples of children raised in orphanages (e.g., during the Romanian communist regime) can subsequently have multiple social, cognitive, and sensorimotor problems (for review see Gunnar, 2001) and smaller cortical gray matter and white matter volumes than children who were placed in foster care or never institutionalized (Sheridan et al., 2012). While this orphanage rearing was variable in general care quality, at a minimum, stable long-term relationships with caregivers are missing (Rutter, 1998). The outcome from "good" orphanages can include problems with executive functions (see Chapter 10) and social cognition, while other aspects of sensorimotor, cognitive, and linguistic development can recover well. At the other extreme, in a sample of children raised in Romanian orphanages for at least the first 12 months, 12% exhibited features of autism, although even here these symptoms tended to diminish over time (Rutter et al., 1999).

> **Further Reading**   Bourne et al. (2022); Maurer et al. (2008); Shackman et al. (2008).

There are several environmental factors that can affect the development of brain and behavior that have recently come under study, including pre- and postnatal parental stress and child nutrition. Another is children who have been raised in low social economic status homes. Children raised in financially poor environments, even in otherwise wealthy countries, have raised risk for a variety of adverse outcomes. The factors that contribute to the low-quality rearing environments of these children are varied and may include the stress and nutritional factors mentioned above, and/or factors including drug abuse or low-quality social interactions. With regards to nutrition, one aspect that has been studied is iron deficiency. Iron is needed for the functioning of all cells and organs, but especially for the developing brain. If there is not enough iron there can be long-term effects on brain and behavior, especially if this occurs in the first 1,000 days of life (McCarthy et al., 2022). Across the world, the nature and effects of the myriad of environmental factors are just beginning to be understood in terms of brain development and subsequent effects on child outcomes. A challenge for developmental cognitive neuroscientists in the future will be to devise theory-driven targeted interventions to help optimize outcomes in different settings (see Chapter 12).

> **Further Reading** Tomalski & Johnson (2010).

Familial Risk Populations

Over the past decade there has been increasing interest in studying infants and toddlers with a different kind of risk factor, one associated with the fact that common developmental disorders like autism and ADHD often run in families. For example, infants who have an older sibling diagnosed with autism are known to be at around 20 times the risk of developing the disorder themselves (Risch et al., 2014). In addition, rates of ADHD are also increased in these autism-risk families, suggesting the presence of some common factors across different disorders. Studying populations of infants at familial risk for conditions such as autism potentially offers powerful leverage on some key issues for common developmental disorders. First, conditions such as autism appear to emerge over the first years of life, at least in terms of their observable behavioral symptoms. Through studying early signs of autism, we can reveal the original core features of the syndrome before they become compounded through years of atypical interactions with the child's social and physical environment—and also understand the potentially multiple pathways that can lead to this outcome. Second, identifying the earlier predictors of a later diagnosis will be important for early intervention. Following from Waddington's epigenetic landscape (Figure 1.2), it is commonly assumed that there is a critical developmental timepoint within which at least some children at risk could be nudged out of the developmental trajectory that results in the autism phenotype, and back into the typical developmental trajectory. Thus, there is increasing interest in prodromal (before the onset of diagnostic symptoms) intervention, which may be able to prevent or alleviate symptoms in some individuals and improve their quality of life (Green et al., 2013). A third reason for the burgeoning interest in studying young children at elevated risk is that it may provide insights into protective factors, or mechanism of adaptation, that enable specific individuals to remain on a typical developmental trajectory in the face of risk factors. Referring to Waddington's epigenetic landscape once again, we may say that these individuals are better buffered against perturbing factors. For example, the chances of girls at familial risk going on to later autism is far lower than the equivalent figure for boys, indicating that, in general, girls may have more protective factors in place. A better understanding of these natural protective and adaptive factors may help in the design of future interventions.

Further Reading   Elsabbagh & Johnson (2010).

Key Issues for Discussion

- What would be an ideal method or technique for studying the development of human brain functions that is not currently available?
- Choose an example of a clear behavioral change during childhood, and discuss what two methods would be most appropriate for revealing the underlying causes and mechanisms of that change.
- Why are populations with particular developmental disorders or with impoverished early environments important for our understanding of the *typical* development of human brain functions?
- To what extent should we seek to change the social environment of infants at risk for later autism to allow them to better adapt for later life?

3

From Gene to Brain

This chapter introduces developmental genetics and outlines contemporary views on the role and structure of genes in building brains and the cognitive processes that they support. The popular 20th-century view of the genome as a direct blueprint for constructing the brain has been superseded by the contemporary view that genes are expressed differently according to their specific temporal and spatial context, and that even the simplest chemical building blocks of the brain (proteins) originate from multiple and variable gene interactions. These considerations make the view that specific genes "code for" particular facets of cognition implausible. Instead, the path from gene to brain to behavior (epigenesis) is both complex and variable. While this complexity may seem daunting, there are several ways to begin to unravel the genetic contribution to developing brain functions. One strategy is to study the role in the brain of relatively simple molecules that are close to being direct products of gene expression. A second strategy is to study variability between typical individuals in genes, brain functions, and cognition. Correlating differences on multiple measures may reveal associations between particular gene variants, aspects of brain function, and behavioral capabilities. A third approach is to study syndromes in which there are known genetic differences to the general population (see also Chapter 2) or specific individuals with rare mutations. By comparing neurocognitive development in these syndromes to the typical trajectory of development we can gain insights into the functional effects and consequences of some genes. A fourth approach involves animal studies to examine particular classes of genes that are expressed very rapidly as a result of learning or developmental plasticity. Finally, we review work on a particular gene, FOX-P2, that illustrates some of the general points made earlier in the chapter and is discussed further in Chapter 9.

The History of the Gene

In Chapter 1 we discussed different historical viewpoints on development. During the 19th and 20th centuries this debate continued, centered on efforts to understand the nature of inheritance. While it was clear that some physical traits of mammals, insects, and plants

Developmental Cognitive Neuroscience: An Introduction, Fifth Edition. Michelle de Haan, Iroise Dumontheil, and Mark H. Johnson.
© 2023 John Wiley & Sons Ltd. Published 2023 by John Wiley & Sons Ltd.
Companion website: www.wiley.com/go/johnson/devneuro5e

are passed from one generation to the next, the mechanisms and substance involved in this inheritance process remained obscure. We will see that the concept of a gene as first hypothesized by Johannsen in 1911 is not the same as the gene described by Watson and Crick in 1953, and their notion of a gene is different from that understood by developmental geneticists today.

The first significant step to localizing the material of inheritance was to identify it as being carried within cells, something that was already recognized by embryologists by the latter half of the 19th century. The next question was whether it was the nucleus of the cell, or the surrounding cytoplasm, that controlled the material responsible for inheritance. In a flurry of exciting research around 1900, several scientists rediscovered the earlier work of Gregor Mendel in which he proposed certain "rules of inheritance" based on his work breeding plants, and they identified the nucleus of the cell as the site of the heritable material. Shortly after, Wilhelm Johannsen (1911) coined the terms *gene, genotype* (the sum total of all genes), and *phenotype* (the end product or result of gene expression). Over the next 40 years these ideas were developed by scientists such as T.H. Morgan who pioneered the use of the fruit fly, Drosophila, for genetic studies. The short breeding cycle of this fly meant that many new experiments could be conducted in a short period of time, resulting in a dramatic increase in our understanding of genetics.

Further Reading   Fox Keller (2002); Stiles (2008).

In the early 1950s the search was on to discover which chemical found in the nucleus was responsible for inheritance. Two scientists in London, Rosalind Franklin and Maurice Wilkins, had been using a technique known as "X-ray diffraction" to reveal the structure of a molecule called DNA (deoxyribonucleic acid). In 1953 James Watson and Francis Crick based nearby in Cambridge used these data to propose the basic "double helix" structure of DNA that allows it to encode information and pass it on to the next generation. Once the chemical structure of DNA was revealed, it turned out to provide an obvious answer to the question of how information was coded and transmitted, that is, the problem of how to translate information to make the proteins that then go on to form complex biological structures such as the brain.

Principles of Gene Function

The beautiful structure of DNA, a right-handed spiral double helix, is importantly related to its function of retaining and transmitting information in at least two ways (see Figure 3.1). First, DNA involves two nucleotide (molecular unit) strands that are usually intertwined, but that can be unraveled when the genetic information needs to be copied during the division of cells to make bodies and brains. Second, each of the strands contains a "code" instantiated as differing sequences of four nucleotide bases (chemical units composed of a sugar-phosphate group and a base compound; they are called adenine, guanine, thymine, and cytosine [A, G, T, and C]). These nucleotide bases act like two

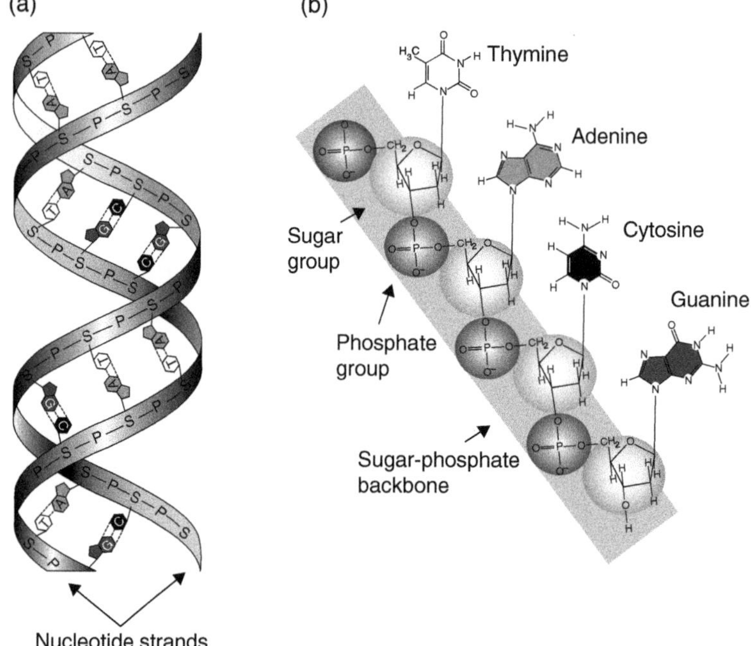

Figure 3.1 (a) The basic double helix structure of DNA in which two nucleotide strands coil around each other. (b) Detail showing how the two strands are linked by chemical bonds between the bases of nucleotides. The four bases are thymine, adenine, cytosine, and guanine. Reprinted by permission of the publisher from *The Fundamentals of Brain Development: Integrating Nature and Nurture* by Joan Stiles, pp. 45, 290, Cambridge, Mass.: Harvard University Press, Copyright © 2008 by the President and Fellows of Harvard College.

sides of a zipper that can be unzipped and a matching other half copied off with the complementary bonding of adenine and thymine and of guanine and cytosine (see Figure 3.1). Genes are specific sequences of nucleotides within a strand of DNA, with the sequences of base compounds providing the basis for gene expression.

The next question to be addressed is how strands of DNA make the basic chemical building blocks of biological tissue, proteins. Surprisingly, it was observed that the proteins are constructed relatively far away within the cell from the actual DNA itself. This means that there has to be some intermediary molecule that transfers the information from the DNA to the cellular machinery for making different proteins. The messenger was discovered to be RNA (ribonucleic acid), a closely related molecule to DNA with a few minor, but critical, differences.

By the early 1960s the identification of DNA and its mechanisms of expression lead to the textbook view that all of the genetic information necessary for the development and functioning of an organism was contained in the sequence of nucleotides within the DNA. The view that genes coded directly for the proteins that compose our bodies (and brains) seemed well established. However, with further research it rapidly became evident that this view was a gross simplification and that, in reality, gene expression is a highly dynamic and context-sensitive phenomenon.

The first place where the simple mapping between genes and proteins breaks down is in the complex steps (involving RNA) between the original unraveling and "reading" of a strand of DNA and the actual construction of a protein. Far from the view that one gene codes for one protein, it turns out that the same sequence of DNA can in many cases generate dozens of different specific proteins. A second consideration is that large segments of DNA (as much as 98%) do not appear to code for proteins. For this reason, these segments have often been referred to as "junk" DNA, but, while they are not directly coding for proteins, sequences of non-coding DNA have been discovered to have a range of functions, such as regulating gene expression, and are not just evolutionary hangovers (Cech & Steitz, 2014). A third caveat to the traditional view is that many genes code for regulatory, rather than structural, proteins. That is, they code for proteins that modulate the expression of other genes, creating complex cascades of interaction that can vary in different locations of the embryo or adult body. These days, genes are also said to be "pleiotropic," i.e., one gene can play multiple different roles at different times of development and in different areas of the developing animal or plant. For example, while around 75% of all of the genes in the human genome are expressed in the brain at some point during development, very few, if any, of these genes are expressed only in the brain. Finally, and perhaps most surprisingly, it has recently become evident that in addition to the DNA inherited at conception there are other differences in the cellular machinery that interact with and influence the expression of genes. This machinery is essential to the functioning of the DNA, which is inert without its presence. Thus, inherited cellular factors other than DNA can influence the genes that get expressed during child development, providing a potential path for inter-generational transfer. In summary, over recent decades it has become clear that while the complex processes associated with gene expression contribute to the emergence of physical and behavioral traits, the relationship between a specific gene and those traits is often very indirect and complex.

Further Reading   Knopik et al. (2016).

Exciting examples are appearing relating to how the life-long expression of genes in individual animals can be regulated by their early environment. For example, a series of elegant studies by Weaver, Meaney, and colleagues has demonstrated that maternal behavior toward newborn rats regulates the expression of genes involved in the same rats' responses to stress later in life (Weaver et al., 2004). Newborn rats whose mother frequently licks and grooms them during the first week grow up to be less fearful, and show less physiological stress reactions, than newborn rats whose mother is less attentive. Cross-fostering studies (with newborns raised by rats other than their biological mother) show that this life-long effect on the young rats is induced by the particular mother's behavior and is not directly genetically inherited. In a series of detailed experiments, it has been established that the maternal licking and grooming of the newborn sets in motion a series of biochemical reactions that regulates the activity of the mechanisms associated with the expression of particular genes (called the "epigenome"). Thus, early sensory experiences can have life-long effects through permanent changes in the timing and amount of different proteins expressed by the genes. Work such as this had led to the new emerging field of "epigenetics" that will clearly have some exciting implications for developmental cognitive neuroscience in the future.

Further Reading   Weaver et al. (2004).

Genetics and Developmental Cognitive Neuroscience

From the perspective of developmental cognitive neuroscience, one of our main challenges is to understand the role of brain function and development in the mapping from genotype to behavioral phenotype (see Pennington, 2001, 2002). Several strategies are currently being employed to understand this mapping.

First, the extent of genetic influences on behavioral or neural phenotypes varies. Twin studies have been used to estimate the heritability of traits. These studies are based on the premise that monozygotic twins are genetically identical, while dizygotic twins are like ordinary siblings and share on average 50% of their genetic variation. However, both twin sets share the same environment, their mother's womb, their family, their neighborhood, and most often their school. If both monozygotic and dizygotic twin pairs resemble each other closely on a trait (there is concordance), it must be due to shared environmental influences. If monozygotic twins resemble each other more than dizygotic twins, i.e., they have a higher concordance rate, then genetic factors must play a role. This approach has shown that all psychological traits show substantial genetic influence, typically in the range of 30–50% but that no traits are 100% heritable (Plomin et al., 2016). Twin studies have further provided evidence that brain structure is under strong genetic control (Blokland et al., 2012) and that heritability is also substantial for brain function measures (e.g., Blokland et al., 2011). While twin studies have limitations, they give evidence that understanding how the genotype leads to observed behavioral phenotype may help us understand the source of individual differences over the course of development.

One direction that has been taken to tackle this question is to focus on cases in which the pathway from gene to brain to behavior is reduced in length and/or complexity. The general strategy here is to study aspects of the brain that are only a few chemical steps away from the direct products of gene expression. Building an organized network of neurons involves a highly complex orchestration of many different genes. However, simple chemicals that modulate neural activity and development, such as neurotransmitters (see Chapter 4), are much closer to the basic chemical building blocks that genes directly produce (proteins and monoamines). A large variety of simple molecules play important roles in brain development and function and these are the subject of much research in developmental neuroscience (see Stanwood & Levitt, 2008, for review). However, we now also have a few examples related to human cognitive development.

A second approach to understanding the pathway to emerging brain function is to study the naturally occurring individual differences among individuals in genetic, brain function, and behavioral measures. The genetic contribution to these differences between individuals comes not from a genetic defect, but from slightly different forms of the same gene (called alleles), whereby single nucleotides (A, C, G, T) differ between individuals. There are often several different variations in the exact coding sequences of genes, called single nucleotide polymorphisms (SNPs). Some of these variants are common in the population, others are rare. Since humans have two copies of every gene, these variations can either be the same for both copies (homozygous) or different (heterozygous). Alleles can also be found in the non-coding (junk) DNA mentioned earlier. The field of behavior(al) genetics is partly based on associations between individual differences in alleles and variations in behavior or personality. The twin studies design described above is an example of behavioral genetics. Another approach is to collect individuals' DNA and test for associations between genetic variation and phenotypic variation. Earlier attempts focused on a few specific SNPs in theoretically driven candidate gene studies. In contrast, more recent genome-wide association (GWA) is an exploratory approach which considers 600,000 to 2 million SNPs covering all common genetic variation. A key finding from this research is that the high heritability of psychological and brain phenotypes observed in twin studies is caused by a combination of a large number of variations in many genes, each having a small effect (polygenicity). Another finding is that phenotypic correlations between cognitive abilities (e.g., between intelligence, mathematics, and reading) are substantially mediated by genetic factors which have been called generalist genes (Plomin et al., 2016), and there is pleiotropy, with no one-to-one mapping between genes and neural systems or behavior (e.g., Meyer-Lindenberg & Weinberger, 2006).

These approaches are merging with developmental cognitive neuroscience as investigators try to correlate allelic variation with variation in cognition, brain structure, and brain function during development. It is assumed that by directly relating genetic variation to brain individual differences we can avoid the potential complicating factor of the different ways in which brain function relates to cognition, and thence to actual behavior. In this way it is hoped that even stronger correlations will be observed than those between genes and measures of behavior (Fan et al., 2003). In later chapters, we will provide an overview of some initial attempts to use the individual differences genetic approach in developmental studies with infants and children.

Further Reading   Plomin et al. (2016); Meyer-Lindenberg & Weinberger (2006).

The third approach to bringing genetics into developmental cognitive neuroscience involves the study of human syndromes and developmental disorders such as those briefly described in the last chapter. While some syndromes, such as autism, have a complex genetic cause involving many genes each having only a small effect (another example of polygenicity), there are other syndromes in which the genetic basis is better defined. One such syndrome is Fragile-X, which involves an allele of the fragile X mental retardation 1 (FMR1) gene on the X chromosome (a singular piece of DNA). Simply put, this gene normally involves between 6 and 55 repeats of a genetic coding sequence. However, in affected families the number of repeats can increase between generations until the point where a baby is born with more than 230 repeats in FMR1. At this point the structure becomes unstable and ceases to function, giving Fragile-X syndrome its name. This impairment of the functioning of the gene results in a lack of production of a particular protein (FMR1 protein).

Since males only have one X chromosome, compared to the two possessed by females, Fragile-X has clearer and more devastating effects in men. In addition to a variety of physical symptoms like an elongated face and flat feet, the syndrome as manifest in males includes some autistic-like symptoms, such as hand flapping and atypical social development, with some Fragile-X individuals meeting the diagnostic criteria for autism. One of the knock-on consequences of the lack of the FMR1 protein is a disturbance in the neurotransmitter called glutamate (see Chapter 4). Current research is attempting to relate the functioning of this neurotransmitter in the brain to some of the cognitive and behavioral differences observed in Fragile-X. Although much research still needs to be done, Fragile-X is perhaps currently the best understood example of the complex pathway from genetic abnormality to cognition (see Figure 3.2). Nevertheless, although Fragile-X involves only one gene, the effects of the defect are widespread and the syndrome involves many different aspects of cognition and behavior.

> **Further Reading**   Cornish & Wilding (2010).

A variant of this general approach is to start with syndromes known to have uneven cognitive or behavioral profiles, before working "backward" to discover their genetic basis. As mentioned earlier, examples of this approach include the search for a genetic basis for autism and for dyslexia. While these approaches have been partially successful, it may be that a closer two-way interaction between geneticists and psychologists that focuses on rare cases will be more successful. For example, as stated in the last chapter, Williams syndrome (WS) results from a missing segment of 23–28 genes on one chromosome. People with this so-called "microdeletion" normally show the full WS physical and cognitive phenotype, including relative strengths in the domains of language and face recognition and serious impairments in visuospatial construction abilities and number skills (Chapters 7, 9, and 11). Research has involved the investigation of individual cases in which not all of the WS genes are deleted: so-called "partial deletion" patients. Some of these cases have the full WS phenotype, some show only a few of the atypicalities, and others show none. By studying in detail the profile of cognitive abilities of such rare cases, a closer understanding of the combinations of genes that contribute to a behavioral phenotype can be reached (Karmiloff-Smith et al., 2003).

> **Further Reading**   Karmiloff-Smith (2008); Welsh et al. (2008).

A fourth approach to beginning to understand the complex role of genes in functional brain development is to engineer animal models, most commonly mice, in which particular genes are knocked out. With regard to these "knockout" mouse models (see Chapter 2), it is important to note that it is not sufficient for understanding the role of a gene(s) in a developmental sequence to demonstrate that knocking out the gene in question eliminates

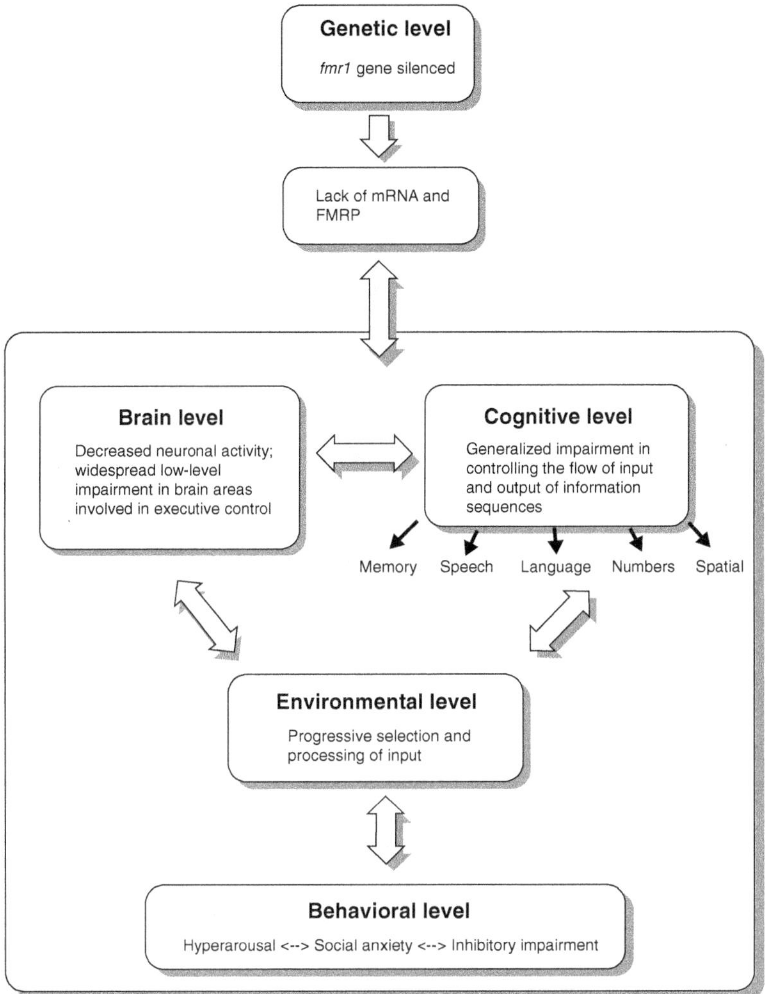

Figure 3.2 An illustration of the complex causal pathway between a genetic level defect and its consequences for behavior from Fragile-X syndrome.

a particular behavior in later life. In these experimental cases, and in human developmental disorders with an identifiable genetic abnormality, we may say that the gene defect *causes* the later deficit. However, this does not allow us to infer that the gene *codes for* the aspect of behavior that is disrupted, or that this aspect of behavior is the "function" of the gene, for all the reasons discussed above.

Finally, it is possible that the genes that play a role in brain plasticity and transmission in synapses may be most amenable to the study of their cognitive and behavioral consequences. In these cases, we can study changes in the expression of genes that take place even in adult animals. One such example is the so-called "immediate early" genes. This special class of genes is involved in rapid plastic changes in the brain that occur during both development and adult learning. In addition to molecular biological studies of these types of genes, in the future one approach to understanding the contribution of the expression of different genes may be through detailed cellular neural network models simulating the effects of gene expression which can then be compared to real neurobiological systems.

The Epigenome

The fact that early sensory experiences can have life-long effects through permanent changes in the timing and amount of different proteins expressed by the genes has led to the newly emerging field of "epigenetics," and attempts to characterize the human epigenome—a dynamic map of how different genes are expressed over developmental time. At first sight, the role of epigenetic processes in the construction of a human brain seems under-constrained given these deep and powerful mechanisms of latent plasticity. This raises the question of what factors constrain patterns of expression of the genome to give sufficient "stability" to reliably result in the typical complexity and structure of the human brain. This key question is currently the focus of much research, and a number of factors that constrain the epigenome are beginning to be identified.

Recent papers have described some of the complexity and dynamics of gene expression derived from studying developing and adult postmortem human brains (Colantuoni et al., 2011; Kang et al., 2011). From these analyses it turns out that 90% of the brain-related genes analyzed were differentially regulated across either different brain regions or different points in developmental time. Interestingly, the majority of this differential expression occurs during prenatal development in humans, with patterns of expression tending to become more fixed with increasing age (Kang et al., 2011).

One of the mechanisms that promote stability within the dynamic epigenome is genomic imprinting. This is a process through which certain genes (less than 1% in mammals) are expressed according to the parent-of-origin of that variant of the gene. Effectively, genomic imprinting silences the allele from either the mother or the father, leaving the remaining one to be exclusively expressed (Keverne et al., 1996). These epigenetic marks are present from the outset of an embryo and can be maintained throughout the life span. Some developmental disorders, such as Angelman syndrome and Prader-Willi syndrome, are associated with deficits in this process.

> **Further Reading**   Zhang & Meaney (2010).

The FOXP2 Gene

One example of a gene that has been related to cognition and development is the FOXP2 gene (officially known as "Forkhead Box P2"). This is perhaps the first celebrity gene as it has been the target of many studies, theoretical discussions, and media reports (see also Chapter 9). The story of research on FOXP2 illustrates several of the points made above, and particularly the need to beware of claims that genes "code for" specific aspects of behavior or cognition. The interest of psychologists in the FOXP2 gene began with the discovery of an unfortunate family (the KE family) in which three generations of family history indicated that about half of its members had an inherited language impairment that was initially suggested to be specific to certain aspects of grammar. When the gene associated with this deficit in the family was discovered to be FOXP2 (Lai et al., 2001), popular science writers and some of the media hailed the discovery of "the gene (that codes) for grammar."

As you will read in Chapter 9, research over the past decade has shown these initial claims to be unfounded for several reasons. First, it quickly became evident that the deficits in the affected family members are much broader than grammar alone, and include coordinating complex mouth movements, timing of rhythmic motor movements, various aspects of language outside grammar, and low general intelligence (Vargha-Khadem et al., 1995). A second reason for questioning the specificity of FOXP2 is that it is also found in many other species that obviously do not possess human language skills, such a mice and birds. Nevertheless, knockout mice lacking FOXP2 function do have significantly reduced vocalizations, and expression of the gene is closely associated with vocal learning in songbirds. When this evidence is taken together with functional imaging studies of affected members of the KE family that show under-activation of language areas of the brain during verb generation and repetition tasks (Liégeois et al., 2003), the overall findings suggest that FOXP2 is one of the genes important for learning to control rapid movement sequences. A final reason for questioning the specificity of the function of FOXP2 in relation to human language is that it is highly conserved in evolution. In addition to being found in mice and birds, the gene can also be found in reptiles. Like all genes, FOXP2 shows minute changes in the amino acids that compose it over evolutionary time. While some of these small changes could potentially affect the functioning of the gene, it seems more likely that any change in the function of the gene arises from it being active at different points in development or in different places in the developing brain (Carroll, 2005).

Further Reading   Marcus & Fisher (2003); Karmiloff-Smith (2008).

Given that FOXP2 appears to be involved in the brain plasticity associated with vocal learning, what exactly is its role? While this is the subject of much current research, it is important to note that FOXP2 is one of a family of so-called "transcription factors." In other words, its molecular function is to transcribe other parts of the DNA into the RNA that makes proteins. This makes the gene well placed to potentially orchestrate the action of a number of other genes. As we will discover in Chapter 9, it is very likely that human language involves many different genes, each of which has only a small effect on the overall outcome. Similarly, FOXP2 has multiple roles in different organs that may differ somewhat between species and also change with developmental time within the same species.

The story of FOXP2 illustrates some of the complexities and challenges that face the researcher interested in integrating genetics into developmental cognitive neuroscience. Since genes are pleiotropic (genes expressed in the brain are nearly always also expressed in other parts of the body), developmental disorders of genetic origin will inevitably be systemic (found throughout the body). For example, in WS heart defects are the most common cause of diagnosis, and in several other syndromes immune system problems have been reported. Thus, while we should not expect to find simple and direct mappings between particular genes and specific aspects of behavior, cognition, or brain function, there is no doubt that the powerful new methods for analyzing segments of the epigenetic pathway from gene to behavior will open new doors of discovery about the emergence of human brain functions.

Key Issues for Discussion

- How is the study of the role of genes in functional brain development limited by the methods that we currently have available?
- Compare and contrast different strategies for relating genes to cognitive development and ability.
- Can genetic evidence be useful for revealing the role of experience in a developmental transition in cognition or behavior?
- Is it useful to associate particular genes with the emergence of specific cognitive functions?

4

Building a Brain

This chapter describes several aspects of the pre- and postnatal development of the brain, with specific reference to humans. We begin with a basic overview of primate brain anatomy, with specific emphasis on arguably the most important structure of the brain for understanding cognitive development, the cerebral neocortex. We then begin our survey of development by outlining some of the key stages of prenatal brain growth, focusing on the birth, migration, and differentiation of cells that subsequently compose particular brain structures. The most obvious manifestation of postnatal development of the human brain is its fourfold increase in volume between birth and the teenage years. We trace the factors that give rise to this dramatic change. We will discover that the change in volume is mainly due to increases in nerve fiber bundles and myelination, rather than due to the addition of new neurons. A surprising aspect of brain development is that some measures of structural and neurophysiological brain development, such as the density of synaptic contacts, show a characteristic "rise and fall" during postnatal life.

The following section addresses the question of the extent to which the differentiation of the neocortex into its well-known areas or regions is prespecified. The "protomap" hypothesis states that the areal differentiation of the cortex is determined by intrinsic molecular markers (and/or prespecification of the proliferative zone in which neurons are born). In contrast, the "protocortex" hypothesis suggests that an initially undifferentiated protocortex is constructed and then divided up largely as a result of input through projections from the thalamus and activity-dependent modifications. Currently available evidence supports a middle-ground view in which large-scale regions are prespecified, while small-scale functional areas can require activity-dependent processes. This implies that the location and connectivity of individual neurons influences their subsequent specialization.

The next section focuses on a clear area of difference between human cortical development and that of other primates: our very extended period of postnatal development. This greatly extended period reveals two differential aspects of cortical development not as clearly evident in other primates: an inside-out pattern of development of layers, and differences in the timing of development across regions. These differential aspects of

Developmental Cognitive Neuroscience: An Introduction, Fifth Edition. Michelle de Haan, Iroise Dumontheil, and Mark H. Johnson.
© 2023 John Wiley & Sons Ltd. Published 2023 by John Wiley & Sons Ltd.
Companion website: www.wiley.com/go/johnson/devneuro5e

human cortical development provide the basis for associations between brain and cognitive development described in later chapters. The chapter concludes with an overview of the postnatal development of some subcortical structures, and our current knowledge of the development of neurotransmitters and modulators. The developmental levels of several neurotransmitters mirror aspects of the differential structural development of cortex, and some neurotransmitters may change their function during development.

An Overview of Primate Brain Anatomy

This book is written with the assumption that the reader has some basic introductory knowledge of the brain. Nevertheless, in order to ensure sufficient knowledge for the reader to follow this and later chapters we need to recap on some basic facts about the brains of primates, including our own species. The brains of all mammals follow a basic vertebrate brain plan that is found even in species such as salamanders, frogs, and birds. The major difference between these species and higher primates is in the dramatic expansion of the overlying cerebral cortex, together with associated structures such as the basal ganglia. Human brain development follows closely the sequence of events observed in other primates, albeit on a slower time schedule (something we will return to later in this chapter).

The neocortex of all mammals, including humans, is basically a thin (about 3–4 mm) flat sheet. Although complex, its general layered structure is relatively constant throughout its extent (see Figure 4.1). The rapid expansion in the overall size of the cortex during evolution has resulted in it becoming increasingly intricately folded with various indentations (sulci) and lobes (gyri). For example, the area of the cortex in the cat is about $100\,\text{cm}^2$, whereas that of the human is about $2400\,\text{cm}^2$. This suggests that the extra cortex possessed by primates, and especially humans, is related to the higher cognitive functions they possess. However, the basic relations between principal structures of the brain remain similar from mouse to human.

Most of the sensory inputs to the cortex pass through a structure known as the thalamus. Each type of sensory input has its own particular nuclei within this region. For example, the lateral geniculate nucleus (LGN) carries visual input to the cortex, while the medial geniculate nucleus (MGN) carries information from the auditory modality. Because of the crucial role of the thalamus in mediating inputs to the cortex, some have hypothesized that it also plays a crucial role in cortical development—an idea that will be discussed at greater length later. The flow of information between thalamus and cortex is not unidirectional, however, since most of the projections from lower regions into the cortex are matched by projections from the cortex back down. Some output projections from the cortex pass to regions that are believed to be involved in motor control, such as the basal ganglia and the cerebellum. However, most of the projections from the cortex to other brain regions terminate in roughly the same regions from which projections arrived (such as the thalamus). In other words, the flow of information to and from the cortex is largely bidirectional. For this reason, it is important not to confuse the terms "input" and "output" with "sensory" and "motor." All sensory and motor systems make extensive use of both input and output fibers, with information passing rapidly in both directions along collateral pathways.

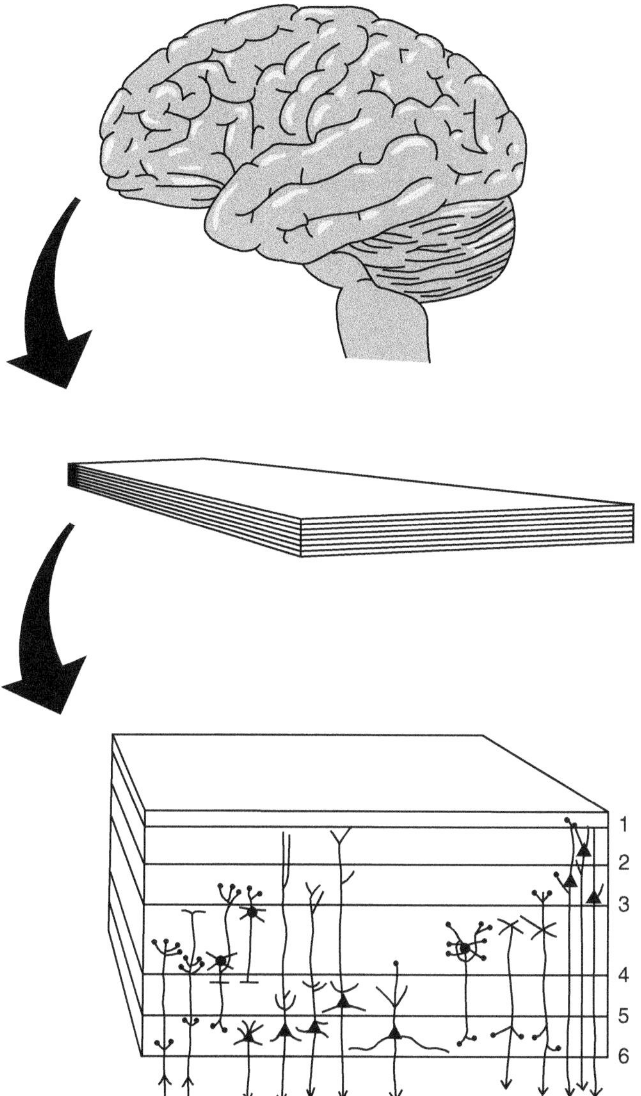

Figure 4.1 A simplified schematic diagram which illustrates that, despite its convoluted surface appearance (top), the cerebral cortex is a thin sheet (middle) composed of six layers (bottom). The convolutions in the cortex arise from a combination of growth patterns and the restricted space inside the skull. In general, differences between mammals involve the total area of the cortical sheet, and not its layered structure. Each of the layers possesses certain neuron types and characteristic input and projection patterns (see text).

The brain has two general types of cells, neurons and glial cells. Glial cells are more common than neurons, but are generally assumed to play no direct role in cognition and computation. However, as we shall see later, they play a very important role in the development of the cortex. The computational unit of the brain has long been thought to be the neuron (Shepherd, 1972). Neurons come in many shapes, sizes, and types, each of which presumably

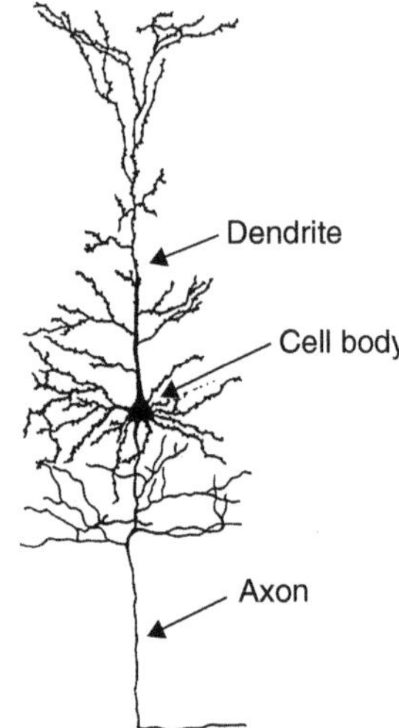

Figure 4.2 A typical cortical pyramidal cell. The apical dendrite is the long process that extends to the upper layers and may allow the cell to be influenced by other neurons. An axon projects to subcortical regions.

reflects their particular function. There are at least 25 different neuronal types within the cortex, although several of these types are relatively rare, and some are restricted to particular layers. About 80% of neurons found in the cortex are *pyramidal cells*, so called because of the distinctive pyramid shape of the cell body produced by the very large apical dendrite (input process), which generally runs tangential to the surface of the cortex (Figure 4.2). These are the neurons whose long axons (output processes) are so often found in the fibers feeding into other cortical and subcortical regions. While pyramidal cells are found in many cortical layers (generally they are larger in the lower layers and smaller in the upper layers), their apical dendrites often reach into its top layer, layer 1 (see below). This long apical dendrite allows the cell to be influenced by large numbers of cells from other (more superficial) layers and regions. This may be a useful feature as the pyramidal cell is a very stable and inflexible class of cell whose output is modulated by groups of more plastic and flexible inhibitory regulatory neurons. Figure 4.1 also shows a schematic section through an area of primate cortex cut at right angles to the surface of the cortex, revealing the layered structure. We will refer to this as the *laminar* structure of the cortex. Each of the laminae has particular cell types within it, and each layer has typical patterns of inputs and outputs.

Most areas of the neocortex (in all mammals) are made up of six layers. The basic characteristics that define each layer appear to hold in most regions of the cortical sheet. Layer 1 has few cell bodies. It is made up primarily of long white fibers running along the horizontal surface, linking one area of cortex to others some distance away. Layers 2 and 3 also contain horizontal connections, often projecting forward from small pyramidal cells to neighboring areas of cortex. Layer 4 is the layer where most of the input fibers terminate and it contains a high proportion of spiny stellate (star-shaped) cells on which these projections terminate. Layers 5 and 6 have the major outputs to subcortical regions of the brain. These layers contain a particularly high proportion of large pyramidal cells, with long descending axons. There are also many neurons involved in intrinsic cortical circuits.

Although this basic laminar structure holds throughout most of the neocortex, there are some variations. For example, the input layer (layer 4) is particularly thick and well developed in primary sensory cortex. Indeed, in the visual system, it is possible to distinguish at least four "sublayers" within layer 4. Conversely, layer 5 (one of the output layers) is particularly well developed in motor cortex, presumably due to its importance in sending output signals from the cortex. It is also clear that different parts of the cortex have different projection patterns from other parts of the cortex. While there may only be a small number of these characteristic projection patterns from one region of cortex to another, there is no single pattern that can be said to be characteristic of all cortical regions. Hence this is another dimension of variation that can contribute to regional specialization within the cortex. Further dimensions of difference include the presence of particular neurotransmitters, the chemicals that transmit across the synapse, and the relative contribution of excitatory compared to inhibitory neurotransmitters. Finally, regions may vary in the timing of key developmental events, such as the postnatal reduction in the number of synapses (Huttenlocher, 1990).

Prenatal Brain Development

The sequence of events during the prenatal development of the human brain closely resembles that of many other vertebrates. Shortly after conception a fertilized cell undergoes a rapid process of cell division, resulting in a cluster of proliferating cells (called the blastocyst) that somewhat resembles a bunch of grapes. Within a few days, the blastocyst differentiates into a three-layered structure (the embryonic disk). Each of these layers will further differentiate into a major organ system. The endoderm (inner layer) turns into the set of internal organs (digestive, respiratory, etc.), the mesoderm (middle layer) turns into the skeletal and muscular structures, and the ectoderm (outer layer) gives rise to the skin surface and the nervous system (including the perceptual organs).

The nervous system itself begins with a process known as *neurulation*. A portion of the ectoderm begins to fold in on itself to form a hollow cylinder called the *neural tube*. The neural tube differentiates along three dimensions: length, circumference, and radius. The length dimension gives rise to the major subdivisions of the central nervous system, with the forebrain and midbrain arising at one end and the spinal cord at the other. The end which will become the spinal cord differentiates into a series of repeated units or segments, while the front end of the neural tube organizes differently with a series of bulges and convolutions forming (see Figure 4.3). By around 5 weeks after conception these bulges can be identified as protoforms for major components of the mammalian brain. Proceeding from front to back: the first bulge gives rise to the cortex (telencephalon), the second gives rise to the thalamus and hypothalamus (diencephalon), the third turns into the midbrain (mesencephalon), and others to the cerebellum (metencephalon) and to the medulla (myelencephalon).

The circumferential dimension (tangential to the surface) in the neural tube is critical, because the distinction between sensory and motor systems develops along this dimension: dorsal (top-side) corresponds roughly to sensory cortex, ventral (bottom-side) corresponds to motor cortex, with the various association cortices and "higher" sensory and motor cortices aligned somewhere in between. Within the brain stem and the spinal cord, the corresponding alar (dorsal) and basal (ventral) plates play a major role in the organization of nerve pathways into the rest of the body.

Differentiation along the radial dimension gives rise to the complex layering patterns and cell types found in the adult brain. Across the radial dimension of the neural tube the bulges grow larger and become further differentiated. Within these bulges cells *proliferate* (are born), *migrate* (travel), and *differentiate* (change form) into particular types. The vast majority of the cells that will compose the brain are born in the so-called *proliferative zones*. These zones are close to the hollow portion of the neural tube (which subsequently becomes the ventricles of the brain). The first of these proliferation sites, the *ventricular zone*, may

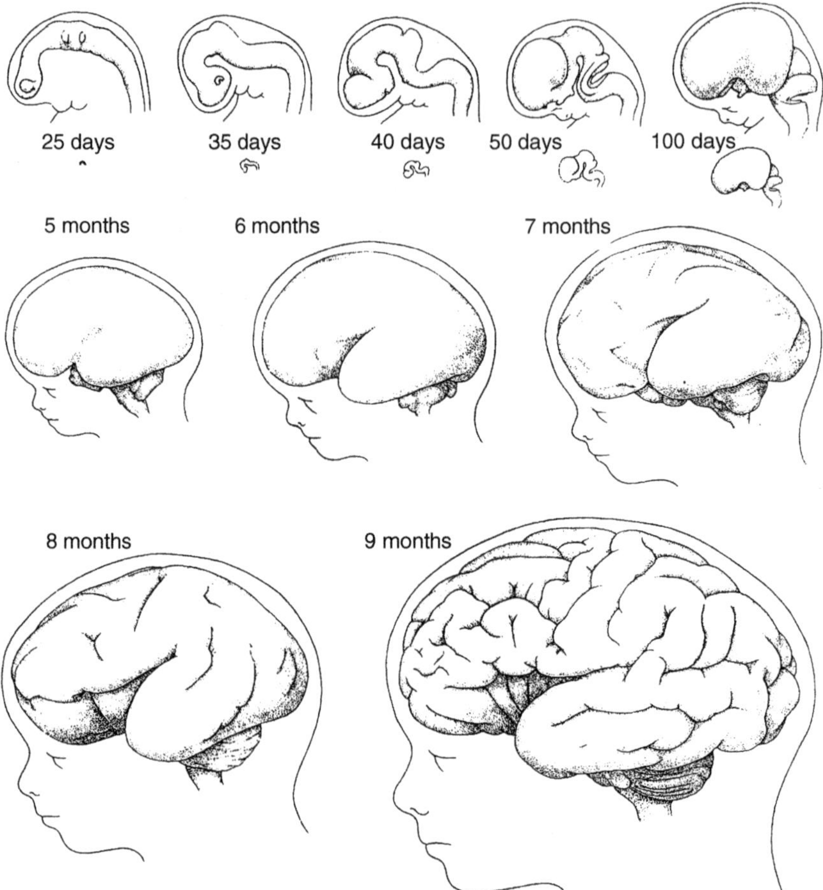

Figure 4.3 A sequence of drawings of the embryonic and fetal development of the human brain. The drawings of brains beneath those of 25–100 days are the same images but drawn to the same scale as those in the row below. The forebrain, midbrain, and hindbrain originate as swellings at the head end of the neural tube. In primates, the convoluted cortex grows to cover the midbrain, hindbrain, and parts of the cerebellum. Prior to birth, neurons are generated in the developing brain at a rate of more than 250,000 per minute.

be phylogenetically older (Nowakowski, 1987). The second, the *subventricular zone*, only contributes significantly to phylogenetically recent brain structures such as the neocortex (i.e., "new" cortex). These two zones yield separate glial (support and supply cells) and neuron cell lines and give rise to different forms of migration. But first we will consider how young neurons are formed within these zones.

Further Reading   Stiles et al. (2015); White & Hilgetag (2008); Sanes et al. (2006).

Neurons and glial cells are produced by division of proliferating cells within the proliferative zones to produce clones (a clone is a group of cells which are produced by division of a single precursor cell—such a precursor cell is said to give rise to a lineage). *Neuroblasts* produce neurons, and *glioblasts* produce glial cells. Each of the neuroblasts gives rise to a definite and limited number of neurons, a point to which we will return later. In at least some cases, particular neuroblasts also give rise to particular types of cells. For example, less than a dozen proliferating cells produce all the Purkinje cells of the cerebellar cortex, with each producing about 10,000 cells (Nowakowski, 1987).

After young neurons are born, they have to travel or *migrate* from the proliferative zones to the particular region where they will be employed in the mature brain. There are two forms of migration observed during brain development. The first, and more common, is *passive cell displacement*. This occurs when cells that have been generated are simply pushed farther away from the proliferative zones by more recently born cells. This form of migration gives rise to an "outside-in" pattern. That is, the oldest cells are pushed toward the surface of the brain, while the most recently produced cells remain closer to their place of birth. Passive migration gives rise to brain structures such as the thalamus, the dentate gyrus of the hippocampus, and many regions of the brain stem. The second form of migration is more active and involves the young cell moving past previously generated cells to create an "inside-out" pattern. This pattern is found in the cerebral cortex and in some subcortical areas that have a laminar structure (divided into parallel layers).

It is important to emphasize that prenatal brain development is not just a passive process involving the unfolding of genetic instructions. Rather, from an early stage interactions between cells are critical, including the transmission of electrical signals between neurons. In one example, patterns of spontaneous firing of cells in the eyes (before they have opened in development) transmit signals that appear to specify the layered structure of the lateral geniculate nucleus (see O'Leary & Nakagawa, 2002; Shatz, 2002) and prepare the visual system for the appropriate flows of visual input when the animal later moves through their environment (Ge et al., 2021). Thus, waves of firing neurons intrinsic to the developing organism may play an important role in specifying aspects of brain structure before sensory inputs from the external world have any effect.

Further Reading   Shatz (2002); Ge et al. (2021).

While there is evidence for the emergence of structure within a brain region resulting from intrinsic waves of neural activity, is there evidence that regions begin to talk to each other to start to form coherent networks? Resting-state networks are the spontaneous and intrinsic brain activities that occur in the absence of any overt task, somewhat like the "idle" state of a car engine, and can be studied in infants while they are still, using functional magnetic resonance imaging (MRI). Researchers have studied the very early development of these resting-state networks in babies who were born preterm (Doria et al., 2010). The networks were identifiable but often fragmentary in infants at 30 weeks of gestational age, but by full term (40 weeks gestational age) complete adult-like networks were present (Ball et al., 2014). These results suggest that brain regions already begin to form coherent networks during the period of rapid brain growth in the third trimester.

Postnatal Brain Development

As mentioned earlier, there is a dramatic increase in the total volume of the brain from birth to teenage years (see Figure 4.4). What factors contribute to this developmental change? Using various techniques, this question can be examined at different levels of detail from microscopic (and electron microscope) changes in neurons and synapses, to the larger scale of divisions of the brain into gray (neurons and their local connections) and white matter (myelinated fiber bundles). Beginning at the microscopic scale, a number of measures of brain anatomy and function show a characteristic "rise and fall" developmental pattern during postnatal life. While the progressive and regressive processes should not be viewed as distinct stages, for the purposes of exposition we will discuss them in sequence.

At first consideration, people often assume that the postnatal increase in the size of the brain is due to the addition of new neurons. However, it turns out that with a few

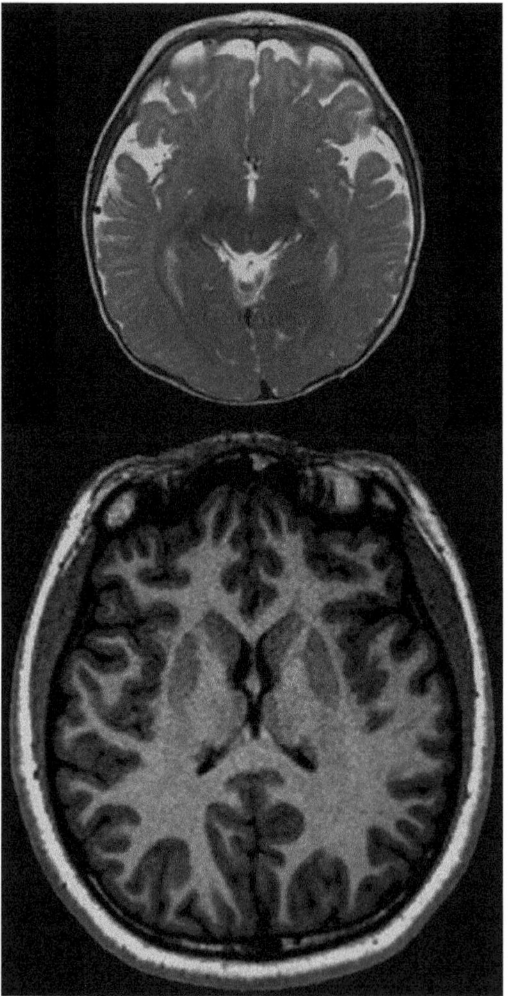

Figure 4.4 MRI structural scans of a 4-month-old infant (top) and a 12-year-old adolescent (below). *Source:* Centre for Neuroimaging Sciences, King's College, London and the Birkbeck-UCL Centre for Neuroimaging.

exceptions this is generally not the case. The formation of neurons and their migration to appropriate brain regions takes place almost entirely within the period of prenatal development in the human. Although there may be some comparatively small-scale addition of neurons in the hippocampus and parts of neocortex (see later), the vast majority of neurons are present by around the seventh month of gestation (Rakic, 1995). In contrast to the lack of new nerve cell bodies, there is, however, a dramatic postnatal growth of synapses, dendrites, and fiber bundles. Further, nerve fibers become covered in a fatty covering (myelin) that considerably adds to the bulk of the brain.

Using autopsy material, an exception to the view that new neurons are generally not added postnatally comes with the discovery of corridors of migrating immature neurons traveling from the subventricular zone to parts of the prefrontal cortex (PFC) in human infants under 18 months postnatal. This migratory pathway significantly decreases in childhood before becoming nearly extinct in adults (Sanai et al., 2011). This developmentally transient pathway of neurons does not follow the same principles as Rakic's radial unit model, potentially because these are inhibitory interneurons and not the more common pyramidal cells discussed earlier. Sanai et al. (2011) speculate that these pathways may provide underpinning mechanisms for delayed plasticity.

Turning back to more major factors in the growth of the brain, perhaps the most obvious manifestation of postnatal neural development as viewed through the standard microscope is the increase in size and complexity of the dendritic tree of most neurons. An example of the dramatic increase in dendritic tree extent during human postnatal development is shown in Figure 4.5. While the extent and reach of a cell's dendritic arbor may increase dramatically, it also often becomes more specific and specialized. Less apparent through standard microscopes, but more evident with electron microscopy, is a corresponding increase in measures of the density of synaptic contacts between cells.

Huttenlocher and colleagues have reported a steady increase in the density of synapses in several regions of the human cerebral cortex (Huttenlocher, 1990, 1994; Huttenlocher et al., 1982; Petanjek et al., 2011). While an increase in synapses (synaptogenesis) begins around the time of birth in humans for all cortical areas studied to date, the most rapid bursts of increase, and the final peak density, occur at different ages in different areas. In the visual cortex there is a rapid burst at 3 to 4 months, and the maximum density of around 150% of adult level is reached between 4 and 12 months. A similar time course is observed in the primary auditory cortex (Heschl's gyrus). In contrast, while synaptogenesis starts at the same time in a region of the PFC, density increases much more slowly and does not reach its peak until after the first year or in later childhood (Petanjek et al., 2011). (It should be noted at this point that there are a variety of possible measures of synaptic density—per cell, per unit dendrite, per unit brain tissue, etc. Careful selection of measures is required so that factors such as increases in dendritic length do not unduly influence the results.)

Further Reading Bourgeois (2001); Huttenlocher (2002); Kostovic et al. (2008); Petanjek et al. (2011).

Another additive process is myelination. Myelination refers to an increase in the fatty sheath that surrounds neuronal pathways, a process that increases the efficiency of

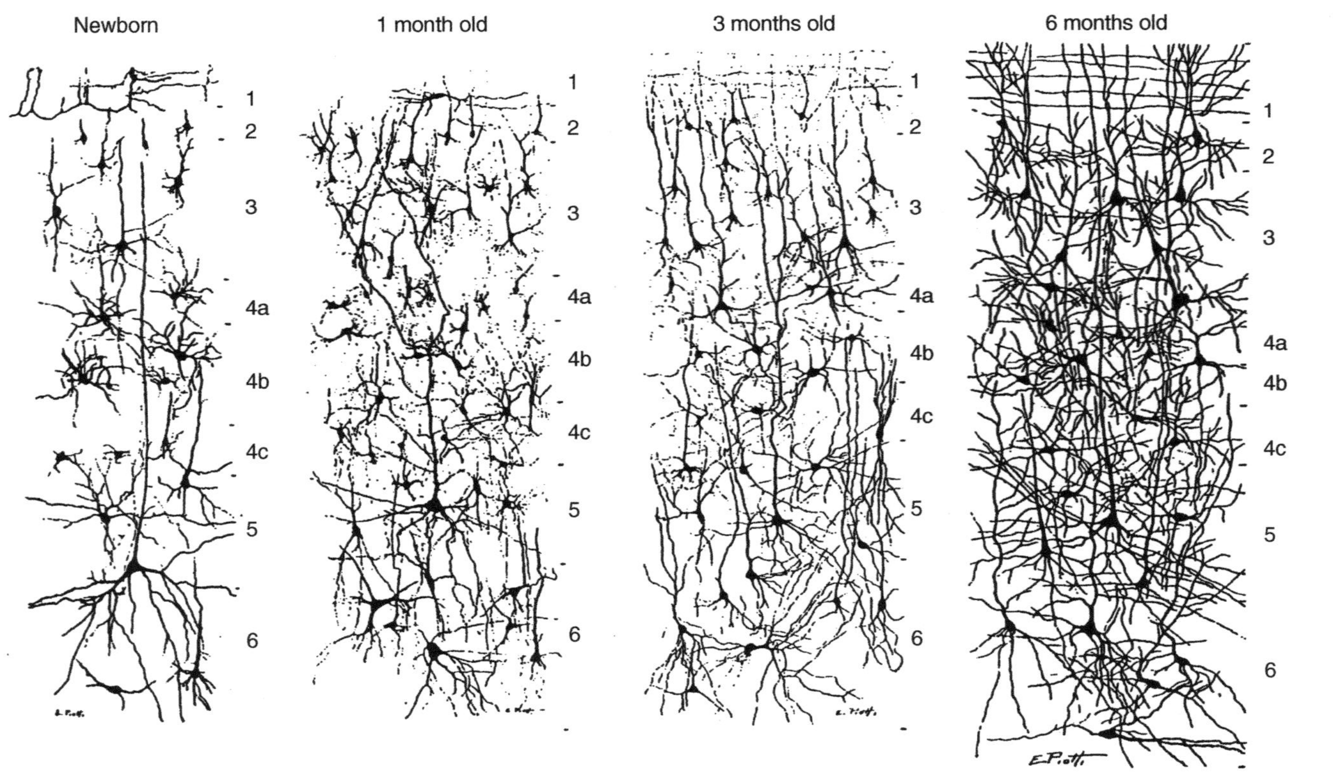

Figure 4.5 A drawing of the cellular structure of the human visual cortex based on Golgi stain preparations from Conel (1939–1967). Reprinted by permission of the publisher from The Postnatal Development of The Human Cerebral Cortex, Volumes I–VIII, by Jesse LeRoy Conel, Cambridge, Mass.: Harvard University Press, Copyright © 1939, 1941, 1947, 1951, 1955, 1959, 1963, 1967 by the President and Fellows of Harvard College. Copyright © renewed 1967, 1969, 1975, 1979, 1983, 1987, 1991.

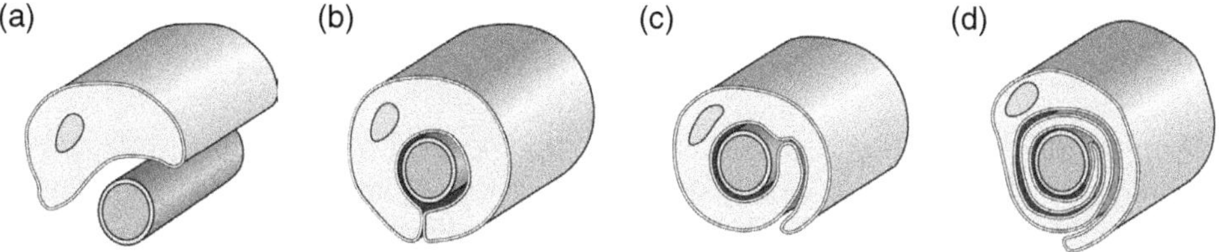

Figure 4.6 The sequence of axon myelination by an oligodendrocyte. (a–d) show the sequence of initial contact, then engulfing and surrounding the axon, followed by spiraling around the axon to form the final myelin sheath. Reprinted by permission of the publisher from *The Fundamentals of Brain Development: Integrating Nature and Nurture* by Joan Stiles, pp. 45, 290, Cambridge, Mass.: Harvard University Press, Copyright © 2008 by the President and Fellows of Harvard College.

information transmission (see Figure 4.6). Myelination greatly increases the speed of transmission of impulses (by as much as 100 times); however, under-myelinated connections in the young human brain are still capable of transmitting signals, and some connections in the adult brain never myelinate. In the central nervous system, sensory areas tend to myelinate earlier than motor areas. Cortical association areas are known to myelinate last, and continue the process into the second decade of life (Yakovlev & Lecours, 1967) (see later). Myelination plays a key role in coordinating the timing of neural processing (e.g., by allowing auditory signals from both ears to be integrated), and recent research suggests that myelination can be dynamically changed by neural activity and participates in nervous system plasticity during and beyond development (Chorghay et al., 2018; Kaller et al., 2017).

> **Further Reading**   Chorghay et al. (2018); Kaller et al. (2017); Klingberg (2008).

Another technique that has sometimes been used to study postnatal development is positron emission tomography (PET). A study of human infants using this technique (Chugani et al., 2002) reported a sharp rise in overall resting brain glucose metabolism (the uptake of glucose from the blood is essential for cell functioning) after the first year of life, with a peak approximately 150% above adult levels achieved somewhere around 4–5 years of age for some cortical areas. While this peak occurred somewhat later than that in synaptic density, an adult-like *distribution* of resting activity within and across brain regions was observed by the end of the first year. This result has been confirmed in functional MRI studies of brain resting-state networks in infants during sleep (Fransson et al., 2007; see also the studies by Ball et al., 2014; Doria et al., 2010). Fransson and colleagues observed several resting-state networks that encompassed cortical regions such as the primary visual cortex, bilateral sensory motor areas, bilateral auditory cortex, parietal cortex, and medial and dorsolateral PFC (see Figure 4.7 in the color plate section). Although these networks differed in some respects from those observed in adults, like the previously discussed research on prenatal resting functional magnetic resonance imaging (fMRI), it suggests that many regions of the infant cortex can be activated within somewhat coordinated networks from early on.

> **Further Reading**   Chugani et al. (2002).

We now turn to regressive events during human postnatal brain development. Such events are commonly observed by those studying the development of nerve cells and their connections in the brains of many animals (for reviews see Sanes et al., 2006). That processes of selective loss have a significant influence on postnatal primate brain development is evident from a number of quantitative measures. For example, in the PET study just mentioned the authors found that the absolute rates of glucose metabolism rise postnatally until they exceed adult levels, before reducing to adult levels after about 9 years of age for most cortical regions.

Consistent with these PET findings, Huttenlocher (1990, 1994) reports quantitative neuro-anatomical evidence from several regions of the human cortex that following the increase in density of synapses described above there is then a period of synaptic loss. Like the timing of bursts of synaptogenesis, and the subsequent peaks of density, the timing of the reduction in synaptic density varies between cortical regions. For example, synaptic density in the visual cortex returns to adult levels between 2 and 4 years, while the same point is not reached until between 10 and 20 years of age for regions of the PFC, or even later (Petanjek et al., 2011). Figure 4.8 in the color plate section illustrates some of these timing differences.

> **Further Reading**   Greenough et al. (2002); Petanjek et al. (2011).

One explanation for the decrease in glucose uptake observed in the PET studies is that it reflects the decrease in synaptic contacts. This hypothesis was investigated in a developmental study conducted with cats (Chugani et al., 1991). In this study, the peak of glucose uptake in cat visual cortex was found to coincide with the peak in overproduction of synapses in this region. However, when similar data from human visual cortex are plotted together (see Figure 4.9), it is apparent that the peak of glucose uptake lags behind synaptic density. An alternative to the hypothesis that reduction of metabolic activity is the result of the elimination of neurons, axons, and synaptic branches is that the same activity may require less "mental effort" once a certain level of skill has been attained.

Most of the developments in the brain discussed so far concern aspects of the structure of the brain. However, there are also developmental changes in what has been known as the "soft soak" aspects of neural function, molecules involved in the transmission and modulation of neural signals. While these will be discussed in more detail in a later section, it is interesting to note at this point that a number of neurotransmitters in rodents and humans also show the rise and fall developmental pattern (see Benes, 1994; Larsen & Luna, 2018, for review). Specifically, the excitatory intrinsic transmitter glutamate, the intrinsic inhibitory transmitter GABA (gamma-aminobutyric acid), and the extrinsic transmitter serotonin all show this same developmental trend.

> **Further Reading**   Berenbaum et al. (2003); Ojeda & Avila (2019).

Thus, the distinctive "rise and fall" developmental sequence is seen in a number of microscopic and metabolic measures of structural and neurophysiological development in

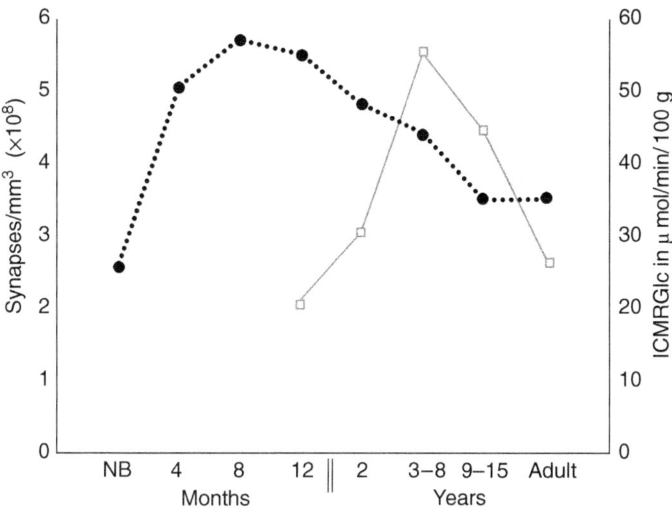

Figure 4.9 Graph showing the development of density of synapses in human primary visual cortex (dotted line: data taken from Huttenlocher, 1990), and resting glucose uptake in the occipital cortex as measured by PET (solid line: data taken from Chugani et al., 1987). ICMRGlc is a measure of the local cerebral metabolic rates for glucose.

the human cortex. Over past decades, a number of laboratories have developed MRI methods to study the structural development of the brain at a larger scale. As discussed in Chapter 2, MRI reveals brain structure at a more gross scale than neurons and synapses, but sufficient to allow the measurement of gray and white matter in different cortical and subcortical regions. Longitudinal studies suggest that cortical gray matter volume or thickness peak in early childhood, around 6 years of age for gray matter volume (Bethlehem et al., 2022; Gogtay et al., 2004; Mills et al., 2016; Zhou et al., 2015). There is considerable heterogeneity between different individuals and between different cortical regions (Brain Development Cooperative Group, 2012; Ojeda & Avila, 2019; Zhou et al., 2015). This "rise and fall" pattern is thought to reflect the pruning or elimination of excess connections between neurons, but could also reflect the cortex being compressed from increasing white matter volumes (Mills et al., 2016). Broadly consistent with earlier reports based on post-mortem neuroanatomical studies, primary sensory areas of cortex, along with the frontal and occipital poles, show the earliest decline in gray matter volume (see Figure 4.10 in the color plate section) (Bethlehem et al., 2022; Gogtay et al., 2004). The remainder of the cortex develops in a broad back-to-front direction, with the parietal lobes showing an earlier peak and decline of gray matter than the PFC and temporal lobes (Brain Development Cooperative Group, 2012; Shaw et al., 2008). The posterior superior temporal cortex, which is a critical part of the social brain network (see Chapter 7) and integrates information from different sensory modalities, could be one of the latest developing cortical regions according to this measure (Gogtay et al., 2004). Whether this sequence reflects the evolutionary sequence in which these structures evolved remains a controversial hypothesis.

Similar MRI data have been collected for the volume of white matter (myelinated fiber bundles), and this shows a general linear increase with age through to early adulthood

(Bethlehem et al., 2022; Mills et al., 2016). This increase in volume likely reflects increases in axon diameter, axonal packing, and myelination (Lebel & Deoni, 2018). Longitudinal diffusion tensor imaging studies have revealed that, as for gray matter, there are tract-specific patterns of developmental changes and significant variability between individuals (Lebel & Beaulieu, 2011). The volumes of subcortical brain regions also change during development; hippocampal and amygdala volumes increase and nucleus accumbens and globus pallidus volumes decrease between mid-childhood and late adolescence, while volumes in the caudate and putamen peak in late childhood/early adolescence before decreasing (Brain Development Cooperative Group, 2012; Goddings et al., 2014). Importantly, the time course of changes in subcortical region volumes differs between genders (Dennison et al., 2013) and reflects both chronological age and puberty stage, suggesting a sensitivity to the hormonal changes associated with puberty (Goddings et al., 2014). There is no consensus regarding whether developmental changes in brain structure are delayed overall in males compared to females, which would be consistent with the later onset of puberty in males but may reflect sex differences in physical size (see Mills et al., 2016, for discussion).

Further Reading   Mills et al. (2016).

Overall, a number of different measures and laboratories have found the rise and fall pattern for neurons and their local connectivity. However, it should be stressed that not all measures show this pattern (e.g., myelination, white matter), and measures such as synaptic density are static snapshots of a dynamic process in which both additive and regressive processes are continually in progress: in other words, there are probably not distinct and separate progressive and regressive phases.

There is considerable variation in brain structure and function in normal adult subjects. This variation at a large scale is driven by genetic influences. A meta-analysis of twin studies has found that total cerebral volume had 84% (confidence interval [CI]: 75–92%) heritability, with estimates of 72% for total gray matter volume and ranging between 60% and 74% for the volumes of each lobe and between 45% and 82% for subcortical volumes (Blokland et al., 2012). Heritability was even higher for total white matter volume (85%, CI: 82–88%); heritability of white matter volumes in each lobe ranged between 62% and 84%. While estimates are less precise because of lower sample sizes, heritability tends to be lower for more local phenotypic measures. For example, the right middle frontal gyrus shows a heritability of 49%. Lowest heritability was found in the uncus (0–2%) and parahippocampal gyri (20–28%) and cingulate cortices (27–28%) (Blokland et al., 2012). In a meta-analysis focusing on developmental populations, Maggioni et al. (2020) found that global brain morphology was heritable in neonates, but with estimates lower than those found in adults, higher heritability of white matter than gray matter, and greater heritability of gray matter volumes in posterior regions and visual and motor regions than frontal regions. A limitation of this work is that twins may have delayed brain maturation compared to singletons, and neonatal MRI is affected by motion artifacts. Studies in childhood and adolescence further show high heritability of global brain structure measures but a complex pattern of changes in genetic influences over the course of development and changes in inter-regional genetic correlations. The heritability of the morphology of frontal and subcortical regions, in particular the caudate, was found to become strong during adolescence and young adulthood (Maggioni et al., 2020).

Building on the evidence for individual variability in the rise and fall in the developmental trajectory of gray matter, Shaw et al. (2006) demonstrated from MRI images that it is the trajectory of change in cortical thickness, and not the thickness itself, that best predicts a measure of intelligence (IQ). In this study, more intelligent children (as assessed by IQ) went through a larger and clearer pattern of rise and fall in their cortical thickness between 7 and 19 years of age than did those with more average intelligence scores (see Figure 4.11 in the color plate section). This groundbreaking study suggested that differences in the dynamic changes that occur during development are critical for our understanding of individual differences in intelligence and cognition in adults.

Further Reading   Maggoni et al. (2020).

The Development of Cortical Areas: Protomap or Protocortex?

A longstanding debate among those who study the developmental neurobiology of the cortex concerns the extent to which its structure and function are prespecified, in the sense that they are the result of genetic, molecular, and cellular level interactions and not determined by the pattern of firing of neurons. As we described earlier, the cortex is a layered structure somewhat reminiscent of a cake composed of different layers of sponge, cream, and jam. Orthogonal to the laminar dimension of cortical structure is its differentiation into regions or areas. Returning to our layered cake, we can consider cortical differentiation in regions as being like the slices of a cake, some of which may contain thicker jam or cream layers. Figure 4.12 illustrates one of the best-known schemes for dividing the cerebral cortex into areas. In the adult primate, most of these cortical areas can be determined by very detailed differences in the laminar structure, such as the precise thickness of certain layers. Often, however, the borders between areas are indistinct and controversial. It is commonly assumed that these anatomically defined areas have particular unique functions. While this has proved to be the case for early sensory and motor areas, there are many cases of functional regions or borders that do not neatly correspond to known neuroanatomical divisions. It should be stressed that the division of the cortex into areas with differing functional specializations is not an exact science in that the detailed features of neuroanatomy relevant for supporting the functions of different areas remain largely unknown.

Despite these caveats, a century of neuropsychology has taught us that the majority of typical adults tend to have similar functions within approximately the same areas of cortex. This observation has led to a common assumption that the division of the cerebral cortex into structural and functional areas is tightly prespecified through genetics. However, as we will see, this assumption is, at best, only partially correct.

In the review that follows we will report evidence that several aspects of the structure of the cerebral cortex, including the general laminar structure and large-scale regions, do not require neural activity to be established. Crucially, however, much of the fine-scale division into functional areas involves activity-dependent processes. We begin by returning to the

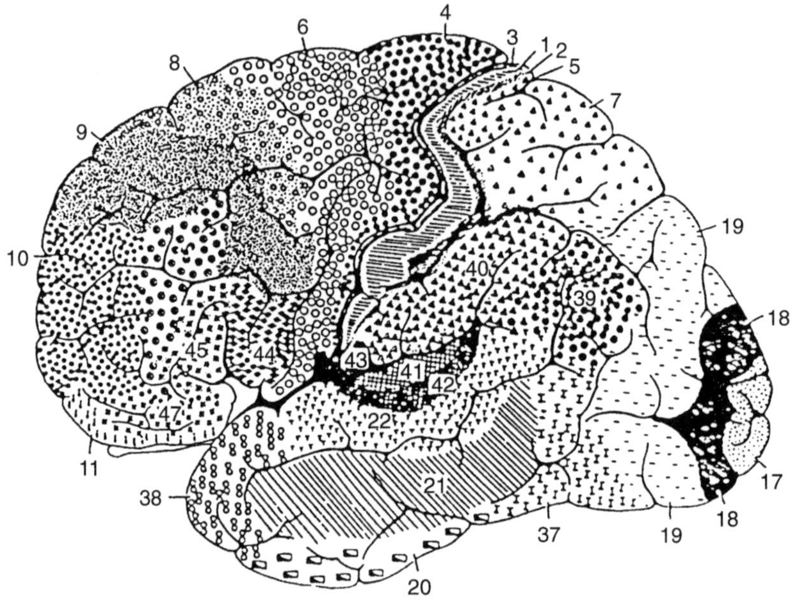

Figure 4.12 Cytoarchitectural map of the cerebral cortex. Some of the most important specific areas are as follows. Motor cortex: motor strip, area 4; pre-motor area, area 6; frontal eye fields, area 8. Somatosensory cortex: areas 3, 1, 2. Visual cortex: areas 17, 18, 19. Auditory cortex: areas 41 and 42. Wernicke's speech area: approximately area 22. Broca's speech area: approximately area 44 (in the left hemisphere).

prenatal development of cortex, and probably the most complete theory of the development of cortical structure: the radial unit model proposed by Pasko Rakic (1988).

As mentioned earlier, most cortical neurons in humans are generated outside the cortex itself in a region just underneath what becomes the cortex, the "proliferative zones." This means that these cells must migrate to take up their locations within the cortex. How is this migration accomplished? Rakic has proposed a "radial unit model" of neocortical differentiation that gives an account of how *both* the areal and the layered structure of the mammalian cerebral cortex arise (Rakic, 1988). According to the model, the laminar organization of the cerebral cortex is determined by the fact that each proliferative unit (in the subventricular zone) gives rise to about one hundred neurons. The progeny from each proliferative unit all migrate up the same radial glial fiber, with the latest to be born traveling past their older relatives. A radial glial fiber is a long process that stretches from top to bottom of the cortex and originates from a glial cell. Thus, radial glial fibers act like a climbing rope to ensure that cells produced by one proliferative unit all contribute to one radial column within the cortex. Rakic's proposed method of migration is illustrated in Figure 4.13.

Further Reading   Rakic (2002).

However, it has become apparent that not all migration within the cortex strictly follows the radial unit model (Polleux et al., 2002). Interspersed between pyramidal cells are a variety of types of interneurons that help balance excitation and inhibition within the cortex.

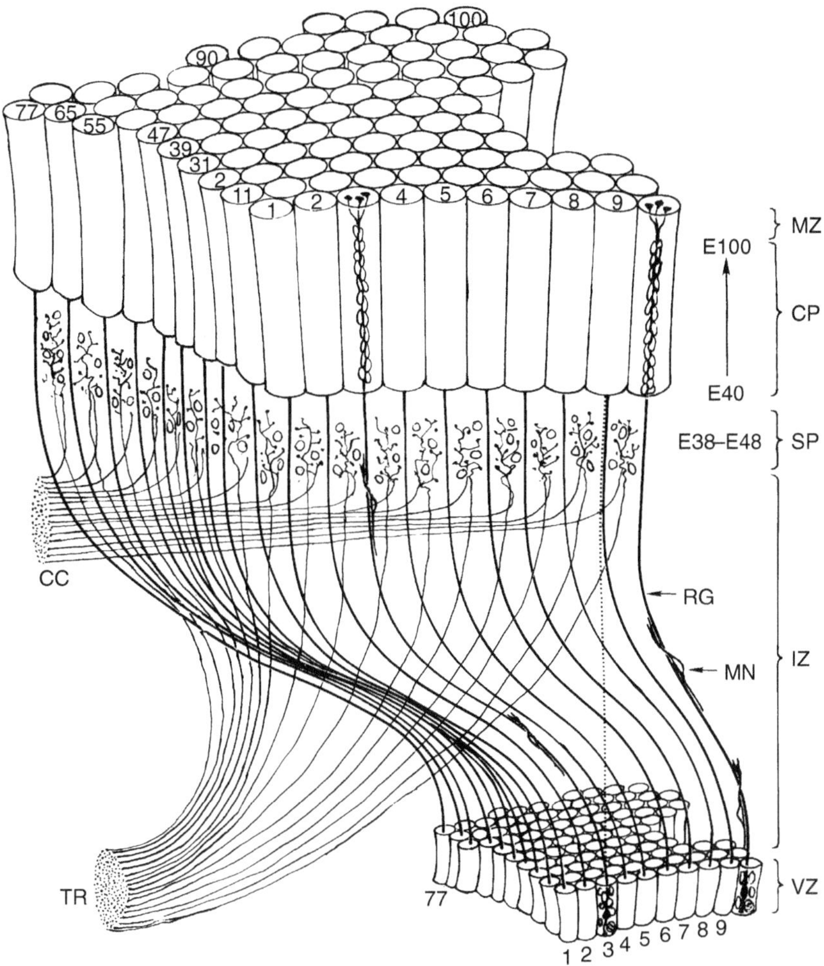

Figure 4.13 The radial unit model of Rakic (1987). Radial glial fibers span from the ventricular zone (VZ) to the cortical plate (CP) via a number of regions: the intermediate zone (IZ) and the subplate zone (SP). RG indicates a radial glial fiber, and MN a migrating neuron. Each MN traverses the IZ and SP zones that contain waiting terminals from the thalamic radiation (TR) and corticocortical afferents (CC). As described in the text, after entering the cortical plate, the neurons migrate past their predecessors to the marginal zone (MZ). Adapted from Goldman-Rakic, P.S. 1987.

Several psychiatric disorders, such as autism and schizophrenia, are thought to involve dysregulation of this balance (LeBlanc & Fagiolini, 2011; Lewis, 2012), and in some cases this may be due to differences in the molecular mechanisms known to control these migration processes (Polleux et al., 2002).

Nevertheless, Rakic's model explains how many cortical cells arrange themselves into the thickness of the cortex; but how does the differentiation into specific layers emerge? While we are unable to answer this question definitively at this point, it is likely that differentiation into particular cell types occurs *before* a neuron reaches its final location. That is, a cell is already developing into the neuron it will become (pyramidal, spiny stellate, etc.)

before it reaches its adult location within the cortex. For example, in genetic mutant "reeler" mice, cells that acquire inappropriate laminar positions within the cortex still differentiate into neuronal types according to their time of origin, rather than the types normally found at their new location. This implies that the information required for differentiation is present at the cell's birth in the proliferative zones; it is not dependent upon their distance from the proliferative zones, or on the characteristics of the neighborhood in which that cell ends up. That is, in the proliferative zones for the neocortex, cell types may be determined at the stage of division.

Although in many cases a cell's identity may be determined before it leaves the proliferative zones, some of the features that differentiate cell types may form later. For example, Marin-Padilla (1990) has proposed that the distinctive apical dendrite of pyramidal cells, which often reaches into layer 1, is a result of the increasing distance between this layer and other layers resulting from the inside-out pattern of growth. Specifically, the increasing separation between layer 1 and the subplate zone which results from young neurons moving into what will become layers 2–6 means that cells which have their processes attached to layer 1 will become increasingly "stretched"—that is, their leading dendrite will become stretched tangential to the surface of the cortex, resulting in the elongated apical dendrite so typical of cortical pyramidal cells. As mentioned earlier, this long apical dendrite allows the cell to be influenced by large numbers of cells from other (more superficial) layers.

Another aspect of the laminar structure of cortex that appears to be regulated by intrinsic cellular and molecular interactions concerns the major connections between cells, in particular the inputs from the thalamus. As mentioned earlier, the main input layer in the cortex is layer 4. A series of experiments by Blakemore and colleagues established that the termination of projections from the thalamus in layer 4 is governed by molecular markers. Slices of brain tissue are able to survive and grow in a Petri dish for several days under the appropriate conditions. Indeed, pieces of thalamus of the appropriate age will actually grow connections to other pieces of brain placed nearby. Molnar and Blakemore (1991) investigated if and how a piece of visual thalamus (LGN) would innervate various types of cortical and non-cortical brain tissue.

In initial experiments they established that when a piece of thalamus (LGN) and a piece of visual cortex were placed close together in the dish, projection fibers from the LGN not only invaded a piece of visual cortex of the appropriate age, but also terminated in the appropriate layer 4. Thus, layer 4 appears to contain some molecular stop signal that tells the fibers to stop growing and form connections at that location. Next, they conducted a series of choice experiments in which the visual thalamus had a piece of visual cortex and some other piece of brain placed nearby. The thalamic fibers turned out to dislike cerebellum, rarely penetrating it, but they did grow into hippocampus. However, the growth into hippocampus (a piece of brain that is closely related to neocortex) was not spatially confined in the way it had been for visual cortex, suggesting that it was just a "growth-permitting substrate."

The evidence discussed so far indicates that the laminar structure of cortex arises from genetic, local cellular, and molecular interactions, rather than it being shaped as a result of thalamic and sensory input. Thus, the identity and location of neurons are determined before birth, and incoming fibers "know" in which layer to stop and make synaptic contacts. We now turn to the question of whether the areal structure of the cortex is determined in a similar way.

Traditionally, two opposing possibilities have been put forward to account for the division of cortex into areas:

1) The areal differentiation of cortex is due to a *protomap* (Rakic, 1988). By this view, differentiation into cortical regions occurs early in the formation of the cortex, and is due to intrinsic molecular and genetic factors. The actual activity of neurons is not required. The cortex is viewed as a mosaic from the start such that at the stage of neural stem cells (before many neurons have been generated) the primitive map has information sufficient for each cortical area to have individually specified features particularly suitable for the input it will receive or the functions it will perform.

2) The different areas of cortex arise out of an undifferentiated *protocortex*. By this view, differentiation into precise areas occurs later in the development of cortex, and it depends on extrinsic factors like input from other parts of the brain or sensory systems. The activity of neurons is required (Killackey, 1990; O'Leary, 2002) for this specification to occur. The division of cortex into areas in the adult brain is influenced by information relayed from the thalamus and from interactions with other areas of cortex via interregional connectivity.

Further Reading   Sur & Rubenstein (2005).

There is a large, complicated, and sometimes apparently conflicting literature on the areal differentiation of neocortex (for reviews see Bystron et al., 2008; Kingsbury & Finlay, 2001; Pallas, 2001; Ragsdale & Grove, 2001; Sur & Rubenstein, 2005). Some experiments appear, at first sight, to be compelling evidence for the protomap view. For example, the newborns of a strain of "knockout" rodent (see Chapter 3) that genetically lack connections between the thalamus and the cortex still have certain well-defined regional gene-expression boundaries within their cortex (Miyashita-Lin et al., 1999; but see later) and some other characteristics of wild-type mice. In another example, *in vitro* studies in which cortical tissue is maintained in culture, and thus isolated from potential extrinsic patterning cues, still show patterns of gene expression consistent with the development of the hippocampus (Tole et al., 2000). Despite these and other studies supporting the idea of genetically specified regionalization of cortex, there are some important caveats, and also evidence in support of the opposing protocortex view.

1) Most of the patterns of gene expression thought to contribute to the differentiation of cortex into areas do not show clearly defined boundaries, but rather show graded expression across large extents of cortex (Sydnor et al., 2021). This suggests that regionalization of cortex could emerge from a combination of different gradients of gene expression. Kingsbury and Finlay (2001) refer to this as a "hyperdimensional plaid" and contrast this with a "mosaic quilt" (protomap) view. Evidence indicates that while more than one hundred genes may show graded differential expression between two different cortical areas during development (Leamey et al., 2008), genetic

influence over the layered structure of cortex is much more direct. Specifically, an autonomous and dissociable genetic pathway coordinates the development of the deeper layers of cortex from that which contributes to the mid and upper layers (Casanova & Trippe, 2006).

2) Despite primary sensory regions being the best candidates for genetic prespecification, we will see in the next section of this chapter that even these regions can have their properties significantly changed through experience. Thus, sensory input may be vital at least for the maintenance of cortical divisions.

3) Evidence for cortical differentiation prior to birth does not allow us to conclude that neuronal activity is not important since spontaneous neural activity within the brain is known to be important for differentiation and subsequent function (Ge et al., 2021; Shatz, 2002), as discussed earlier.

Further Reading   Shatz (2002); Sur & Rubenstein (2005); Sydnor et al. (2021).

Reviews of the current evidence converge on views that are midway between the proto-map and protocortex hypotheses (Bystron et al., 2008; Kingsbury & Finlay, 2001; Pallas, 2001; Ragsdale & Grove, 2001) or suggest that the dichotomy no longer remains as the differentiation of cortex into structural and functional areas involves "an interwoven cascade of developmental events including both intrinsic and extrinsic mechanisms" (Sur & Rubenstein, 2005). Most agree that graded patterns of gene expression create large-scale regions with combinations of properties that may better suit certain computations (similar to a very coarse protomap). It is within these large-scale regions that smaller scale functional areas arise through the neural activity-dependent mechanisms associated with the protocortex view. A hypothetical example is that one region may receive particular thalamic input projections, overlaid with a certain pattern of neurotransmitter expression, and the presence of specific neuromodulators. This combination of circumstances, combined with neural activity, may then induce further unique features such as particular patterns of short-range or long-range connectivity. Differentiation into smaller areas within the larger regions may occur through the selective pruning of connections. Thus, Kingsbury and Finlay (2001) refer to this perspective on cortical differentiation as a "hyperdimensional plaid" because the patterning that emerges in a plaid is the result of small changes in many threads. Similarly, O'Leary and colleagues call this general view the "cooperative concentration" model since some different gradients of gene expression may act as opposing forces in shaping cortical regions (Hamasaki et al., 2004).

Research on the combinations of factors that determine cortical differentiation continues to be very active and is a fast-moving field. Findings on developing mice suggest that while patterns of gene expression start off in the graded way described earlier during prenatal development, in some cases the patterns become more localized to specific areas even when you prevent thalamic input to the cortex (Hamasaki et al., 2004). However, this conclusion may be confined to the primary sensory areas that are best defined and most studied. In a nice example of the interplay between different factors in cortical specialization, Chou et al. (2013) demonstrated with a type of knockout mice that while genetic mechanisms intrinsic to the cortex are important for defining an overall area of visual cortex, the subsequent further division of this region into more specialized areas requires input from the thalamus. More specifically, their work suggests that while factors associated with a protomap view predominate for the specification of primary sensory areas, the specification of higher order areas may depend more on structured sensory input and activity-dependent processes. This is important as work on primary sensory areas has dominated much research in the field, and even in the mouse there are around 10 times more higher order areas than primary sensory ones.

Another line of work investigates a key element of Rakic's protomap hypothesis. According to this, regional differentiation occurs largely through the different numbers of cells generated in different layers in different regions of the cortex. The clearest example of this comes from examination of the primary visual cortex (area 17) in primates that has a larger number of neurons than neighboring regions (such as area 18). As mentioned earlier, this difference results from the number of divisions in the progenitor cells that are assumed to constitute part of the protomap. However, recent research shows that fibers from the thalamus enter the zone with the progenitor cells and have "mitogenic" effect (Dehay & Kennedy, 2007). A mitogenic effect means that the rate of cell division to create new cells is enhanced. In other words, fibers from the retina that project only to a specific part of the zone containing cortical progenitor cells may influence the rate (and therefore number) of production of new neurons. Thus, intriguingly, the cortical protomap itself may be shaped by input from other regions known to show spontaneous intrinsic waves of neural activity.

Perhaps the biggest challenge for the future is to understand how the emerging structural differentiation of the cortex relates to the emergence of functions, a question that we will return to many times later in this book. In this regard it is important to remember that there are actually only a few examples of clear structural regionalization that map tightly onto functional areas, and there are good reasons to believe that these cases may be exceptions to the general rule. For example, comparisons across a large number of species have led several experts to argue that primary sensory areas that receive direct input from the primary sensory thalamic nuclei are less susceptible to change between species than most of the rest of the cortex (e.g., Krubitzer, 1998). In particular, the primary visual cortex in primates has unique characteristics that have led some to propose that it is the most recently evolved part of cortex. As just mentioned, in the primary visual cortex, inputs from the visual thalamus may regulate the extent of cell proliferation in the ventricular zone (see Dehay & Kennedy, 2007), ensuring that this area of cortex has a rate of neuron production nearly twice that in neighboring areas. The entorhinal cortex, the region of cortex most closely associated with the hippocampus, shows some differentiation from surrounding cortex as early as 13 weeks after gestation (Kostovic et al., 1993). However, for the majority of cortex in the mouse, and the vast majority in humans, there is currently no evidence for cortical divisions arising through the on/off expression of a single gene rather than gradients of expression across multiple different genes.

To summarize so far, the basic laminar structure of the cerebral cortex in mammals appears to be largely universal. Cellular and molecular level interactions determine many aspects of the layered structure of cortex and its patterns of connectivity. Cortical neurons are often differentiated into specific types before they reach their destination (although some of the characteristic features of cell types are shaped by their journey to that site, e.g.,

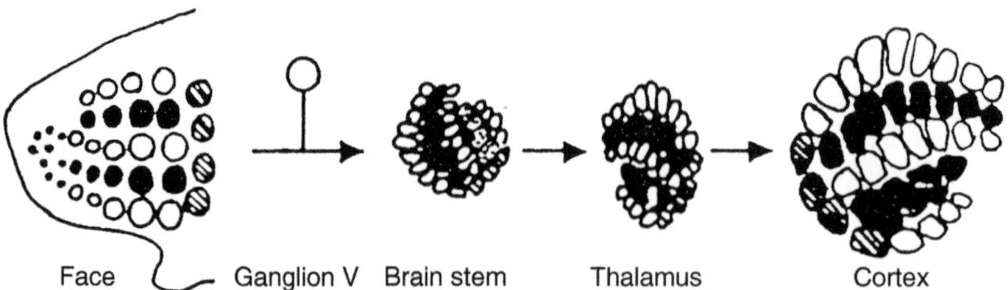

Figure 4.14 Patterning of areal units in somatosensory cortex. The pattern of "barrels" in the somatosensory cortex of rodents is an isomorphic representation of the geometric arrangement of vibrissae found on the animal's face. Similar patterns are present in the brain stem and thalamic nuclei that relay inputs from the face to the barrel cortex.

the long apical dendrite that typifies pyramidal cells reflects a literal "stretching" of processes during the migration process). However, this does not mean that cells in a particular area are necessarily prespecified for processing certain kinds of information.

A good example of how a region of cortex can become differentiated comes from work on the so-called "barrel fields" that develop in the somatosensory cortex of rodents. Each barrel field is an anatomically definable functional grouping of cells that responds to a particular whisker on the animal's snout (see Figure 4.14). Barrel fields are an aspect of the areal structure of cortex that emerges postnatally and are sensitive to whisker-related experience over the first days of life. For example, if a whisker is removed, then the barrel field that normally corresponds to that whisker does not emerge, and neighboring ones may occupy some of the cortical space normally occupied by it (for review see Schlaggar & O'Leary, 1993). Figure 4.14 illustrates how the areal divisions of the cortex arise as a result of similar divisions in structures closer to the sensory surface. In this case, it is almost as if the sensory surface *imposes* itself on to the brain stem, thence to the thalamus, and finally on to the cortex itself. The barrel field compartments emerge in sequence in these areas of the brain, with those closest to the sensory surface forming first, and the cortex patterns emerging last. While there is some contribution to the formation of barrel fields from intrinsic factors (Cooper & Steindler, 1986), a map of sensory space comes to occupy the somatosensory cortex in a reliable and replicable way largely driven by the sensory surface.

As quantity of cortex expands between species in phylogeny, it is likely we will discover that differences in the timing of neurodevelopmental events between areas of cortex also influence the factors that shape their specialization (Cahalane et al., 2011). As just discussed in the example of barrel fields, projections from the thalamus can enter the developing cortex to "capture" their target cortical areas. The sensory input or motor output via the thalamus entrains the still developing cortical neuronal morphology and connectivity in order to fine-tune it for processing specific kinds of information (Cahalane et al., 2011; Sur & Leamey, 2001). It is no surprise, therefore, that in mammals primary sensory thalamic afferents always innervate the cortex first to initially form primary sensory areas of cortex. Less obvious is that this approximately posterior-to-anterior innervation by thalamic afferents runs counter to the timetable of development of the cortical neurons themselves, as cortical neurogenesis proceeds in a rostro-caudal progression with frontal areas of cortex being differentiated before more posterior regions (Cahalane et al., 2011). Thus, there are diametrically opposed gradients of development across cortex that ensure that while the differentiation in some (generally more posterior) regions is heavily dominated by sensory input, the differentiation of more anterior regions is not so constrained. In these anterior areas the early morphological development of neurons and their connectivity is likely to be more shaped by their intrinsic (spontaneous) activity and patterns of interconnectivity with other cortical areas. This happy accident of developmental timing may result in an anterior portion of the cortex that is detached from being directly constrained by sensory input, and therefore becomes generally biased to processing information that needs to be integrated across other regions.

In this section we have reviewed some of the literature on laminar and areal specification of cortex during normal development, presenting evidence for both the protomap and pro-tocortex accounts. This apparently conflicting evidence can be reconciled by the notion that large-scale regions of cortex have particular combinations of graded gene expression that are then refined into smaller functional areas through activity-dependent processes (Chou et al., 2013).

Differential Development of Human Cortex

The main phylogenetic changes in cortical development in primates are in the extent of cortical tissue, and the more prolonged time course of development in primates and humans (see Section 0). In this section we overview some of the differential development of regions that occurs over the first decade.

With regard to the development of the laminar organization of the cortex (the layers of the cake), although most cortical neurons are in their appropriate locations by the time of birth in the primate, the inside-out pattern of growth observed in prenatal cortical development extends into postnatal life. Extensive descriptive neuroanatomical studies of cortical development in the human infant by Conel over a 30-year period led him to the conclusion that the postnatal growth of cortex proceeds in an inside-out pattern with regard to the extent of dendrites, dendritic trees, and myelination (Conel, 1939–1967). Conel's general conclusions have since been validated with modern neuroanatomical methods (e.g., Becker et al., 1984; Purpura, 1975; Rabinowicz, 1979), and a modern rea-nalysis of Conel's original data has deemed it to be highly consistent and reliable (Shankle et al., 1998). In particular, the maturation of layer 5 (deeper) in advance of layers 2 and 3 (superficial) seems to be a very reliably observed sequence for many cortical regions in the human infant (Becker et al., 1984; Rabinowicz, 1979). For example, the dendritic trees of cells in layer 5 of primary visual cortex are already at about 60% of their maxi-mum extent at birth. In contrast, the mean total length for dendrites in layer 3 is only at about 30% of their maximum at birth. Furthermore, higher orders of branching in den-dritic trees are observed in layer 5 than in layer 3 at birth (Becker et al., 1984; Huttenlocher, 1990). Interestingly, this inside-out pattern of growth is not evident in the later occurring rise and fall in synaptic density.

Differential development in human postnatal cortical growth is also evident in the areal dimension (the slices of the cake), and we will return to this topic in several later chapters. Even in mid-gestation, there is more evidence for differential gene expression in different cortical areas in humans than in other species studied (Dehay & Kennedy, 2009; Johnson et al., 2009). In postnatal development, Huttenlocher (1990, 1994; Huttenlocher & Dabholkar, 1997a) reports clear evidence of a difference in the timing of postnatal neuro-anatomical events between the primary visual cortex, the primary auditory cortex, and the frontal cortex in human infants, with the latter reaching the same developmental land-marks considerably later in postnatal life than the first two. It is worth noting that this differential development within the cerebral cortex has not been reported in other primate species (Bourgeois, 2001; Rakic et al., 1986). For example, Rakic and colleagues reported that all areas of cortex appear to reach their peak in synaptic density about the same time— around 2–4 months in the rhesus macaque (alternatively called rhesus monkey), roughly corresponding to 7–12 months in the human child. Contrary to Huttenlocher's findings in humans, this suggests that there may be a common genetic signal to increase connectivity across all brain regions, simultaneously, regardless of their current maturational state. A sudden event of this kind stands in marked contrast to known region-by-region differences in the time course of cell formation, migration, myelination, and metabolism in the human cortex (Conel, 1939–1967; Yakovlev & Lecours, 1967). The most likely difference between the human results and those from the macaques is that the protracted postnatal develop-ment of humans means that regional differences are more evident. In the macaque, how-ever, possible regional differences are compressed into a much shorter time, making them harder to detect (Huttenlocher, 1994). However, Goldman-Rakic (1994) suggests that the differences between the macaque and human results may be due to the neuroanatomical techniques used.

Further Reading   Huttenlocher (2002).

Consistent with the reports from human postmortem tissue, a study in which the functional development of the human brain was investigated by PET also found differ-ential development between regions of cortex (Chugani & Phelps, 1986; Chugani et al., 2002). In infants under 5 weeks of age, glucose uptake was highest in sensorimo-tor cortex, thalamus, brain stem, and the cerebellar vermis. By 3 months of age there were considerable rises in the parietal, temporal, and occipital cortices, basal ganglia, and cerebellar cortex. Maturational rises were not found in the frontal and dorsolateral occipital cortex until approximately 6–8 months. These developments are shown in Figure 4.15.

In addition to the formation of dendritic trees and their associated synapses, most fibers become myelinated during postnatal development. As described earlier, myelin is a mem-brane wrapping around axons that improves conduction. Owing to the increased fat con-tent of the brain caused by myelination of fibers, structural MRI images can reveal a clear gray–white matter contrast, and this allows quantitative volume measurements to be made during development (see Sampaio & Truwit, 2001). While some controversy remains about the interpretation of images from infants under 6 months (due to a higher than adult water content in both gray and white matter at this age), there is consensus that the appearance of brain structures is similar to that of adults by 2 years of age, and that all major fiber tracts can be observed by 3 years of age (Bourgeois, 2001; Huttenlocher &

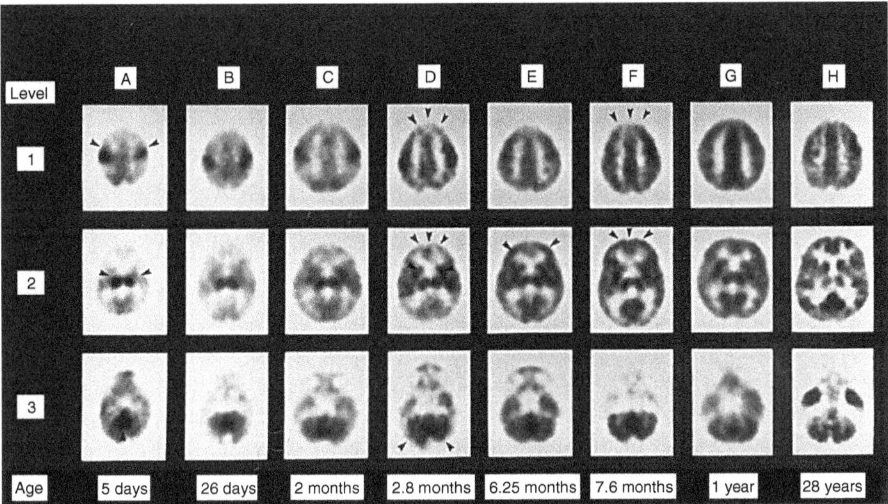

Figure 4.15 PET images illustrating developmental changes in local cerebral metabolic rates for glucose (lCMRGlc) in the normal human infant with increasing age. Level 1 is a superior section, at the level of the cingulate gyrus. Level 2 is more inferior, at the level of caudate, putamen, and thalamus. Level 3 is an inferior section of the brain, at the level of cerebellum and inferior position of the temporal lobes. Gray scale is proportional to lCMRGlc, with black being highest. Images from all subjects are not shown on the same absolute gray scale of lCMRGlc; instead, images of each subject are shown with the full gray scale to maximize gray scale display of lCMRGlc at each age. (A) In the 5-day-old, lCMRGlc is highest in sensorimotor cortex, thalamus, cerebellar vermis (arrows), and brain stem (not shown). (B, C, D) lCMRGlc gradually increases in parietal, temporal, and calcarine cortices; basal ganglia; and cerebellar cortex (arrows), particularly during the second and third months. (E) In the frontal cortex, lCMRGlc increases first in the lateral prefrontal regions by approximately 6 months. (F) By approximately 8 months, lCMRGlc also increases in the medial aspects of the frontal cortex (arrows), as well as the dorsolateral occipital cortex. (G) By 1 year, the lCMRGlc pattern resembles that of adults (H). Wiley.

Dabholkar, 1997a, b). Some reports suggest that after a rapid increase in gray matter volume up to about 4 years of age, there is then a prolonged period of slight decline that extends into adult years (Chugani et al., 2002; but see Huttenlocher & Dabholkar, 1997a, b). Whether this decline in gray matter is due to dendritic and synaptic pruning remains unknown, although in some studies the time course of the rise and fall coincides (Huttenlocher & Dabholkar, 1997a, b). Changes in the extent of white matter are of interest since they presumably reflect inter-regional communication in the developing brain. While increases in white matter extend through adolescence into adulthood, particularly in frontal brain regions (Huttenlocher et al., 1982), the most rapid changes occur during the first two years. Myelination appears to begin at birth in the pons and cerebellar peduncles, and by 3 months has extended to the optic radiation and splenium of the corpus callosum. Around 8–12 months the white matter associated with the frontal, parietal, and occipital lobes becomes apparent.

Further Reading Klingberg (2008).

The differential laminar and regional development observed in the human, as described in this section, provides the basis for many of the associations between brain growth and cognitive change to be described in the following chapters. But first we need to review some other aspects of postnatal brain development.

Postnatal Brain Development: Adolescence

In previous sections we have learned that the trajectory of the development of the brain is not that of a simple linear increase. Rather, it is characterized by more complex patterns involving both rises and falls in different measures. One period of later development that has commonly been associated with "dips" in development is adolescence. While we will discuss changes in adolescence in several of the domain-based chapters that follow, it also worth considering adolescence as a specific phase of human brain development in its own right.

At the onset of puberty, significant changes begin to occur in brain structure and chemistry. These changes involve continuing myelination of connections, and changes in the density of synapses, particularly within the prefrontal region of the cortex. There is some evidence that adolescence is associated with the fastest rate of cortical thinning (Zhou et al., 2015). The role of the hormonal changes associated with puberty in structural and functional brain development is still unclear. Results from studies assessing gender differences in brain development around puberty suggest that puberty may contribute to the reorganization of brain structures and maturational changes, and that sex differences emerge in puberty. As mentioned above, the changes in the volume of subcortical regions have been found to vary as a function of age, puberty stage, and sex (Goddings et al., 2014). However, there are still mixed results regarding the impact and direction of pubertal influences, therefore more large-scale longitudinal studies are required (Blakemore & Choudhury, 2006; Goddings et al., 2019).

In terms of behavior, adolescence is commonly described as a period of increased impulsive and risk-taking behavior. Scientists have investigated the hypothesis that this is related to a lack of inhibitory control, possibly mediated by a prolonged maturation of the functioning of PFC and changes in the brain's "reward" network leading to a different balance of these neural systems in adolescence compared to childhood and adulthood (see Chapter 10). Specifically, using longitudinal structural MRI data, regions associated with risk-taking and sensation-seeking behaviors, the amygdala and nucleus accumbens, showed relatively earlier structural maturation of gray matter volumes than the prefrontal cortex (Mills et al., 2016). One study investigated whether individual differences in the likelihood of engaging in risky behaviors is associated with activity in the brain circuitry that anticipates a reward in adults, adolescents, and children (Galvan et al., 2006). Results from this and related studies suggest that while both impulsive behavior and risk-taking behavior are prevalent in adolescents, they show different developmental trajectories and have partially different brain bases. Impulsive behavior (lack of inhibition) is related to PFC development and gradually diminishes from childhood to adulthood. In contrast, individuals prone to risky behavior (reward network) are at further risk during adolescence when the brain systems involved in the anticipation of reward are undergoing developmental changes.

During adolescence many other "executive functions" such as selective attention, working memory, problem solving, and multi-tasking, but also aspects of social cognition, improve steadily (see Chapter 10). While such executive functions are commonly related to the PFC, fMRI studies indicate that broad networks of cortical regions are involved in these changes (e.g., Crone & Dahl, 2012; Luna et al., 2004).

> **Further Reading**   Olson & Luciana (2008).

Postnatal Brain Development: The Hippocampus and Subcortical Structures

This chapter has focused primarily on the cerebral neocortex since this is the part of the brain that shows most protracted postnatal development. However, other brain structures such as the hippocampus and cerebellum also show some postnatal development and, as we will see, have been associated with cognitive changes in infancy and childhood. The postnatal development of some subcortical structures (such as the hippocampus, cerebellum, and thalamus) poses something of a paradox: on the one hand, there is much behavioral and neural evidence to indicate that these structures are functioning at birth, while, on the other, they all show some evidence of postnatal development and/or functional reorganization. One explanation for this is that as the neocortex develops postnatally, its interactions with subcortical regions undergo certain changes. Evidence from assessments of functional connectivity during resting states suggest that subcortical regions may influence cortical processing more heavily early in development, with cortical networks gradually gaining their independence from subcortical influence during the course of childhood (Supekar et al., 2009).

The limbic system is normally taken to include the amygdala, the hippocampus, and the limbic regions of cortex (cingulate gyrus and parahippocampal gyrus [entorhinal cortex]). While these latter cortical regions follow the same developmental timetable as other regions of the cortex, they are differentiated from the rest of the cortex at an early stage, and are therefore unlikely to show the same degree of plasticity. As discussed earlier, gyral development (folding) does not necessarily indicate architectural specificity. Nevertheless, the gyral folding associated with the cingulate region is discernible as early as 16–19 weeks of gestational age in humans and the parahippocampal gyrus within the temporal lobe at 20–23 weeks gestational age (Gilles et al., 1983). In contrast, other prominent gyri in the cortex do not emerge until 24–31 weeks. The major nuclear components of the limbic system, such as the hippocampus, start to differentiate from the developing temporal lobe around the third and fourth months of fetal development. After this, further differentiation of the hippocampus takes place, resulting in it becoming a rolled structure tucked inside the temporal lobe and surrounded by tissue known as the dentate gyrus (see Seress, 2001, for review). It has been known for some time that in rodents neurogenesis continues into postnatal life in the dentate gyrus region (Wallace et al., 1977). More recently, it has been confirmed in humans that granule neurons continue to be produced throughout adulthood (for review see Kozorovitskiy & Gould, 2008). This production of new neurons is influenced by hormones, and, at least in rats, some types of learning enhance the number of new neurons produced. The functional importance of adult neurogenesis in this region remains to be determined.

> **Further Reading**   Kozorovitskiy & Gould (2008); Seress & Abraham (2008).

The cerebellum is a brain structure thought to be involved in motor control, but which probably also plays a role in some aspects of higher cognitive functioning. Within two months after conception the cerebellum has formed its three primary layers, the ventricular (V), intermediate (I), and marginal (M) layers. However, its development is prolonged and neurogenesis in this region continues postnatally with only about 17% of the final number of granule cells present at birth, and neurogenesis possibly continuing until 18 months (Spreen et al., 1995). Despite being one of the few regions of the human brain to show postnatal neurogenesis, cerebellar functional development as measured by resting PET shows high glucose metabolic activity as early as 5 days old (postnatal), the same schedule as other sensorimotor regions such as the thalamus, brain stem, and sensorimotor cortex (Chugani, 1994).

Some subcortical regions also change their response properties during adolescence, though this may be a reflection of their interactions with the cortex. For example, the amygdala is an important part of the social brain (Chapter 7) that has been consistently related with emotion processing. In adults, the amygdala is activated by the perception of fearful facial expressions. While this structure also responds to fearful faces in children of 11 years, it responds equally to neutral faces at this age, suggesting a less finely tuned function (Thomas et al., 2001). Other studies of amygdala function in development have revealed gender differences over the adolescent period: in females the response of the amygdala to fearful faces decreased during adolescence whereas it did not in males (Killgore et al., 2001). The opposite pattern was observed for a region of the PFC. Killgore and colleagues interpret these findings in terms of the greater increase in the regulation of emotion by females (mediated by prefrontal systems).

Neurotransmitters and Neuromodulators

The aspects of brain development discussed so far have mainly concerned its neurons and "wiring." However, there are also developmental changes in what has been referred to as soft soak aspects of neural function. Soft soak refers specifically to the chemicals involved in the transmission and modulation of neural signals. Neurons and their dendrites can be thought of as lying in a bath composed of various chemicals that modulate their functioning. In addition, other chemicals play a vital role in the transmission of signals from one cell to another. Neurotransmitters in the cerebral cortex may be classified into those that arise within the cortex (intrinsic), and those that arise from outside the cortex (extrinsic) (see Benes, 1994). The intrinsic transmitters can be further divided according to whether they have an excitatory effect or inhibitory effect on post-synaptic sites.

The intrinsic excitatory transmitter glutamate is thought to play an important role in the axons of pyramidal cells that project to intrinsic cortical microcircuits, other cortical regions, and subcortical regions (Streit, 1984). In rats, the developmental time course of different glutamatergic pathways varies considerably. In general, however, it is the receptors for the transmitter, rather than the quantity of the transmitter, that increase with postnatal age. This development seems to follow the rise and fall pattern seen in other aspects of neural development. Specifically, in the rat between postnatal days 10 and 15 the amount of glutamate binding in cortical regions increases rapidly and reaches a peak around 10 times the levels observed in adults (Schliebs et al., 1986). By day 25 these levels have reduced drastically.

GABA is probably the most important intrinsic inhibitory transmitter in the adult mammalian brain. Intriguingly, however, while GABA has a largely excitatory function in early postnatal development (Ben-Ari, 2002; Ben-Ari et al., 2012), there is then an activity-dependent shift to inhibition in early postnatal development that may drive the opening and closing of critical periods in sensory cortex (Hensch, 2005; Hensch et al., 1998). Experiments with knockout mice relevant to human development disorders, such as FMR1 knockout model for Fragile-X syndrome, show that these delicately balanced development shifts may be disrupted (Harlow et al., 2010). While there are a variety of ways to measure GABA activity which sometimes give differing results (see Benes, 1994), in the typically developing human the same overall pattern of rise and fall seen for glutamate is also observed for GABA. Specifically, the density of GABA receptors increases rapidly in the perinatal period and doubles over the first few weeks before later declining (Brooksbank et al., 1981). The extent to which the rise and fall in these intrinsic neurotransmitters mirrors that observed in the structural measures discussed earlier, such as glucose uptake and synaptic density, is currently unclear and requires further research. It is clear, however, that the levels of GABA can be influenced by the extent of sensory experience (Fosse et al., 1989).

Work has begun to reveal the importance of an appropriate balance between glutamate and GABA, and thus excitatory and inhibitory influences, during development, and the influence this balance may have on connectivity changes and synaptic pruning. For example, one type of knockout mouse has reduced GABAergic neurons in the cortex and hippocampus by adulthood (Cobos et al., 2005). The resulting reduction in inhibition leads to the withdrawal of glutamate receptors from excitatory neurons in order to maintain an appropriate excitation/inhibition balance (Howard et al., 2013). Consequently, these mice show reduced connectivity in the auditory cortex (Seybold et al., 2012). Seybold and colleagues note that when faced with a lack of inhibition and potential overactivity, "the nervous system may sacrifice connectivity and computational power for stability" (p. 13833). Such examples of brain adaptation (homeostasis) may be relevant for understanding human developmental disorders (Johnson et al., 2015).

Extrinsic neurotransmitters arise from a number of different subcortical locations. One of these transmitters, acetylcholine, originates mainly from the basal forebrain (Johnston et al., 1979). Interestingly, the innervation of cortex by cholinergic fibers follows the inside-out pattern of growth described earlier, with the deeper cortical layers being innervated before the more superficial ones. In humans, this cholinergic innervation begins prenatally, though adult levels are not reached until about 10 years of age (Diebler et al., 1979). However, the binding sites within the cortex for this transmitter decrease from birth onward, possibly due to synaptic pruning (Ravikumar & Sastry, 1985).

Another neurotransmitter with origins outside the cortex is norepinephrine (or noradrenaline), which originates in a cluster of nuclei called the locus coeruleus. As well as its role as a neurotransmitter, norepinephrine has been associated with cortical plasticity (Kasamatsu & Pettigrew, 1976). In several mammals there is an extensive network of noradrenergic fibers in the cortex at birth which may be more dense than that seen in adults (Coyle & Molliver, 1977).

Serotonin originates in the brain stem raphe nuclei, and in both rats and primates the level increases rapidly over the first few weeks of life (Johnston, 1988). In rhesus monkeys the adult pattern of projection of serotonin fibers is reached by the sixth postnatal week, though levels of serotonin continue to rise after this (Goldman-Rakic & Brown, 1982). There is some evidence (from specific binding sites) of a later decrease in serotonin in human cortex and hippocampus (Marcusson et al., 1984). Like acetylcholine, serotonin is found mainly in the deeper cortical layers at birth, consistent with the structural inside-out gradient of development discussed earlier.

The fourth main extrinsic cortical transmitter, dopamine (which originates in the substantia nigra), likewise shows an inside-out pattern around the time of birth, at least in rats (Kalsbeek et al., 1988). In these animals, dopaminergic fibers show the adult pattern of projection into the frontal and cingulate cortex through extended postnatal development (Bruinink et al., 1983).

Finally, there is increasing study of the effects of hormones on cognitive and brain development. Like most hormones, oxytocin can have multiple effects on body and brain. However, it has become increasingly associated with social behavior in several different species, including humans (Insel, 2010). Oxytocin affects a number of the brain regions involved in processing social information, and its role in social development is now being explored. For example, Krol et al. (2021) found that infants' degree of attention to frowning and smiling faces is affected by a genetic polymorphism associated with oxytocin. Variation in social attention is also linked to a combination of genetic factors and breastfeeding experience likely to modulate oxytocin levels (Krol et al., 2015).

In summary, it appears that:

- Most intrinsic and extrinsic transmitters are present in the cortex at birth—at least in rats and probably also in humans, but these show changes in distribution and overall levels for some time after birth.
- Several transmitters of both intrinsic and extrinsic origins show the characteristic rise and fall evident in some measures of structural neuroanatomical development.
- Several transmitters of extrinsic origin show the same inside-out gradient of cortical development observed in structural measures.
- Neurotransmitters may play multiple roles during development. For example, noradrenaline may also regulate cortical plasticity.
- Some transmitters show a differential distribution throughout the cortex. This differential distribution may play some role in the subsequent specialization of regions of cortex for certain functions.

Further Reading   Benes (2001); Cameron (2001); Berenbaum et al. (2003); Richards (2003); Stanwood & Levitt (2008); Insel (2010).

What Makes a Brain Human?

An important issue for the work described in this chapter is to what extent human brain development is similar to that of other species. This issue is of the upmost theoretical and practical importance for several reasons. First and foremost, most researchers are ultimately interested in how the human mind arises from its underlying brain development, and several of the topics of later chapters—such as number and language—are assumed to be, to at least some degree, unique to humans. This raises the question of what is unique about the human brain and the developmental processes that give rise to it. A related question is how applicable studies with other species are to human brain development.

As mentioned earlier, primates generally have a much more prolonged timetable for brain development than other mammals. Even between *Homo sapiens* and other primates there is a wide difference in timing, with our postnatal cortical development being extended by roughly a factor of four longer than other primates. We discussed earlier that this prolonged postnatal development in humans stretches out differential laminar and regional cortical differentiation that are more compressed in time in other species. What is the significance of this stretched out timetable of brain development? Finlay and Darlington (1995) compared data on the size of brain structures from 131 mammalian species, and concluded that the order of landmarks of brain development is conserved across a wide range of species. Further, they noticed that, controlling for overall brain and body size, the time course of these landmarks was related to the relative size of structures of the brain in a systematic way. Specifically, disproportionately large growth occurs in the late-generated structures such as the neocortex when the overall timetable is slowed. By their analysis, the structure most likely to differ in size in the relatively slowed neurogenesis of primates is therefore the neocortex.

Finlay and Darlington extended their model of the evolution of the brain to human prenatal development (Clancy et al., 2000). The model predicts that the more delayed the general time course of brain development in a species, the larger the relative volume of the later developing structures (such as the cerebral cortex, and particularly the frontal cortex) will be. In accordance with this general prediction, the slowed rate of development in humans is associated with a relatively larger volume of cortex, and an especially large frontal cortex.

Why would allowing more time for the cortex to develop have the effect of increasing its size? Casting our minds back to the radial unit model proposed by Rakic (1988), a single round of additional symmetric cell division at the proliferative unit formation stage would have the effect of doubling the number of ontogenetic columns, and hence expanding the area of cortex. In contrast, an additional single round of division at a later stage, from the proliferative zones, would only increase the size of a column by one cell (about 1%). There is very little variation between mammalian species in the layered structure of the cortex, while the total surface area of the cortex can vary by a factor of 100 or more between different species of mammal. It seems likely, therefore, that species differences originate (at least in part) in the timing of cell development (i.e., the number of rounds of cell division that are allowed to take place within and across regions of the proliferative zones).

Thus, it is likely that the increased extent of cortex in our brains, and particularly PFC, is at least in part a happy by-product of slowing down the overall timetable of brain development. This suggests that evidence from other mammals will be highly relevant to the study of human brain development since we are looking at fundamentally the same process.

One caveat to this conclusion, however, must be that the point on the timetable of brain development at which different animals are born may differ greatly. In this regard, the relatively delayed time course of human brain development also has another very important benefit. It allows a prolonged postnatal period during which interaction with the environment can contribute to the tuning and shaping of circuitry.

Whether it is just the slowed timing of brain development that produces the unique human brain remains controversial. Subtle differences in the steps of cortical development between primates and rodents are already known, and some reports of possibly species-specific progenitor cells or neurons in humans merit further research (Bystron et al., 2008).

General Summary and Conclusions

We have reviewed some of the highlights of pre- and postnatal development of the brain. While many of these landmarks of development are similar between humans and other mammals, the timing of human brain development is characterized by being slower and more protracted. According to some theories, this slowed development allows the building of relatively more cortex, and particularly frontal cortex. One major feature of human postnatal development is that brain volume quadruples between birth and adulthood, mainly due to increases in nerve fiber bundles, dendrites, and myelination. Another major feature is that several measures of structure and neurophysiology, such as the density of synaptic contacts, show a characteristic rise and fall during postnatal development.

The issue was raised as to whether the differentiation of the neocortex into anatomical and functional areas is prespecified. The protomap hypothesis states that the differentiation of the cortex into areas is determined by intrinsic molecular markers or prespecification of the proliferative zones. The protocortex hypothesis states that an initially undifferentiated protocortex is divided up largely as a result of input through projections from the thalamus and is activity-dependent. A review of currently available evidence supports a middle-ground view in which large-scale regions are prespecified, while the establishment of small-scale higher order areas require activity-dependent processes.

The very extended period of postnatal development seen in the human brain reveals two differential aspects of cortical development not as clearly evident in other primates: an inside-out pattern of development of layers, and differences in the timing of development across regions. These differential aspects of human cortical development will provide the basis for associations between brain and cognitive development described in later chapters.

Key Issues for Discussion

- How essential are animal studies for understanding human brain development?
- What, if any, lines of evidence could definitively decide between the protomap and protocortex accounts of cortical differentiation?
- What might be the functional consequences of the "rise and fall" pattern observed for several measures (like the density of synapses) of postnatal development?

5

Vision, Orienting, and Attention

This chapter reviews selected topics in vision, visual orienting, and visual attention for which attempts have been made to relate brain development to behavioral change. The contribution of peripheral systems (retinal) development in the emergence of basic visual functions is briefly alluded to before discussion of the neural basis of the development of binocular vision. Neuroanatomical and computational modeling evidence highlights the importance of the segregation of inputs to specific cells in the primary visual cortex. A compelling behavioral test of the increased segregation hypothesis is discussed. Moving on from sensory processing to sensorimotor integration, the following section describes an attempt to use the developmental neuroanatomy of the cortex to make predictions about transitions in orienting behavior in infants. A number of behavioral marker tasks for cortical regions involved in eye movement control are reviewed and discussed in relation to this maturational model. Recent work has shown that infant eye movement measures associate with attention measures later in development, and that some of these abilities may be amenable to training. Finally, experiments concerned with the development of covert (internal) attention shifts in infants and children are described. A number of changes in the flexibility and speed of shifting attention are shown to originate in infancy but continue into childhood. Differences in attention attributable to either typical genetic variation, or that associated with developmental disorders, are described.

The Development of Vision

Visual pathways, and especially the visual cortex, are among the most studied regions of the brain. More than 25 visual areas of the primate cortex have been identified, and, by means of single-cell recording, neuroimaging, and neuropsychological studies, attempts have been made to understand the functions of these regions. On the behavioral side, much is known about visual psychophysics, and our knowledge about visual cognitive abilities is rapidly expanding. Vision would therefore seem to be a good starting place for studying the functional consequences of the developing brain.

Developmental Cognitive Neuroscience: An Introduction, Fifth Edition. Michelle de Haan, Iroise Dumontheil, and Mark H. Johnson.
© 2023 John Wiley & Sons Ltd. Published 2023 by John Wiley & Sons Ltd.
Companion website: www.wiley.com/go/johnson/devneuro5e

Immediately as we begin to consider this issue it becomes obvious that it is hard to determine whether changes in visual abilities during development are due to limitations in the periphery, such as the structure of the eye, lens, or eye muscles, or whether they are due to changes within the brain. It is clear that immaturities in peripheral sensory systems place limits on the perceptual capacities of young infants. For example, it is well known that immaturity of the retina limits spatial acuity. Some have even argued that such limitations are necessary to prevent overwhelming developing visual circuits with too much data too soon (Turkewitz & Kenny, 1982). The issue remains, however, whether peripheral limitations provide the major constraints on the development of perception, or whether development of visual pathways within the brain is the primary limiting factor. While this issue continues to be debated (Iliescu & Dannemiller, 2008), Banks and colleagues (e.g., Banks & Shannon, 1993) conducted an "ideal observer" analysis in which the morphology of neonatal photoreceptors and optics was compared to that of adults. This analysis generated estimates of the contribution of optical and receptor immaturity, as opposed to central immaturities, to deficits in infant spatial and chromatic vision. Observed differences in these aspects of vision between adults and infants turn out to be significantly greater than would be predicted on the basis of peripheral limitations only, indicating that central nervous system pathway development is an important contributing factor in the development of vision. Given that central nervous system factors play an important role even in the development of spatial acuity, we can inquire further into the nature and source of these constraints. Figure 5.1 summarizes some of the steps in visually guided behavior that will be discussed in the next few chapters. It is important to note too that developing visual pathways are affected by the quality of visual input they receive (e.g., Bathelt et al., 2020, and below). Processing related to perceiving and acting on objects is discussed in the next chapter. The processing of social visual stimuli, such as faces, is discussed in Chapter 7. In the present chapter we focus on two aspects of visual processing: orienting and attention.

> **Further Reading** Iliescu & Dannemiller (2008); Maurer et al. (2008); Atkinson & Braddick (2003).

Alongside evidence from behavioral and psychophysiological measures of visual abilities, in recent years investigators have begun to use functional imaging to see which visual pathways and structures can be activated at different ages. Owing to the difficulties of keeping young infants still for long enough, functional magnetic resonance imaging (MRI) studies with this population typically involve sedated infants being scanned for clinical reasons. Nevertheless, the studies have demonstrated that sedated and sleeping infants respond to visual stimulation in some of the same visual cortical regions as adults (Born et al., 1996, 2002; Yamada et al., 1997, 2000). These findings have been confirmed in healthy awake infants using functional near-infrared spectroscopy (fNIRS) (see Chapter 2). Following on from an earlier study with a single recording site (Meek et al., 1998), Taga and colleagues used a multichannel NIRS system to study blood oxygenation in occipital and frontal sites while 2 to 4-month-olds viewed dynamic schematic face-like patterns (Taga et al., 2003). The results demonstrated that a localized area of the occipital cortex responded

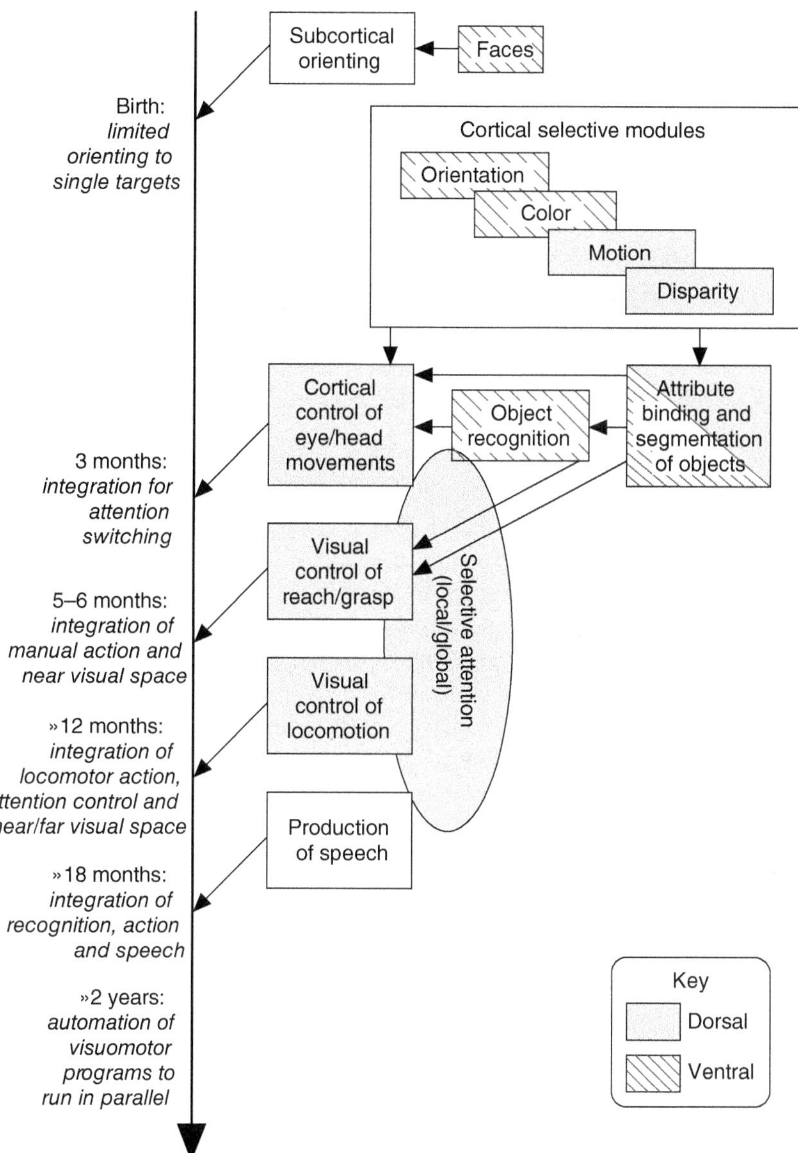

Figure 5.1 Diagram of the developmental sequence of visual behavior (left of vertical line) and ventral- and dorsal-stream neural systems contributing to this (right of vertical line).

to brief changes in luminance contrasts in the form of event-related changes in blood oxygenation similar to those seen in adults. More importantly, this study has helped establish the use of this technique to explore visual function during early infancy. For example, NIRS has been used to demonstrate activation in the prefrontal cortex in 4-month- to 2-year-olds in visual exploratory tasks (Delgado-Rayes et al., 2020). This may reflect activation of frontal areas involved in control of eye movements (see below).

The central visual field of most primates is binocular, requiring integration of information between the two eyes. This integration is achieved in the primary visual cortex. Functional and anatomical structures observed in layer 4 of the primary visual cortex, so-called "ocular dominance columns," are thought to be important for binocular vision (see Figure 5.2). These columns arise from the segregation of inputs from the two eyes. In other words, neurons in a single ocular dominance column are dominated by input from one eye in adult mammals. Ocular dominance columns are thought to be the stage of processing necessary to achieve binocular vision and, subsequently, detection of disparities between the two retinal images. Owing to the fact that ocular dominance columns are known to have a sensitive period for their formation, and to be sensitive to the differential extent of input from the two eyes, they have become a popular model system for developmental neurobiology (e.g., Bear & Singer, 1986; Kasamatsu & Pettigrew, 1976; Rauschecker & Singer, 1981).

Held (1985) reviewed converging evidence that binocular vision develops at approximately the end of the fourth month of life in human infants. Visually evoked potential measures (short latency event-related potential [ERPs] elicited by a conspicuous visual stimulus) with dynamic correlograms find evidence for binocularity around 3 months (see Atkinson & Braddick, 2003, for review). One of the abilities associated with binocular vision, stereoacuity, increases very rapidly from the onset of stereopsis, such that it reaches adult levels within a few weeks. This is in contrast to other measures of acuity, such as grating acuity, which increase much more gradually. Held suggested that this very rapid spurt in stereoacuity requires some equally rapid change in the neural substrate supporting it. On the basis of evidence from animal studies, he proposed that this substrate is the development of ocular dominance columns found in layer 4 of the primary visual cortex. While Held's proposal was initially based on a simple causal association between the formation of ocular dominance columns and the onset of binocularity, other research in his laboratory has been concerned to describe a link between the process of change at the two levels.

As mentioned in the previous chapter, processes of selective loss commonly contribute to the sculpting of specific pathways in the cortex. Neurophysiological evidence indicates that the inputs from the two eyes to the cortex are initially mixed so that they synapse on common cortical neurons in layer 4 (see Figure 5.3). These layer 4 cells project to

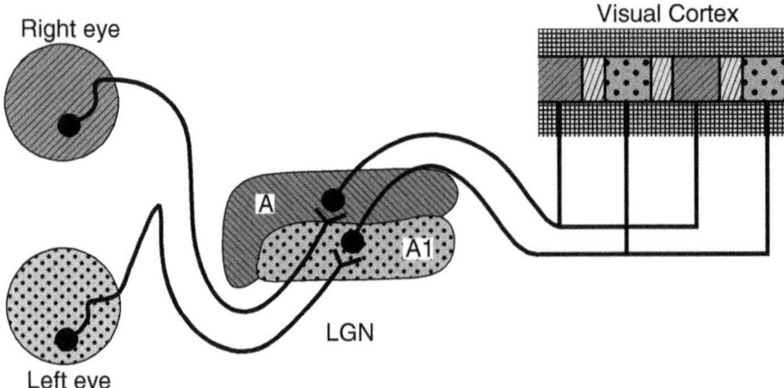

Figure 5.2 Simplified schematic diagram illustrating how projections from the two eyes form ocular dominance columns in the visual cortex. LGN, lateral geniculate nucleus.

Figure 5.3 (a) Afferents from both eyes synapse on the same cells in layer 4, thereby losing information about the eye of origin. (b) Afferents are segregated on the basis of eye origin (R and L), and consequently recipient cells in layer 4 may send their axons to cells outside of that layer so as to synapse on cells that may be disparity-selective.

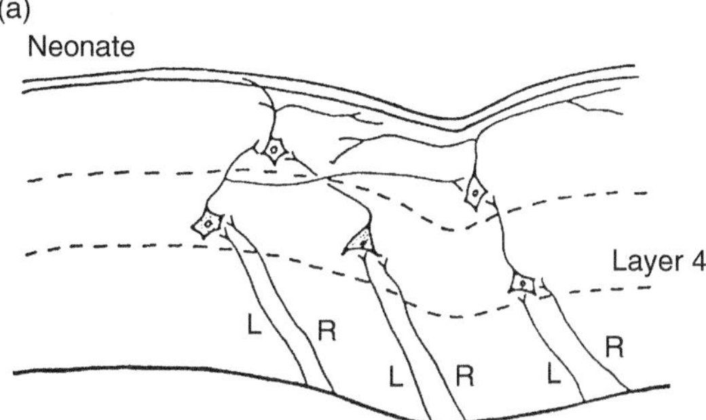

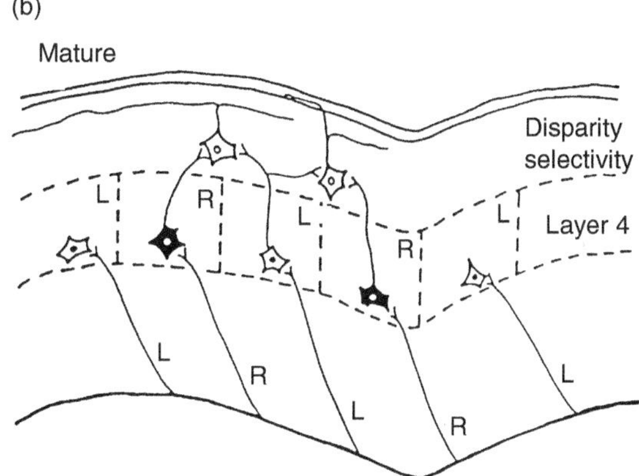

disparity-selective cells (possibly in cortical layers 2 and 3). During ontogeny, geniculate axons originating from one eye withdraw from the region, leaving behind axons from the other eye. Held suggested that it is these events at the neural level that give rise to the sudden increase in stereoacuity observed by behavioral measures in the human infant.

This process of selective loss has the information-processing consequence that information from the two eyes which was previously combined in layer 4 of the primary visual cortex becomes segregated (Held, 1993). Specifically, there will be a certain degree of integration between the eyes that will decline once each neuron receives innervation from only one eye. Held and colleagues elegantly demonstrated this increasing segregation of information from the two eyes by showing that infants under 4 months can perform certain types of integration between the two eyes that older infants cannot. In this experiment, Held and his colleagues presented a grating to one eye of an infant, and an orthogonal grating to the other eye (Shimojo et al., 1983). Infants under 4 months perceived a single grid-like representation instead of two sets of gratings that were orthogonal to each other. This is because, presumably, synaptic inputs from the individual eyes have not segregated into ocular dominance columns. As a result, a given cortical layer 4 neuron may have synaptic inputs from each eye and will effectively "see" the image from each eye simultaneously. That is, information from the two eyes would be summed in layer 4, resulting in an averaging of the two signals. Since the inputs from each eye summate, in the case of the orthogonal gratings a grid was perceived. Older infants (older than 4 months of age) do not perceive this grating because neurons in cortical layer 4 of striate cortex only receive inputs from one eye. The inputs to layer 4 have been segregated from one another, and a given layer 4 neuron will receive input from one eye or the other.

The loss of these connections is probably due to the refinement of synapses by selective loss. This refinement most likely occurs through activity-dependent neural mechanisms since it has been shown that the formation of ocular dominance columns can be experimentally blocked by reducing neuronal activity (Stryker & Harris, 1986). However, while these processes may strengthen and maintain ocular dominance columns, some research indicates that the initial formation of the columns can occur without structured visual experience in animals (see Iliescu & Dannemiller, 2008, for review). It is possible that the prenatal intrinsic spontaneous waves of retinal activity discussed in Chapter 4 structure the lateral geniculate nucleus (LGN) into eye-specific layers. This would happen because neighboring cells in one retina (left or right) will fire at nearly the same time, but not be correlated with activity in cells from the other retina. Since "cells that fire together wire together," eye-specific layers could form spontaneously. These eye-specific layers in the LGN could then potentially impose their structure on the developing visual cortex.

Evidence from human infants has been brought to bear on the question of the importance (or otherwise) of experience for the onset of binocular vision. Jandó et al. (2012) compared results from a group of babies born at term (close to nine months post conception) to a group of babies born prematurely (two months early). They monitored infants' brain responses to random patterns matched across the two eyes, as compared to their brain responses to mismatching patterns (where each eye received different information). Their results showed clearly that the onset of the binocular response occurred four months after birth for both term and preterm infants, showing a strong role for visual experience in guiding the development of binocularity (Jandó et al., 2012).

In several parts of this book we will discuss different streams of visual processing in the brain. Visual processing in the primate brain is initially divided into a subcortical route involving structures such as the superior colliculus, pulvinar, and amygdala, and several cortical routes that spread out from the LGN and primary visual cortex. In Chapter 7, we will discuss relations between the subcortical route and activation of cortical circuits in the processing of social stimuli. In Chapter 6, we will examine the differential processing of object and number information in two cortical streams; the dorsal (*where* or *action*) pathway, and the ventral (*what* or *perception*) pathway. In the next section of this chapter we focus on cortical and subcortical routes that are involved in the control of eye movements and visually guided actions.

The Development of Visual Orienting

In the previous section attempts to relate brain development to an aspect of sensory processing were discussed. In this section we move to a domain, visual orienting, which allows the study of the effects of brain development on the integration between sensory input and motor output. Visual orienting involves moving the eyes and head in response to, or in anticipation of, a new sensory stimulus. Most of the tasks to be described involve stimuli that are well within the visual capacities of infants, and involve a form of action (movements of the eyes; saccades) that infants can readily accomplish. Thus, while there is continuing development of both sensory and motor processing throughout infancy, through carefully designed experiments we are able to focus more on the integration between sensory input and motor output.

There are also other reasons for studying the development of visual orienting. One of these is that over the first year of life it is the human infant's primary method of gathering information from its environment and being an active explorer of the environment from early on (Aslin, 2007). These shifts of gaze allow the infant to select particular aspects of the external world for further study and learning. For example, as we will see in this chapter, simply by shifting the head and eyes infants can ensure that they are exposed to faces more than to other stimuli. In addition, most of what we have learned about mental processes in infants has come from tasks in which some measure of looking behavior, such as preferential looking or habituation to a repeatedly presented stimulus, is used (see Chapter 2).

Despite the importance of shifts of eye gaze in early infancy, however, until the past decades very little was known about how the development of the brain relates to changes in visual orienting abilities. This was despite a substantial literature on the neural basis of saccades which comes from neuropsychological and neuroimaging studies of human adults, as well as single-cell recording and lesion studies in non-human primates (see Andersen et al., 2000, for review).

In one of the first attempts to relate brain development to behavioral change in the human infant, Gordon Bronson (1974, 1982) argued that the early development of vision and visual orienting could be attributed to a shift from subcortical visual processing to processing in cortical visual pathways over the first six months of life. Specifically, Bronson cited evidence from electrophysiological, neuroanatomical, and behavioral studies that the primary (cortical) visual pathway is not fully functioning until around 3 months postnatal. As discussed earlier, it has more recently become evident that there is some, albeit limited, cortical activity in newborns, and that the onset of cortical functioning probably proceeds by a series of graded steps, rather than in an all-or-none manner. Meanwhile, neurophysiological research on monkeys and neuropsychological research with human adults has revealed that there are multiple pathways involved in oculomotor (eye movement) control and attention shifts in the primate brain. A number of the structures and pathways involved in oculomotor control in primates are illustrated in Figure 5.4.

Further Reading   Atkinson & Braddick (2003); Iliescu & Dannemiller (2008); Johnson (2002); Richards (2001, 2003).

Most of the pathways and structures illustrated in Figure 5.4 are known to be involved in particular types of information processing related to the execution and planning of eye movements. When considering the integration between sensory inputs and motor outputs it is important that the pathways discussed can be traced from the source of the input (the eye) to the muscles that shift the eyes. A diagram of such pathways may look different from

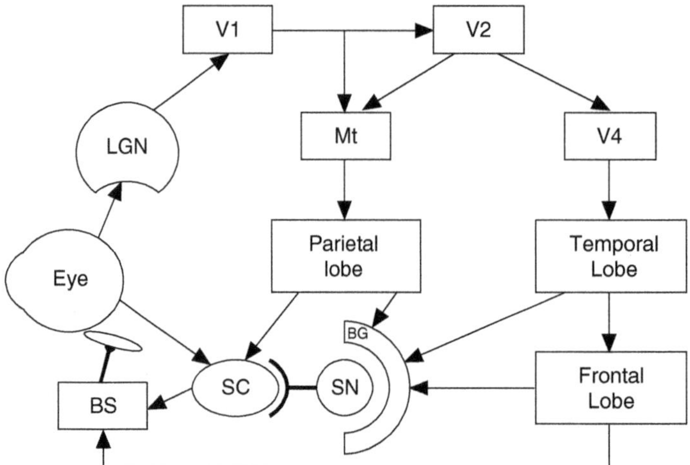

Figure 5.4 Diagram representing some of the main neural pathways and structures involved in visual orienting and attention. BS, brain stem; LGN, lateral geniculate nucleus; V1, V2, and V4, visual cortical areas; MT, middle temporal area; SC, superior colliculus; SN, substantia nigra; BG, basal ganglia.

one that just shows pathways for sensory processing. In Chapter 4 we discussed the dissociation between visual pathways for action and perception.

Four brain pathways will be discussed here. The first is the pathway from the eye to the superior colliculus. This subcortical pathway has input mainly from the temporal visual field (the peripheral or outer half of the field of each eye) and is involved in the generation of rapid reflexive eye movements to easily discriminable stimuli. The other three pathways to be discussed here share a projection from the eye to cortical structures via the midbrain (thalamic) visual relay, the LGN, and the primary visual cortex (V1). The first of these goes both directly to the superior colliculus from the primary visual cortex and also via the middle temporal area (MT). Some of the structures on this pathway are thought to play an important role in the detection of motion and the smooth tracking of moving objects. The next pathway proceeds from V1 to other parts of the visual cortex and thence to the frontal eye fields (FEFs). As we will see later, this FEF pathway is thought to be involved in more complex aspects of eye movement planning such as anticipatory saccades and learning sequences of scanning patterns. Finally, there is a more complex and less well understood pathway that involves tonic (continual) inhibition of the superior colliculus (via subcortical structures called the substantia nigra and basal ganglia). Schiller (1985) proposed that this pathway ensures that the activity of the colliculus can be regulated. Others suggest that this oculomotor pathway forms an integrated system with the FEFs and parietal lobes (e.g., Alexander et al., 1986) and that it plays some role in the regulation of the subcortical oculomotor pathway by the other cortical pathways.

The challenge has been to relate the development of these various pathways to the visuomotor competence of the infant at different ages. The three perspectives on the development of human brain function offer different views on this. According to the maturational view, we need to relate the sequential development of the different pathways to the onset of new functions as assessed by marker tasks. According to the skill learning view, the infant's brain needs to acquire the sensorimotor skill of generating accurate and informative saccades, and so areas involved in skill acquisition will be important. Finally, according to the interactive specialization view, some of the pathways will initially have poorly defined borders and functions (a lack of specialization), and only with experience will they become dissociable.

Of these three approaches, the maturational view has largely held sway with regard to visual orienting until recently. To date, this has involved two complementary approaches: predictions about the sequence of development of the pathways from developmental neuroanatomy (Atkinson, 1984; Johnson, 1990), and the administration of marker tasks (see Chapter 2) to ascertain the functional development of particular structures or pathways. An example of the first approach was originally presented by Johnson (1990). In this analysis he proposed that, first, the characteristics of visually guided behavior of the infant at particular ages are determined by which of the pathways (shown in Figure 5.4) are functional, and, second, which of the cortical pathways are functional is influenced by the developmental state of the primary visual cortex. The basis of this claim at the neuroanatomical level lies in three sets of observations. First, the primary visual cortex is the major (though not exclusive) "gateway" for input to most of the cortical pathways involved in oculomotor control (Schiller, 1985). Second, the primary visual cortex shows a postnatal continuation of the prenatal "inside-out" pattern of cortical growth described in Chapter 4, with the deeper layers (5 and 6) showing greater dendritic branching, length, and extent of myelination than more superficial layers (2 and 3) around the time of birth. Third, there is a restricted pattern of inputs and outputs from the primary visual cortex (e.g., the projections to V2 depart from the upper layers; see Chapter 4). By combining these observations with information about the developmental neuroanatomy of the human primary visual cortex, Johnson hypothesized the following sequence of development of cortical pathways underlying oculomotor control: the subcortical pathway from the eye directly to the superior colliculus (probably including cortical projections from the deeper layers of V1 to superior colliculus), followed by the cortical projection which inhibits the superior colliculus pathway, followed by the pathway through cortical structure MT, and finally the pathway involving the FEFs and related structures.

Following these predictions derived from developmental neuroanatomy, we can return to behavioral experiments to see if the transitions here support the predicted sequence of pathway development. Beginning with the newborn infant, evidence from measures of the extent of dendritic arborization and myelination indicate that only the deeper layers of the primary visual cortex are likely to be capable of supporting organized information-processing activity in the human newborn. Since the majority of feed-forward intracortical projections depart from outside the deeper layers (5 and 6), most of the cortical pathways involved in oculomotor control will only be receiving weak or disorganized input at this stage. However, evidence from various sources, such as visually evoked potentials, indicates that information from the eye is entering the primary visual cortex in the newborn. Thus, while some of the newborn's visual behavior can be accounted for in terms of processing in the subcortical pathway, Johnson (1990) proposed that there is also information processing occurring in the deeper cortical layers at birth. At least two characteristics of visually guided behavior in the human newborn are consistent with predominantly subcortical control such as saccadic pursuit tracking and preferential orienting to the temporal field. To go through these in more detail:

- The ability of infants to track a moving stimulus in the first few months of life has two characteristics (Aslin, 1981). The first is that the eye movements follow the stimulus in a saccadic or step-like manner, as opposed to the smooth pursuit found in adults and older infants. The second characteristic is that the eye movements tend to lag behind the movement of the stimulus, rather than predicting its trajectory. Therefore, when a newborn infant visually tracks a moving stimulus, it could be described as performing a series of saccadic eye movements. Such behavior is consistent with subcortical control of orienting.

- Newborns much more readily orient toward stimuli in the temporal (the half of the visual field of each eye which is furthest from the nose), as opposed to the nasal (the half of the visual field of each eye closer to the nose), visual field (e.g., Lewis et al., 1979). Posner and Rothbart (1981) suggest that midbrain structures such as the colliculus can be driven most readily by temporal field input. This proposal has been confirmed in studies of adult "blindsight" patients by Rafal et al. (1990), who established that distracter stimuli placed in the temporal "blindfield" had an effect on orienting into the good field, whereas distracter stimuli in the nasal blindfield did not. Evidence from studies of infants in which a complete cerebral hemisphere has been removed (to alleviate epilepsy) indicates that the subcortical (collicular) pathway alone is capable of generating saccades toward a peripheral target in the cortically "blind" field (Braddick et al., 1992).

Further Reading   Isa & Yoshida (2021).

Around 1 month of age infants show "obligatory attention" (also known as "sticky fixation") (Hood, 1995; Johnson, Posner, & Rothbart, 1991; Stechler & Latz, 1966). That is to say, they have great difficulty in disengaging their gaze from a stimulus in order to make a saccade to another location. Sometimes an infant around 1 month of age can spend as long as several minutes fixedly gazing at a seemingly uninteresting aspect of the environment, such as a section of carpet, before bursting into tears! Although this phenomenon is still poorly understood, Johnson suggested that it was due to the development of tonic inhibition of the colliculus via the substantia nigra (see Figure 5.4). Since this pathway projects from the deeper layers of the primary visual cortex to the colliculus, it is hypothesized to be the first strong cortical influence on oculomotor control. This (as yet) unregulated tonic inhibition of the colliculus has the consequence that stimuli impinging on the peripheral visual field no longer elicit an automatic exogenous saccade as readily as in newborns. Work using computational modeling has suggested that infants' fixations reflect on-line perceptual and cognitive activity similar to that seen in adults. However, the developmental state of the oculomotor system affects this relationship at 6 months (Saez de Urabain et al., 2017).

By around 2 months of age infants begin to show periods of smooth visual tracking, although their eye movements still lag behind the movement of the stimulus. At this age they also become more sensitive to stimuli placed in the nasal visual field (Aslin, 1981) and to coherent motion (Wattam-Bell, 1990). Johnson proposed that the onset of these behaviors coincides with the functioning of the pathway involving structure MT. The enabling of this route of eye movement control may provide this cortical stream with the ability to regulate activity in the superior colliculus.

Associated with further dendritic growth and myelination within the upper layers of the primary visual cortex strengthening the projections from V1 to other cortical areas, around 3 months of age the pathways involving the FEFs may become functional. This development may greatly increase the infant's ability to make "anticipatory" eye movements and to learn sequences of looking patterns, both functions associated with the FEFs. With regard to the visual tracking of a moving object, now not only do infants show periods of smooth tracking, but their eye movements often predict the movement of the stimulus in an anticipatory manner. A number of experiments by Haith and colleagues have demonstrated that anticipatory eye movements can be readily elicited from infants by this age. For example, Haith et al. (1988) exposed 3½-month-old infants to a series of picture slides which appeared either on the right- or on the left-hand side of the infant. These stimuli were either presented in an alternating sequence with fixed inter-stimulus interval (ISI), or with an irregular alternation pattern and ISI. It was observed that the regular alternation pattern produced more stimulus anticipations and faster reaction times to make an eye movement than in the irregular series. Haith and colleagues concluded from these results that infants of this age are able to develop expectancies for non-controllable spatio-temporal events. Canfield and Haith (1991) tested 2- and 3-month-old infants in an experiment which included more complex sequences (such as left–left–right, left–left–right, etc.). Consistent with the predictions from developmental neuroanatomy, while they failed to find significant effects with 2-month-olds, 3-month-olds were able to acquire at least some of the more complex sequences.

Given that this survey of the behavioral evidence pertaining to the development of visual orienting was broadly consistent with the predictions from developmental neurobiology (see Table 5.1 for a summary), the next step was to develop and consider marker tasks for

Table 5.1 Summary of the Relation between Developing Oculomotor Pathways and Behavior

Age	Functional anatomy	Behavior
Newborn	SC pathway + layer 5 and 6 pyramidal output to LGN and SC	Saccadic pursuit tracking Preferential orienting to temporal visual field "Externality effect"
1 month	As above + inhibitory pathway to SC via BG	As above + "obligatory" attention
2 months	As above + MT (magnocellular) pathway to SC	Onset of smooth pursuit tracking and increased sensitivity to nasal visual field
3 months and over	As above + FEF (parvocellular) pathway to SC and BS	Increase in "anticipatory" tracking and sequential scanning patterns

SC = superior colliculus; LGN = lateral geniculate nucleus; BG = basal ganglia; MT = middle temporal area; FEF = frontal eye fields; BS = brain stem.

Table 5.2 Marker Tasks for the Development of Visual Orienting and Attention

Brain region	Marker task	Studies
Superior colliculus	Inhibition of return	Clohessy et al. (1991); Simion et al. (1995)
	Vector summation saccades	Johnson et al. (1995)
Middle temporal area	Coherent motion detection; structure from motion	Wattam-Bell (1991)
	Smooth tracking	Aslin (1981)
Parietal cortex	Spatial cueing task	Hood & Atkinson (1991); Hood (1993); Johnson (1994); Johnson & Tucker (1996)
	Eye-centered saccade planning	Gilmore & Johnson (1997)
Frontal eye fields	Inhibition of automatic saccades	Johnson (1995)
	Anticipatory saccades	Haith et al. (1988)
Dorsolateral prefrontal cortex	Oculomotor delayed response task	Gilmore & Johnson (1995)

some of these and other parts of the cortex with a role in oculomotor control. Table 5.2 shows a number of marker tasks that have been developed for the functioning of structures involved in oculomotor control and visual attention shifts.

Marker tasks for several cortical regions thought to play a role in oculomotor control, the parietal cortex, FEFs, and dorsolateral prefrontal cortex (DLPFC, Figure 5.5), have been developed, and show rapid development between 2 and 6 months of age in human infants. Starting with marker tasks for the FEFs, frontal cortex damage in humans results in an

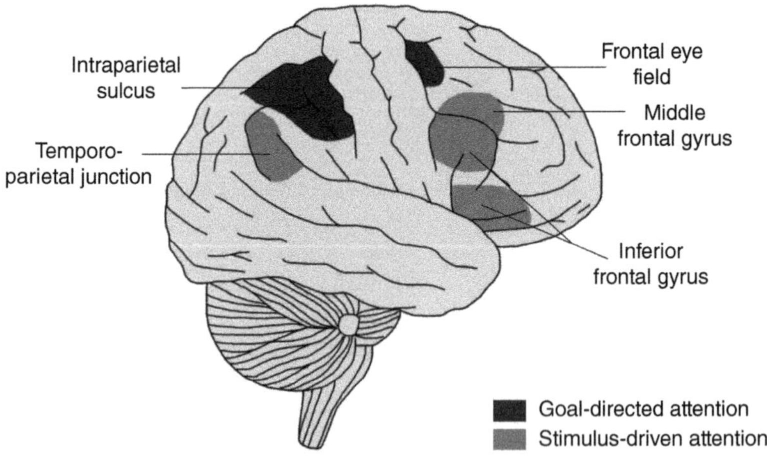

Figure 5.5 Brain areas involved in stimulus-driven and goal-driven attention. Figure provide by Iroise Dumontheil.

inability to suppress involuntary automatic saccades toward targets, and an apparent inability to control volitional saccades (Fischer & Breitmeyer, 1987; Guitton et al., 1985). For example, Guitton et al. (1985) studied normal participants and patients with frontal lobe lesions or temporal lobe lesions in a so-called "anti-saccade" task. In this task participants are instructed to *not* look at a briefly flashed cue, but rather to make a saccade in the opposite direction (Hallett, 1978). Guitton et al. (1985) reported that while normal participants and patients with temporal lobe damage could do this task with relative ease, patients with frontal damage, in particular those with damage around the FEFs, were severely impaired. Patients with frontal damage had particular difficulty in suppressing unwanted saccades toward the cue stimulus.

Johnson (1995) developed a version of the anti-saccade task for use with infants. In an adult anti-saccade task participants are told to make a saccade in the opposite direction to that in which a target appears. Clearly, one cannot give verbal instruction to a young infant to look to the opposite side from where the cue stimulus appears. Instead, in this version of the task infants are motivated by making a second target stimulus reliably more dynamic and colorful than the first. Thus, after a number of such trials, infants may learn to inhibit their tendency to make a saccade to the first stimulus (the cue) when it appears in order to respond as rapidly as possible to the more attractive second stimulus (the target). A group of 4-month-old infants showed a significant decrease in their frequency of looking to the first (cue) stimulus over a number of such trials (Johnson, 1995). A second experiment demonstrated that this decrement was not due to differential habituation to the simpler stimulus. A more refined version of this saccade inhibition task tailored to individual differences between infants has since been published (Holmboe et al., 2008). Since 4-month-olds are able to inhibit saccades to a peripheral stimulus, it is reasonable to infer that their FEF circuit is functioning by this age.

Other tasks conducted with infants are consistent with an increasing prefrontal cortex endogenous control over shifts of attention and saccades around 6 months of age. For example, Funahashi et al. (1989, 1990) devised an oculomotor delayed response paradigm to study the properties of neurons in the DLPFC of macaque monkeys. In this task the monkey plans a saccade toward a particular spatial location, but has to wait for a period (usually between 2 and 5 seconds) before actually executing the saccade. Single-unit recording in the macaque indicates that some cells in the DLPFC code for the direction of the saccade during the delay. Further, reversible microlesions to the area result in selective amnesia for saccades to a localized part of the visual field. A subsequent positron emission tomography (PET) study on human participants has confirmed the involvement of DLPFC (and parietal cortex) in this task (Jonides et al., 1993).

Gilmore and Johnson (1995) devised an infant version of the oculomotor delayed response task: a marker task for DLPFC (see Figure 5.6). The results obtained to date indicate that 6-month-old human infants can perform delayed saccades successfully with delays of up to 5 seconds, suggesting some influence of the prefrontal cortex on eye movement control by this age.

In terms of the three viewpoints on human functional brain development outlined earlier, the Johnson (1990) model of the development of visual orienting could be described as a maturational hypothesis. The emphasis of this model was on how maturational changes in neuroanatomy cause or allow new brain pathways to become active.

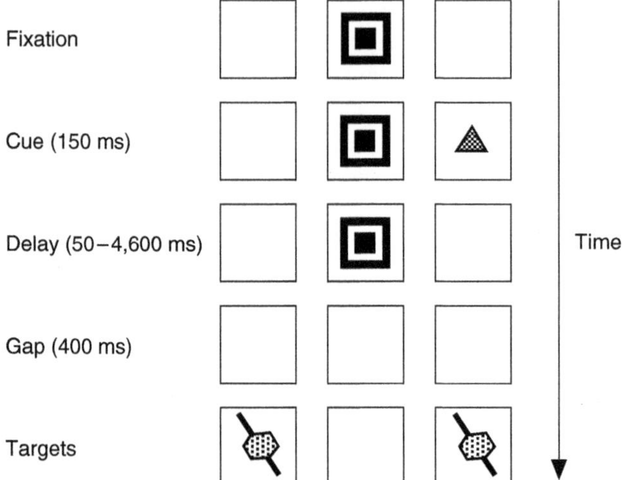

Figure 5.6 The oculomotor delayed response task as designed for use with infants. Infant subjects face three computer screens on which brightly colored moving stimuli appear. At the start of each trial, a fixation stimulus appears on the central screen. Once the infant is looking at this stimulus, a cue is briefly flashed up on one of the two side screens. Following the briefly flashed cue, the central stimulus stays on for between 1 and 5 seconds, before presentation of two targets on the side screens. By measuring delayed looks to the cued location prior to the target onset, Gilmore and Johnson (1995) established that infants can retain information about the cued location for several seconds.

Further Reading   Richards (2008); Johnson (2002).

Data collected more recently suggest that the original model requires some modification, and that it may be beneficial also to consider skill learning perspectives. A more direct assessment of the hypotheses advanced in the Johnson model can be provided by event-related potential measures (time-locked to the initiation of the eye movements: Balaban & Weinstein, 1985; Csibra et al., 1997). By time-locking to the onset of eye movements we can examine the brain events that precede the production of this simple action. In adults such experiments reveal characteristic pre-saccadic components recorded over the parietal cortex prior to the execution of saccades. The clearest of these components is the pre-saccadic spike potential (SP), a sharp positive-going deflection which precedes the saccade by 8–20 ms (Csibra et al., 1997). The spike potential is observed in most saccade tasks in adults and is therefore thought to represent an important stage of cortical processing required to generate a saccade.

Csibra et al. (1998) investigated whether there are pre-saccadic potentials recordable over parietal leads in 6-month-old infants. Given the prediction that by this age infants have essentially the same pathways active for saccade planning as do adults, the authors were surprised to find no evidence of this component (see Figure 5.7) in their infant participants (see also Vaughan & Kurtzberg, 1989). This finding suggests that the target-driven saccades performed by 6-month-olds in their study were controlled largely by subcortical routes for visually guided responses mediated by the superior colliculus.

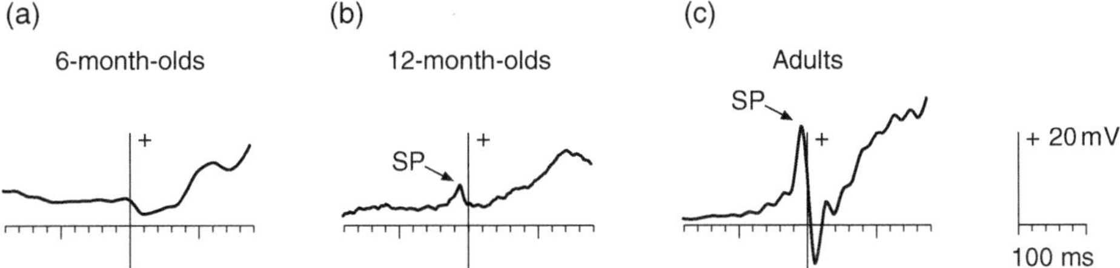

Figure 5.7 Grand-average saccade-locked potentials at the midline, parietal electrode in (a) 6-month-old infants, (b) 12-month-old infants, and (c) adults. The vertical bar marks the saccade onset, a spike potential (SP) is evident in adults and 12-month-olds, but not at 6 months.

Because this result was surprising, the authors decided to conduct two follow-up studies. In one of these they tested 12-month-olds with the same procedure. These older infants did show a spike potential like that observed in adults, though somewhat smaller in amplitude (see Figure 5.6). The other study explored whether the dorsal pathway could be activated in very young infants through a more demanding saccade task. Specifically, they compared ERPs before reactive (target-elicited) and anticipatory (endogenous) saccades in 4-month-old infants (Csibra et al., 2001). They were not able to record any reliable posterior activity prior to either reactive or anticipatory eye movements. Thus, even when the saccade is generated by cortical computation of the likely location of the next stimulus, as in the case of anticipatory eye movements, posterior cortical structures do not seem to be involved in the planning of this action.

In all these infant ERP studies, while there was a lack of evidence for posterior cortical control over eye movements in the experiments with 6-month-olds, the authors observed effects recorded over frontal leads. These saccade-related effects were consistent with the FEF disinhibition of subcortical (collicular) circuits when a central foveated stimulus is removed (Csibra et al., 1998, 2001). In brief, these findings were interpreted in terms of the FEFs helping to maintain fixation on to foveated stimuli by inhibiting collicular circuits. This is consistent with the predictions of the Johnson (1990) model. However, when saccades to peripheral stimuli are made, the ERP evidence indicated that these are largely initiated by collicular circuits, sometimes as a consequence of inhibition being released by the FEFs. In a converging line of thinking, Canfield and colleagues argued on the basis of behavioral and further neuroanatomical evidence that the FEF pathway would precede the more posterior pathways developmentally (Canfield et al., 1997). This early involvement of the FEF pathway could be consistent with a skill learning hypothesis in which more anterior structures get activated earlier than posterior circuits. By this interpretation, there is greater involvement of frontal cortex circuitry in infants because they are still acquiring the skill of planning and executing eye movements.

The third perspective on functional brain development, interactive specialization, could also be brought to bear on this data. From this perspective, the pathways involved in visual orienting are initially less distinct from both surrounding tissue and each other (see Iliescu & Dannemiller, 2008). Further, rather than the onset of a new ability being associated with the onset of functioning in a "new" area, the interactive specialization approach encourages the view that there will be widespread changes across one or more pathways associated with new abilities. In a study where functional activity in children and adults was assessed by fMRI) during the onset of demanding saccade tasks, multiple cortical and sub-cortical areas appeared to change their response pattern (Luna et al., 2001), rather than one or two previously silent regions becoming active (mature).

Saccade Planning

One cortical area involved in oculomotor control for which a more experience-dependent approach has been taken is the parietal cortex. This is a region of the primate cortex which has been implicated in aspects of saccade planning in monkey cellular recording studies, human functional neuroimaging studies, and neuropsychological studies of brain-damaged patients. It is also a region of cortex that undergoes marked developmental changes between 3 and 6 months of age, as demonstrated in both neuroanatomical studies of post-mortem brains (Conel, 1939–1967) and PET (Chugani et al., 1987). Andersen and col-leagues have recorded from single cells in parts of the macaque monkey parietal cortex, and many of these cells code for saccades within an eye- or head-centered frame of refer-ence. In other words, their receptive fields respond to combinations of eye or head position, on the one hand, and retinal distance from the fovea to the target, on the other. This is in contrast to parts of the superior colliculus in which cells commonly respond according to the retinal distance and direction of the target from the fovea.

Zipser and Andersen (1988) and Andersen and Zipser (1988) constructed a computer model in which the some of the units ("hidden layer"; see Chapter 1) developed response properties which closely resembled those observed within regions of the primate parietal cortex. While the details of this model need not concern us here, the fundamental point to note is that the representations for generating saccades within an eye- or head-centered frame of reference emerged as a result of training, and did not require to be "hard-wired" into the network. An important question that remains is whether infants only develop the ability to use extra-retinal coordinates to plan saccades during postnatal life. If they do, this would be consistent with the assumption underlying the model that representations con-trolling eye- or head-centered action need to be constructed postnatally and result from constraints imposed by the network structure and its interaction with the external environment.

Gilmore and Johnson (1997) conducted several experiments designed to ascertain whether the ability of infants to use extra-retinal frames of reference to plan saccades emerges over the first few months of life. In one of these experiments 4- and 6-month-old infants were exposed to two simultaneously flashed targets on a large monitor screen. The targets were flashed so briefly that they were gone before the infant started to make a sac-cade to them. The authors then studied the saccades which infants made in response to these targets (see Figure 5.8). In many trials they made two saccades, the first of these being to the location of one of the two targets. The infants having made a saccade to one of the two targets, Gilmore and Johnson examined whether the second saccade that they made was to the actual location of the second target or whether it was to the retinal loca-tion (the location on the retina at which that target had originally appeared). To make the second saccade to the correct spatial location requires infants to be able to take into account the fact that their eyes had shifted position and then compute the saccade necessary given

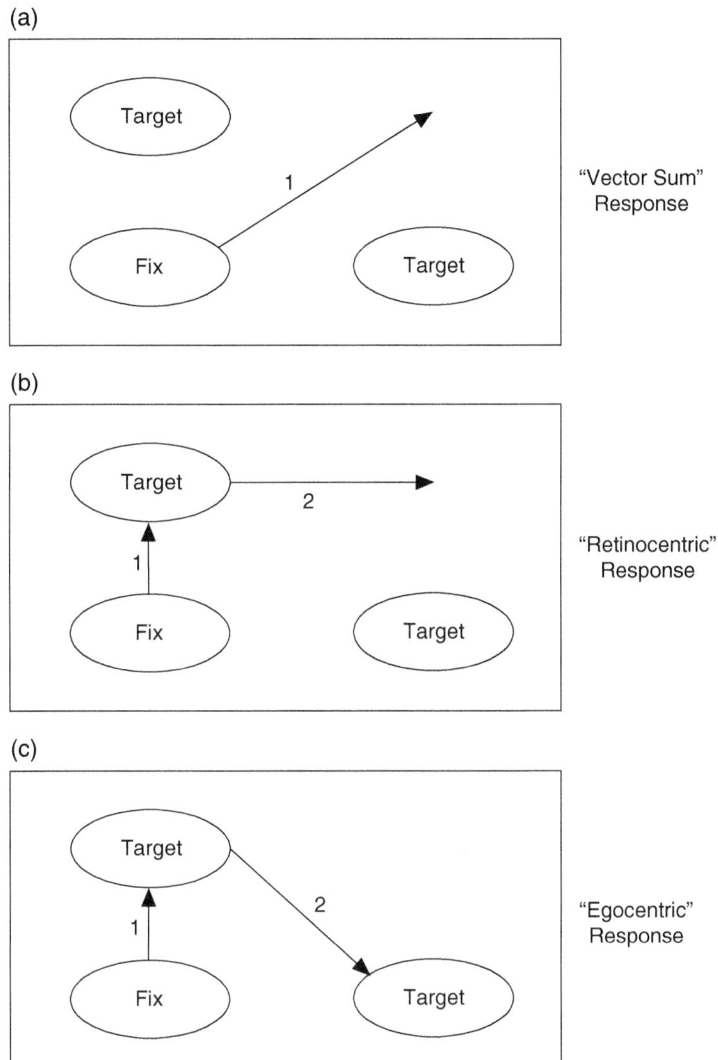

Figure 5.8 Three types of saccades made by young infants in response to two targets briefly flashed as shown. (a) A "vector summation" saccade in which eye movement is directed between the two targets. (b) A "retinocentric" saccade in which the second saccade is directed to the location corresponding to the retinal error when the flash occurred. (c) An "egocentric" saccade which corresponds to the use of extra-retinal information to plan the second saccade. Between birth and 6 months, infants shift from the first two types of response to the third.

the new eye position. The results indicated that for 4-month-olds the majority of second saccades were directed to the retinal location in which the target had appeared. In contrast, for the 6-month-olds, the majority of second saccades were made to the correct spatial location for the other target. These results suggest that the ability to use extra-retinal cues to plan saccades emerges through the first six months of life. However, saccades based on retinal location (and thought to be subcortical in origin) are probably present from birth (Gilmore & Johnson, 1997).

Further Reading   Johnson et al. (2001).

Most of the research described so far has entailed the construction of marker tasks associated with particular brain regions, and therefore entail controlled experimental conditions with stimuli simplified in relation to the real world. Use of eye tracking with infants and toddlers (see Chapter 2) has allowed investigators to explore visual orienting and saccade planning in more details and in semi-naturalistic contexts, such as during the free viewing of natural visual scenes presented on a large screen (Aslin, 2007). One of the variables that can be measured using an eye tracker is fixation duration—the short time period for which the eyes are static in between saccades. During a fixation several processes are known to occur, such as the foveal visual information being processed and encoded in working memory, the next saccade target being selected from available peripheral visual stimuli, and preparing the oculomotor program to bring the next target into foveal vision (Rayner, 1998). One study used head-mounted eye tracking to investigate the dynamics of attention in 14-month-old infants during the "A not B" task, an early measure of working memory and attention (Mulder et al., 2020). This study indicated that flexible attention during the task and attentional focus were related to optimal performance.

Fixation duration in adults is known to be a stable measure of individual differences in attention even across different tasks (Castelhano & Henderson, 2008), and thus it may prove a useful infant predictor of later behavioral and cognitive skills. To test this idea, Papageorgiou and colleagues utilized existing eye-tracker data from infants and found that fixation duration recorded during the first year associated with measures of temperament and attention skills around 3–4 years of age (Papageorgiou et al., 2014). For example, mean fixation duration in infancy correlated with later childhood measures of "effortful control" (the ability to regulate emotions and to inhibit a dominant response) and symptoms of hyperactivity and inattention.

The use of eye trackers has also opened the door to the early training of attention skills, as we can set up situations in which the child's own looking patterns on a screen determine what they see (gaze contingency). Gaze-contingent eye tracking has recently been used to see whether some oculomotor and attention skills can be influenced by training in early development. Reviews of the effects of cognitive and attention training regimes have concluded that the younger the age at which such training occurs, the more likely it is to result in effects that transfer to other tasks (distal transfer) (Wass, 2018). Wass and colleagues developed a number of eye-tracker "baby games" in which 11-month-old infants engaged in gaze-contingent baby games that encouraged them to, for instance, track a slowly moving object across a field of view that included distractors and occluding surfaces (Wass et al., 2011). After several training sessions on different games targeted at attentional control, assessments of oculomotor skills and attention showed reduced saccadic and disengagement reaction times, and improvements in cognitive control. These changes were not seen in a matched group of babies who simply passively watched videos for the same period of time. Currently, it remains unknown whether these effects persist over a longer period of time, or whether they are transient shifts that quickly return to the pre-training baseline. One small-scale randomized control trial using a similar approach with infants at risk for attention deficit/hyperactivity disorder (ADHD) did not show reliable short-term effects of gaze-based training, but further work needs to be done to establish the usefulness of this approach (Goodwin et al., 2021; see below for further discussion of ADHD).

> **Further Reading**   Scerif (2010); Wass (2018).

In this section we have seen that while even newborns are capable of simple target-driven saccades, oculomotor skills continue to develop throughout the first year. Influential maturational models have been only partially supported by recent ERP and neuroimaging evidence. In particular, the finding that frontal regions may be more important than more posterior areas presents a challenge to the maturational approach. It is possible that in early development these pathways are initially less segregated than is observed in adults, and therefore may interact and process information in quite different ways (Iliescu & Dannemiller, 2008). Studies have shown how measures of individual differences in eye movements in infants associate with later attention-related behaviors, and that gaze-contingent paradigms can be used to influence aspects of visual orienting and attention early in life.

Visual Attention

Our discussion so far has mainly concerned *overt* shifts of attention due to eye and head movements. However, adults are also capable of shifting their attention *covertly* (i.e., without moving the eyes or other sensory receptors) and of sustaining their attention for prolonged periods of time. This allows us to enhance our processing of some spatial locations or objects within our visual field, to the exclusion of others.

One way in which evidence for covert attention has been provided in adults is by studying the effect on detection of cueing saccades to a particular spatial location. A briefly presented cue serves to draw covert attention to the location, resulting in the subsequent facilitation of detection of targets at that location (Maylor, 1985; Posner & Cohen, 1980). While detection of and responses to a covertly attended location are *facilitated* if the target stimulus appears very shortly after the cue offset, with longer latencies between cue and target, saccades toward that location are *inhibited*. This latter phenomenon, referred to as "inhibition of return" (Posner et al., 1985), may reflect an evolutionarily important mechanism which prevents attention returning to a recently processed spatial location. In adults, facilitation is reliably observed when targets appeared at the cued location within about 150 ms of the cue, whereas targets appearing between 300 and 1300 ms after a peripheral (exogenous) cue result in longer detection latencies (e.g., Maylor, 1985; Posner & Cohen, 1980, 1984).

Following lesions to the posterior parietal lobe, adults show severe neglect of the contralateral visual field. According to Posner and colleagues, this neglect is due to damage to the "posterior attention network." This refers to a brain circuit which includes not only the posterior parietal lobe, but also the pulvinar and superior colliculus (Posner, 1988; Posner & Petersen, 1990; see Figure 5.4 for all but the pulvinar). Damage to this circuit is postulated to impair participants' ability to shift covert attention to a cued spatial location. The involvement of these regions in shifts of visual attention has been confirmed by PET studies. As mentioned above, both neuroanatomical (Conel, 1939–1967) and PET (Chugani et al., 1987) evidence from the human infant indicate that the parietal lobe is undergoing substantive and rapid development between three and six months after birth. The question arises, therefore, as to whether infants become capable of covert shifts of attention during this time.

Since infants do not accept verbal instruction and are poor at motor responses used to study spatial attention in adults, such as a key press, the only behavioral response available to demonstrate facilitation and inhibition of a cued location is eye movements. That is, overt shifts are used to study *covert* shifts of attention by examining the influence of a cue stimulus (which is presented so briefly that it does not normally elicit an eye movement) on infants' subsequent saccades toward conspicuous target stimuli. Using these methods, Hood and Atkinson (1991, see Hood, 1995) reported that 6-month-old infants have faster reaction times to make a saccade to a target when it appears immediately after a brief (100 ms) cue stimulus than when it appears in an uncued location. A group of three-month-old infants did not show this effect. Johnson (1994; Johnson & Tucker, 1996) employed a similar procedure in which a brief (100 ms) cue was presented on one of two side screens, before bilateral targets were presented either 100 or 600 ms later. It was hypothesized on the basis of the adult findings that the 200 ms stimulus onset asynchrony (SOA) would be short enough to produce facilitation, while the long SOA trials should result in preferential orienting toward the opposite side (inhibition of return). This result was obtained with a group of 4-month-old infants, suggesting the possibility that infants are capable of covert shifts of attention at this age. Consistent with the previous findings, these effects were not observed in a group of 2-month-old infants.

Another manifestation of covert attention concerns so-called "sustained" attention. Sustained attention refers to the ability of participants to maintain the direction of their attention toward a stimulus even in the presence of distracters. Richards (2001, 2003) has developed a heart rate marker for sustained attention in infants. The heart-rate-defined period of sustained attention usually lasts for between 5 and 15 s after the onset of a complex stimulus (see Figure 5.9).

In order to investigate the effect of sustained attention on the response to exogenous cues, Richards (2001, 2003) used an "interrupted stimulus method" in which a peripheral stimulus (a flashing light) is presented while the infant is gazing at a central stimulus (a TV screen with a complex visual pattern). By varying the length of time between the onset of the TV image and the onset of the peripheral stimulus, he was able to present the peripheral stimulus either within the period of sustained covert attention or outside it. Richards found that during the periods when heart rate was decreased (sustained endogenous attention) it took twice as long for the infant to shift their gaze toward the peripheral stimulus as when heart rate had returned to pre-stimulus levels (attention termination). Further, those

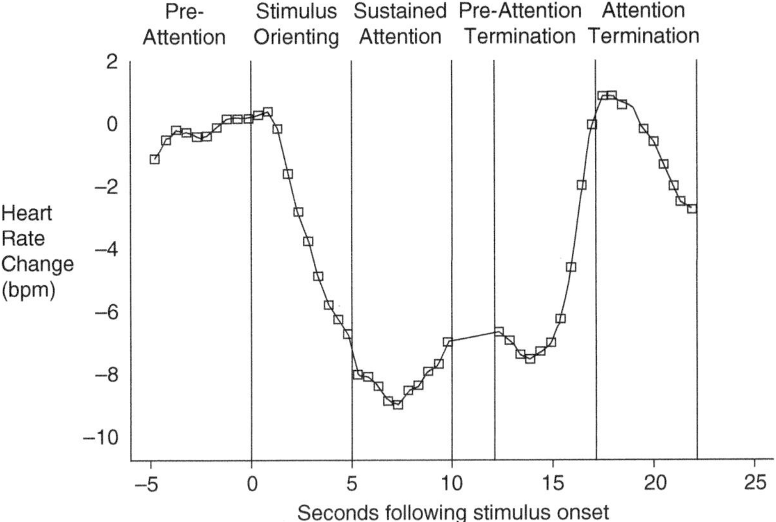

Figure 5.9 Heart-rate-defined phases of sustained attention.

saccades that are made to a peripheral stimulus during sustained attention are less accurate than normal and involve multiple hypometric saccades, a characteristic of collicular-generated saccades (Richards, 1991). Thus, the lack of distractibility during periods of sustained attention is likely to be due to cortically mediated pathways inhibiting collicular mechanisms. This may also be related to connectivity within key cortical attention networks, which was found to develop rapidly in electroencephalography (EEG) measures from 6 to 12 months (Xie et al., 2019).

> **Further Reading** Richards (2008), and see Figure 5.5 for details of brain areas involved in different types of attention.

A number of studies have traced developments in visual attention during childhood using purely cognitive methods, but have described the continuation of similar transitions to those observed in early infancy. Three transitions that have been described are the greater ability to expand or constrict a field of attention (e.g., Chapman, 1981; Enns & Girgus, 1985), the greater ability to disengage attention from distracting information or invalid cueing (Akhtar & Enns, 1989; Enns & Brodeur, 1989), and a faster speed of shifting attention (Pearson & Lane, 1990).

Enns and Girgus (1985) tested school-age children and adults in speeded classification tasks involving a stimulus composed of two elements that varied in distance (visual angle). Participants had to classify stimuli on the basis of one of the two elements. Younger children (6–8 years) experienced more interference when the elements were closely spaced than did older children (9–11 years) and adults. The same stimuli were used for a second task in which both of the elements had to be taken into account. In this task the younger children had difficulty when the elements were separated by large visual angles.

The authors conclude that younger children have problems in contracting and expanding the size of attentional focus. An ERP study of auditory attention also concluded that there is a development in the ability to narrow attentional focus during childhood (Berman & Friedman, 1995).

An observation that may be related is that older children and adults are often reported as being able to shift their attention more rapidly than younger ones. For example, Pearson and Lane (1990), using a spatial cueing paradigm, observed that younger infants took longer to covertly shift their attention to more peripheral targets, whereas they were almost as fast as adults to shift to targets very close to fixation. This indicates that it is the speed of shifting, rather than the latency to elicit a covert shift, which improves with age. The increasing speed of shifting attention with age has also been reported during infancy (Johnson & Tucker, 1996), indicating that this developmental transition may be a gradual one that begins early in life.

In several studies, younger children and infants have been argued to have greater difficulty in disengaging from distracting stimuli or invalid spatial cues. For example, Enns and Brodeur (1989) used a spatial cueing paradigm to cue either neutrally (all locations cued), unpredictably (random cueing), or predictably (cue predicts target presentation). The results from participants aged 6, 8, and 20 years indicated that while all age groups automatically oriented attention to the cued location, the children processed targets in non-cued locations more slowly than did adults, and did not take advantage of the predictability of the cues. Thus the costs and benefits of cueing were greater in the younger participants due to an increased cost imposed by invalid cueing. In an fMRI study, Konrad and colleagues (Konrad et al. 2005) found that 8–12-year-old boys showed greater activation in the superior frontal gyrus and putamen/insula and less activation in the temporo-parietal junction than adults (20–24-year-old males) when reorienting their attention after an invalid visual cue, compared to the valid cue condition, as well as being slower to reorient than adults. This is another example of a shift from frontal to more posterior brain activation over the course of development. Tipper et al. (1989) suggest that these deficits are due to the relative inability to inhibit irrelevant stimuli, a conclusion that is consistent with more recent findings (Brodeur & Boden, 2000; Wainwright & Bryson, 2002). Once again, similar developmental trends have been observed in infancy experiments where younger infants are more likely to fail to disengage from competing stimuli, such as the obligatory attention described earlier.

Research on the neural basis of the development of attention in childhood has been based on four topics: (a) ERP studies, (b) studies of typical genetic variation, (c) effects of early cortical damage, and (d) developmental disorders. Richards (2003) describes several experiments in which he has used the spatial cueing procedure while recording ERPs from infants in order to detect neural signatures of covert attention. In one study he examined the "P1 validity effect" in young infants. The P1 is a large positive ERP component that occurs around 100 ms after stimulus presentation. Studies with adult participants have shown that the P1 is enhanced in scale in valid trials (where the cue correctly predicts the target) (Hillyard et al., 1995). This is of interest since this short-latency component reflects early stages of visual processing, demonstrating that shifts of covert attention modulate early sensory processing of the target. Richards (2003) reports that while there was little ERP evidence for covert attention shifts in 3-month-old infants, by 5 months the pattern of

ERP data resembled that in adults, indicating that infants at this age were shifting attention to the cued location covertly.

Some recent studies have examined how commonly occurring genetic variants (allelic variation; see Chapter 3) influence aspects of attention in the typical population during development. Based on the idea that mappings between individual variation in genetics and behavior may be more direct in infants than in adults (Johnson & Fearon, 2011), several teams have begun to explore how common genetic variation in babies may relate to the development of their attention skills. For example, Leppanen and colleagues attempted to associate genetic polymorphisms linked to the production of serotonin in the central nervous system to variations in attention skills in a group of typically developing infants (Leppanen et al., 2011). They found that babies with one variant (the T-carrier genotype of the tryptophan hydroxylase gene 2 [TPH2]-703 G/T polymorphism) had greater difficulty in shifting their attention from a central stimulus to a target presented in the periphery than babies with a different variant (TPH2 G/G homozygotes). Although one of the simplest of attention paradigms, the ability to disengage attention from a central stimulus and orient to a peripheral one turns out to be a sensitive marker of typical developmental stages and atypicality. As discussed above, in typical development infants go through a transient stage of sticky fixation in which they seem unable to make such shifts of attention. Another recent study of this kind shows that polymorphisms in dopaminergic system genes also affect components of infant attention (Holmboe et al., 2010). This study investigated variants of four dopamine system genes on different aspects of attention, including the "freeze frame" task discussed earlier. The authors found that when in the presence of an engaging central stimulus, infants with one variant (the Met/Met variant of the catechol-O-methyltransferase gene [COMT] valine 158 methionine polymorphism) were significantly less distractable than those with another genotype (the Val/Val genotype). In addition, this effect was only present in infants that did not have a particular variant of another gene (two copies of the dopamine active transporter 1 gene [DAT1] 10-repeat allele), suggesting an interaction between two dopaminergic genes in influencing this aspect of visual orienting. Interestingly, in both of these studies the association with the polymorphisms in question were clearer in trials in which the central stimulus was more dynamic and colorful (Holmboe et al., 2010) or more affectively salient (Leppanen et al., 2011), suggesting that the degree of engagement in processing the central stimulus is critical for the associations observed. Further work building on this has indicated that dopaminergic, rather than cholinergic, systems are key in this process (Markant et al., 2014). Exploring further possible interactions between dopamine and other neurotransmitter polymorphisms in early attentional development remains an exciting direction for future research.

Another way to examine the neural basis of covert orienting in development has been to assess the consequences of perinatal damage to the cerebral cortex. Joan Stiles and colleagues examined spatial cueing in infants who had unfortunately suffered perinatal damage to one of four quadrants of the cortex (Johnson et al., 1998). The results were somewhat surprising in that the posterior lesions that would normally cause deficits in adults had no effect on the infants. In contrast, frontal damage did have a measurable effect on spatial cueing. While surprising, these results fit well with work from other laboratories. For example, Craft, Schatz, and colleagues have studied the consequences of perinatal brain injury (sometimes associated with sickle cell anemia) on performance in spatial cueing tasks during childhood. In several studies deficits were observed following anterior (frontal) damage, and not (or less) with posterior damage (Craft et al., 1994; Schatz et al., 2000, 2001).

A third way to address the neurodevelopment of covert attention is to study disorders of this process in groups of atypically developing children. ADHD is characterized by inattention, hyperactivity, and impulsivity beginning before 7 years of age (Karatekin, 2001). Estimates for its prevalence run as high as 3–5% of school children in the United States, although this figure varies somewhat across cultures. Another condition that has been linked to deficits in attention is autism (see Chapter 2 for an introduction to autism). Two components of attention that have been commonly studied in both autism and ADHD are orienting and reflexive attention-shifting (as described above), and sustained attention (associated with frontal areas such as the anterior cingulate, FEFs, and DLPFC; for review see Petersen & Posner, 2012). Using methods similar to those described above, atypicalities in visual orienting have been well characterized in autism, with several studies showing that children diagnosed with the condition show relatively specific problems in disengaging and shifting attention under competition conditions (e.g., Keehn et al., 2010; Landry & Bryson, 2004). A recent meta-analysis found generally slowed orienting in autism that was not modulated by the presence of a central stimulus and that increased in magnitude with age (Landry & Parker, 2013). In contrast, children with ADHD show generally slowed reaction times to respond to a peripheral stimulus but this difficulty is not increased in competition conditions (Tajik-Parvinchi & Sandor, 2013). Thus, this literature indicates that disengagement (sticky fixation) is particularly problematic for children with autism, whilst children with ADHD may show a more general slowing of orienting.

Children with ADHD show well-documented deficits in sustained attention across a variety of contexts (e.g., Loo et al., 2009; Schoechlin & Engel, 2005). Slower reaction times are often accompanied by increased intra-individual variability, which is generally assumed to reflect occasional lapses of attention (for review see Tamm et al., 2012). The literature on sustained attention in autism is less clear, and direct comparisons of sustained attention in autism and ADHD have variously indicated greater impairments in ADHD (e.g., Johnson et al., 2007); similar impairments but with a greater decrease in vigilance in ADHD over time (Swaab-Barneveld et al., 2000); similar deficits but more impulsive behavior in autism (Riccio & Reynolds, 2001); or broadly similar deficits across autism, ADHD, and comorbid groups (Nydén et al., 2010).

As with other symptoms of autism that are evident following diagnosis, the question arises as to whether deviant patterns of attention are a compounded symptom of original deficits in other domains, such as in social cognition (see Chapter 7), or whether initial problems with attention actually cause some of the other symptoms of the condition (Elsabbagh & Johnson, 2007). One way to address this issue is to study babies at risk for a later diagnosis of autism or ADHD by virtue of having an older sibling already diagnosed. Consistent with evidence from older children diagnosed with autism, by around 12 months of age slowed disengagement from a central to a peripheral stimulus is an early marker of infants who go on to diagnosis (Elison et al., 2013; Elsabbagh, Fernandes, et al., 2013; Zwaigenbaum et al., 2005). Similar atypical behaviors are also seen during object exploration (Sacrey et al., 2013) and in response to social stimuli such as a name call (Nadig et al., 2007). These differences are typically not present at 6 months (Elsabbagh, Fernandes, et al., 2013; Nadig et al., 2007; Sacrey et al., 2013; Zwaigenbaum et al., 2005; though see Elison et al., 2013), suggesting that they emerge on an accelerated timescale relative to other early behavioral symptoms of autism. A meta-analysis of early life (0–5 years) precursors of ADHD found, across 149 studies, that children with current or later emerging ADHD were experiencing difficulties in a broad range of domains, with largest effect sizes, including attention, intra-individual variability, inhibition, and also language ability, sensory processing, and motor ability (Shephard et al., 2022). This meta-analysis further investigated the impact of early interventions on these precursors across 22 studies. Most studies were carried out with preschool-age children (3–5-year-olds) and the interventions mostly directly or indirectly targeted self-regulation through e.g., parent-training programs, cognitive training, neurofeedback, or yoga. The results indicated that when pooled, these interventions improved ADHD symptoms and working memory measures but not other measures, including attention.

> **Further Reading**   Cornish & Wilding (2010).

As discussed in Chapter 2, ADHD is associated with structural brain differences; it is also associated with differences in brain activation during attention tasks. In a meta-analysis of 13 studies of visuospatial selective, sustained, and flexible attention tasks, participants with ADHD were found to show lower activation than typically developing participants in the right DLPFC, left putamen and globus pallidus, right posterior thalamus and tail of the caudate, and right inferior parietal lobe, precuneus, and superior temporal lobe, as well as increased activation in the left cuneus and right cerebellum (Hart et al., 2013). While the number of studies was insufficient to compare these patterns of activation at different ages, a metaregression analysis indicated that a greater percentage of patients receiving long-term psychostimulant treatment was associated with more typical right caudate activation during attention tasks, a similar effect to that observed for right basal ganglia structural differences (Nakao et al., 2011; see Chapter 2).

General Summary and Conclusions

This chapter began with a discussion of the contribution of peripheral systems (retinal) development in the emergence of basic visual functions. Retinal development can only partially explain improvements in basic visual functions, indicating that brain changes are important also. With regard to binocular vision, neuroanatomical and computational modeling evidence highlights the importance of the segregation of inputs to layer 4 cells in the primary visual cortex. This increased segregation of inputs from the two eyes can be viewed as the result of a self-organizing neural network constrained by certain intrinsic and extrinsic factors.

Moving from sensory processing to sensorimotor integration, we proceeded to describe an attempt to use the developmental neuroanatomy of the cortex to make predictions about developmental changes in visual orienting in infants. A number of behavioral marker tasks for cortical regions involved in oculomotor control were reviewed and discussed in relation to the maturational model. Neuroimaging and behavioral studies require modifications to the original model, and suggest that an interactive specialization view may eventually prove more fruitful. The advent of eye-tracker technology suitable for use in early development has allowed for new types of investigations of eye movement control. Fixation duration in infancy has been associated with later measures of attention in some studies, while other aspects of visual orienting proved amenable to training via eye tracking.

Finally, experiments concerned with the development of covert (internal) attention shifts in infants and children were described. A number of changes in the flexibility and speed of shifting attention were shown to originate in infancy but continue into childhood. Some of these abilities have been related to typical genetic variants and may be compromised in developmental disorders.

Key Issues for Discussion

- To what extent do maturational models explain changes in the development of visual orienting abilities over the first year?
- What methods are best used to assess covert attention processes in infants, toddlers, or children?
- Are attention problems in autism likely to be a cause or a consequence of difficulties in other domains?

6

Perceiving and Acting in a World of Objects

Objects are special in our sensory world since they are both recognized and manipulated. It is important to know what objects are, and where they are, in order to act on them. Neuroscience evidence suggests that these two "what and where" functions are computed by two different pathways: a ventral recognition pathway and a dorsal sensorimotor action pathway. Research examining the development and relationship between these two pathways is discussed, within the context of brain-inspired models. Williams syndrome is discussed as an example of possible dissociation between the pathways, and the concept of "dorsal stream vulnerability" is introduced and illustrated with the example of developmental coordination disorder. Bursts of high-frequency neural oscillations have been related to the binding of features to compose objects and to the retention of objects following occlusion, and are a useful measure to understand these processes in infants and during development.

Physical objects are special in our sensory world since—unlike landscapes, faces, and sounds—they are not only recognized and categorized but are also often manipulated using our hands and feet. As adults, perceiving and acting on everyday objects seems easy. However, further consideration reveals that complex computations and rich representations need to underlie these processes. For example, objects have to be recognized from multiple different viewpoints, and under conditions where they are partially obscured (partial occlusion) or in front of complex backgrounds (object parsing). Our fingers and hands have to be adjusted according to the anticipated size and weight of the object, and our wrists oriented appropriately to let us grasp the object in the appropriate place and correct orientation. There have been many behavioral studies conducted on object processing in infants and young children, however, evidence from cognitive neuroscience is also important to inform and constrain theories of this domain of cognitive development.

Developmental Cognitive Neuroscience: An Introduction, Fifth Edition. Michelle de Haan, Iroise Dumontheil, and Mark H. Johnson.
© 2023 John Wiley & Sons Ltd. Published 2023 by John Wiley & Sons Ltd.
Companion website: www.wiley.com/go/johnson/devneuro5e

The Dorsal and Ventral Visual Pathways

In Chapter 5 we discussed several pathways that underlie visual orienting and attention. There is now a substantive body of evidence that visual information processing about objects is divided into two relatively separate streams in the brain. The detailed connectivity of these cortical routes is very complex (see van Essen et al., 1992), but, in simplified terms, one pathway (the ventral route) extends from the primary visual cortex through to parts of the temporal lobe, while the other pathway (the dorsal route) goes from the primary visual cortex to the parietal cortex (see Figure 6.1). The exact point at which the routes separate is still debated. For reasons we will discuss below, the ventral route is sometimes called the *What* or *Perception* pathway, while the dorsal route is often called the *Where* or *Action* pathway. We will start our discussion with the dorsal route.

All visually guided actions take place in space, but the spatial processing required will differ according to the action to be performed. Evidence for multiple spatial systems can be seen in the spatial coding shown by neurons in the dorsal stream. For example, as we heard in Chapter 5, some cells in the parietal cortex anticipate the retinal consequences of saccadic eye movements and update the cortical (body-centered) representation of visual space to provide continuously accurate coding of the location of objects in space. Other cells have gaze-dependent responses. That is, they mark where the individual is looking with respect to eye-centered coordinates systems. Both of these provide egocentric spatial codings that are only useful over very short periods of time since every time the animal moves the coordinates have to be recomputed.

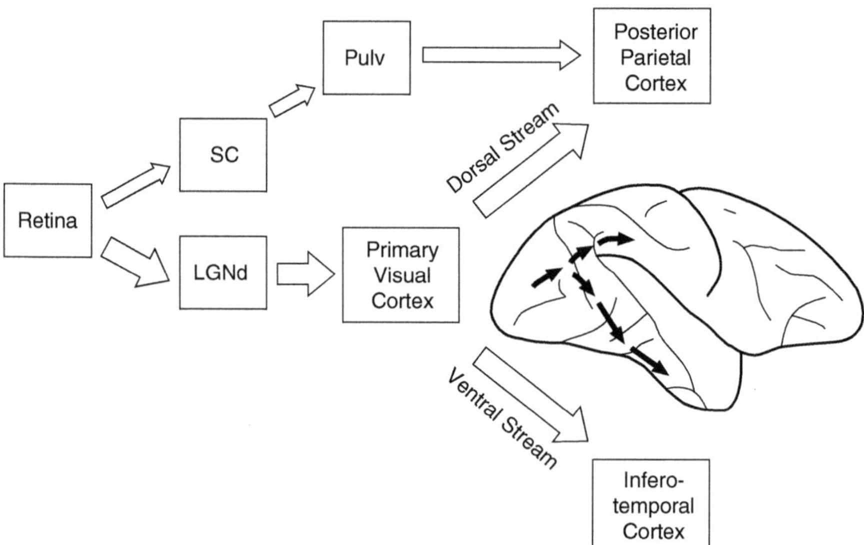

Figure 6.1 Major routes whereby retinal input reaches the dorsal and ventral streams. The diagram of the brain on the right of the figure shows the approximate routes of the projections from primary visual cortex to posterior parietal and the inferotemporal cortex, respectively. LGNd, lateral geniculate nucleus, pars dorsalis; Pulv, pulvinar; SC, superior colliculus.

In the real world, target objects are often moving. Hence, it is not only necessary for us to track that motion in order to localize the object in space, but it is also necessary to anticipate the object's movement. Some cells in the parietal cortex appear to be involved in the tracking of moving objects. Moreover, many of these cells continue to respond after the stimulus has disappeared (Newsome et al., 1988). In addition, there is selectivity in other parts of the dorsal stream for relative motion and size changes when an object moves toward or away from the viewer. Many neurons on the dorsal pathway are also driven by large-scale optical flow fields, suggesting that self-motion is being computed. Cells in the dorsal pathway also code size, shape, and orientation, which are necessary for the proper reaching for and grasping of an object.

There is also evidence suggesting that the different spatial-temporal systems described above are partially segregated into different regions and routes within the dorsal pathway. Thus, the dorsal stream could be viewed as a pathway with many parallel computations of different spatial-temporal properties occurring at once. Different streams compute different spatial-temporal analyses in different coordinate systems, possibly with different effector systems in mind. Milner and Goodale (1995) argue that the cells of the parietal cortex are neither sensory nor motor, but rather sensorimotor. They are involved in transforming retinal information (sensory) into motor coordinates (motor) and in transducing perceptual input into motor actions.

The properties of neurons in the adult ventral stream seem complementary to those in the dorsal stream. As one proceeds along the ventral stream, cells respond to more and more complex clusters of features. At the higher levels, the complex cells show remarkable selectivity in their firing. These neurons are all selective to the figural and surface properties of objects (i.e., the internal object features). More importantly, many of the cells have very large receptive fields on the retina. That means that, although they can process feature information, they lose much of their spatial resolution on the retina. In effect, these cells develop spatially invariant representations of objects by responding to the presence of a consistent feature cluster independently of its position. Some cells seem to respond maximally to a preferred object orientation (independently of position) thereby computing a "view-centred representation." Other cells respond equally to an object in any orientation. That is, they have developed a transformation-invariant representation. Thus, the properties of neurons along the ventral pathway are entirely consistent with what would be expected of a system concerned with recognizing objects, scenes, and individuals with enduring characteristics rather than the moment-to-moment changes in the visual array that occur in a natural setting. Transformation-invariant representations could provide the basic raw material for recognition memory and other long-term representations of the visual world.

Why should such different streams of processing be present in adult primates? If the representations in the dorsal stream are closely linked to the functions of the motor system, then it is not surprising that spatial-temporal information is at a premium down this pathway. Motor actions involve localizing targets within a three-dimensional spatial-temporal world. In contrast, recognition or identification of objects requires that spatial-temporal variability be minimized. Early work in machine vision found that view-invariant recognition (i.e., the ability to recognize an object as the same independently of orientation and location) was a very difficult computational problem. One of the most efficient ways of doing this is to factor out spatial variability. However, removing spatial information from the object representation is completely at odds with the requirements of the motor system. Hence the need for two distinct object classes of object representations.

Further Reading   Johnson et al. (2008); Atkinson & Braddick (2020).

The three approaches to human functional brain development discussed in Chapter 1 cast a different light on developmental hypotheses about the dorsal and ventral pathways. According to a maturational viewpoint, we need to consider which of the pathways develops first, and whether this can explain aspects of behavioral development of object processing. For example, Atkinson (1998) argued, on the basis that infants and children can perceive coherent forms prior to judging motion coherence, that the dorsal pathway develops later than the ventral. However, evidence from developmental neuroanatomy in human infants is not compelling in this regard. For example, in resting (with no task) positron emission tomography (PET) studies of glucose uptake, virtually identical overall patterns of developmental change are seen in the temporal and parietal cortex (Chugani et al., 2002). Resting blood flow measurements and structural neuroanatomy studies cannot inform us directly about function. In Chapter 5 we reviewed several functional imaging studies with young infants showing activity in primary visual and some ventral pathway structures. However, these tasks involved passive viewing of two-dimensional visual patterns or faces and were thus unlikely to evoke dorsal pathway activity even in adults. In non-human primates, there is evidence of ventral pathway functioning from as young as 6 weeks of age. Rodman and colleagues (Rodman, et al., 1991) established that neurons within the superior temporal sulcus were activated by complex visual stimuli, including faces, from the earliest age at which they could record—6 weeks. Unfortunately, equivalent data are not available for dorsal pathway neurofunctions, and so no comparison is possible. Thus, there is currently very little direct evidence relevant to the question available (but see Iliescu & Dannemiller, 2008).

From a skill learning perspective, we need to inquire into the acquisition of perceptual recognition skills in the ventral pathway and the emergence of sensorimotor integration skill in the dorsal pathway. Further, we might investigate whether interactions between processing in the two pathways is acquired during development. The third perspective outlined in Chapter 1, interactive specialization, suggests yet another theoretical possibility for the dorsal and ventral visual pathways: that as the two pathways become more specialized and acquire representations appropriate for either recognition or action, they become less interactive and tend to be co-activated less often. This view would also predict that the two pathways initially begin intermixed and hard to dissociate, but with development the dissociation becomes more complete. Further, from this perspective, the complementary specialization of the two pathways emerges during the development of the individual. This issue will be taken up further below, when brain pathways involved in processing of tools are discussed.

Hidden Objects

It was Jean Piaget who originally noticed some unusual "errors" in the way that young children thought about, and acted on, objects that were out of view. Specifically, Piaget (1954) reported that infants fail to retrieve objects partly occluded by a surface until 7–8 months, fail to search manually for fully occluded objects until several months later, and fail to trace the spatial-temporal trajectory of a hidden object to its final destination until 1½ years of age. Piaget proposed that these developmental changes reflect a conceptual revolution in the first years of life that results in the construction of the first true object representation: an "object concept."

Subsequently, studies of the early development of object perception have qualified Piaget's original account. Experiments that rely on measures of looking, rather than manual reaching, have shown that even 4-month-olds perceive that a partly occluded object continues behind its occluder. For example, after familiarization with an object moving behind a central occluder, infants generalized to (i.e., looked less at) a fully visible complete object relative to a fully visible display with a gap where the occluder had been (Johnson & Aslin, 1996; Johnson & Nanez, 1995; Kellman & Spelke, 1983; Slater et al., 1990). Two-month-old human infants were not found to perceive complete objects over occlusion when tested with the displays used with 4-month-olds, but they performed like the older infants when tested with enhanced displays in which the occlusion and motion relationships were easier to detect (Johnson & Aslin, 1995). In studies using even more enhanced displays, abilities to perceive the complete shapes of partly occluded displays have been extended to 3-week-old human infants (Kawabata et al., 1999).

These findings provide evidence that human perceptual systems rapidly come to detect one kind of invariance in natural scenes: the invariant shape of an object over changing patterns of occlusion and background. Experiments in psychophysics and in cognitive neuroscience provide evidence that adults detect many other invariant properties of objects as well, including the invariant view-dependent shape of an object over changes in object size and position (e.g., Grill-Spector et al., 1998). These abilities also appear to have their roots in infancy, where some evidence for size and shape constancy has been obtained even with newborn infants (Slater et al., 1982). All these findings cast doubt on Piaget's claim that the ability to perceive objects depends on a process akin to scientific reasoning. They raise the question of whether infants might also be able to represent objects that they cannot see, when they are tested with looking methods. Evidence from several laboratories indicate that in looking tasks infants represent fully occluded objects long before they pass Piaget's object permanence tasks (e.g., Baillargeon, 1993; Spelke et al., 1992), although scattered negative findings also have been reported using this method (see Haith & Benson, 1998, for review). Given that different results are obtained with looking methods compared to other infant behavioral testing paradigms, several groups have turned to cognitive neuroscience theories and methods to resolve this paradox. Studies in cognitive neuroscience suggest several potential answers to this issue that are currently under active investigation: (a) the extent of integration of the dorsal and ventral visual pathways, (b) the changing strength of representations of objects in the brain, and (c) the possible inability of infants to plan the necessary actions to retrieve hidden objects. Since the second of these possibilities is discussed in Chapter 13, and the third in Chapter 10, we will focus here on explanations associated with the dorsal and ventral visual pathways.

Mareschal and colleagues chose to explore the hypothesis that the discrepancy between looking and reaching tasks is due to a relative lack of integration between the dorsal and ventral visual pathways in early infancy (Mareschal et al., 1999). The logic behind this hypothesis is that object-directed action toward hidden objects requires a degree of interaction between the two pathways. This idea was initially explored through a connectionist model designed to simulate some of the computational properties and development of the dorsal and ventral pathways. Figure 6.2 shows a schematic outline of this "dual route" processing model. Consistent with the neurophysiology reviewed above, the model has an object recognition route (equivalent to the ventral pathway) and a trajectory prediction route (dorsal pathway). The former route develops a spatially invariant representation of an object to which it is exposed (using the unsupervised learning algorithm developed by Foldiak, 1996), while the latter learns to predict the next "retinal" position of the object. Through recurrent connections, both pathways have some degree of memory. A third component of the model is a response integration network that corresponds to the infant's ability to coordinate and use the information it has about object position and object identity. This network integrates the internal representations generated by the two pathways as required by a retrieval response task.

The model embodies the basic architectural constraints on visual cortical pathways revealed by neuroscience: an object recognition network that develops spatially invariant feature representations of objects; a trajectory prediction network that is blind to surface features and computes appropriate spatial-temporal properties even if no actions are undertaken toward the object; and a response module that integrates information from these two networks for use in voluntary actions.

A prediction of the above model has been tested (Mareschal & Johnson, 2003). In this behavioral experiment infants viewed an array presented on a video monitor in which two different objects moved behind, and then reappeared from behind, an occluding surface.

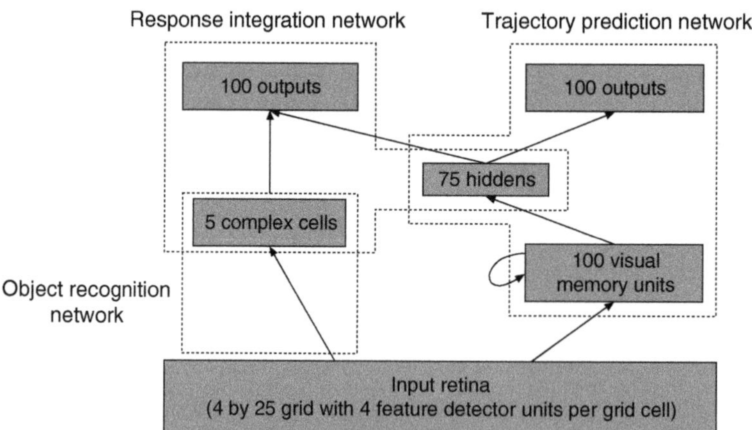

Figure 6.2 A diagram showing the object processing model of Mareschal et al. Mareschal et al. 1999/John Wiley & Sons.

In some conditions these objects were potentially graspable toys, while other conditions involved stimuli more likely to engage ventral pathway function such as faces (see Chapter 7). When the objects reappeared following removal of an occluder, the objects concerned could have either their features switched (e.g., color, or the identity of a face), or their spatial location, or both. If infants were surprised (looked longer) when just features were changed, this was taken to indicate ventral (recognition) pathway processing, whereas if surprise was shown when only the location of the objects changed, this was taken to indicate dorsal pathway retention. The results indicated that infants were able to encode either the surface features, or the spatial location, but not both. This suggests that infants are only capable of activating either the dorsal or the ventral pathway, but, unlike adults, not both at the same time. As yet, behavioral evidence such as this does not conclusively decide between the view that there is increasing integration between the two visual pathways, or the contrary idea that there may be increasing dissociation between the two pathways (Jacobs et al., 1991; O'Reilly, 1998).

The ideas behind the Mareschal model have also been extended by Kaufman, Mareschal, and Johnson (2003), who reviewed some of the apparently inconsistent findings from behavioral experiments on object perception in infants in terms of the potential graspability or other properties of the objects involved. The logic behind this analysis was that behavioral tasks that involve objects that are the appropriate size, shape, and distance to be grasped are more likely to activate the dorsal pathway. Thus, the information that infants encode about these objects relates to their spatial location and shape, and not their color or other surface features. In contrast, tasks that involve objects that are too large, or too far away, to be grasped by the infant may lead to object processing centered on the surface features important for visual recognition. Future work to explore these ideas will benefit from electroencephalography (EEG)/ event-related potential (ERP), event-related oscillations (see below), or other forms of functional imaging.

Not all studies support the idea that there is increasing integration between the pathways. Chinello et al. (2013) assessed children and adults on tests of dorsal stream function, including finger gnosis (the ability to distinguish which fingers are lightly touched without seeing them), non-symbolic numerical abilities and spatial abilities, and ventral stream function, including object and face recognition. They found that dorsal- and ventral-related functions follow two uncorrelated developmental trajectories, and that there was a pattern of clear within-stream cross-task correlation in children that seemed to be lost in adults, with two notable exceptions: performance in face and object recognition on one side, and in symbolic and non-symbolic comparison on the other, remain correlated. Overall, they conclude that development entails a progressive *decorrelation* process, possibly due to greater specialization and cortical segregation, even among functions which are more correlated early in life.

Investigators have employed functional imaging to more directly examine the relatedness or independence of the dorsal and ventral pathways in older children. One study examined dorsal and ventral stream activation in response to viewing tools—these are interesting objects of study, because tools are objects, and thus expected to activate the ventral stream, but also invite actions, and thus also may activate the dorsal stream. Dekker et al. (2011) found that 6-year-old children showed adult-like activation of dorsal and ventral stream pathways in response to viewing tools. This is in line with a subsequent study in which children as young as 4 years old showed a similar pattern of activation in the tool-processing network as adults, although the researchers also note there is refinement in this network in subsequent years (Kersey et al., 2016).

What is the picture very early in development? There are difficulties in studying the tool-use network in young infants, as they do not yet have the capabilities for tool use. One approach to this has been to use resting state (no task) functional magnetic resonance imaging (fMRI) to look at the dorsal and ventral pathways. Wen et al. (2022) took this approach and, using dorsal (tool) and ventral (face) processing resting state networks, as defined in human adults, examined whether similar functional networks were active in human neonates (with no experience with tool use), adolescent macaque monkeys (with experience with basic but not complex tool use), and human adults (with experience with complex tool use). They found that human neonates showed similar patterns of activation in the dorsal and ventral pathways as human adults, and that these pathways showed some differentiation even in the neonates. Interestingly, the monkeys showed little differentiation between the pathways, and a different response to the humans. While these findings indicate that certain resting state pathways may operate already in human newborns, it is important to keep in mind a few things when interpreting these findings: (a) the study looked at networks defined based on adult networks and did not explore the neonate brain from a "neutral" point of view; and (b) this study does not test which brain networks are actually activated in tool use in early life, therefore it is possible that in early stages (e.g., prior to 4 years, based on the Kersey et al. study) a more widespread network is activated which becomes increasingly specialized/refined.

Future studies can explore the developmental relationship between the dorsal and ventral visual streams in children less than 4 years of age in order to establish whether category specialization in the two streams develops at the same time, or whether (a) development of organization-by-category in the dorsal stream precedes and perhaps even drives the development of the ventral stream (Mahon et al., 2007; Mounoud et al., 2007), or (b) development of organization-by-category in the ventral stream precedes and perhaps drives the development of the dorsal stream (Braddick et al., 2003; Klaver et al., 2008).

A different approach to understanding the relation between dorsal and ventral stream processing is to examine developmental disorders in which one component is particularly affected. This approach has led to the proposal that there is a "dorsal stream vulnerability" across several developmental disorders (Atkinson & Braddick, 2020). This has come from the findings that maturation of sensitivity to global visual motion (dorsal stream) is more variable and is delayed compared to that of global visual form (ventral stream) in many disorders. In this view, the compromised dorsal stream function also leads to difficulties in visuomotor and visual attention abilities.

Micheletti et al. (2021) investigated dorsal stream vulnerability in a group of children with developmental coordination disorder (DCD), a condition that affects physical coordination and is characterized by moving clumsily and carrying out daily activities less well than expected. As the disorder affects sensorimotor integration, it has been interpreted as a disorder related primarily to atypical development of the dorsal system, with the ventral system spared. Michelletti and colleagues used a task called "Ball in the Grass," in which children were presented with an array of seemingly random short lines, in which a "ball" could appear either defined by the coherent movement of a subset of the lines (dorsal stream) or the coherent orientation of a subset of the lines (ventral stream). They found that 5–12-year-olds with DCD performed more poorly on the dorsal stream task than children without DCD, but performed equally as well on the ventral stream task. Interestingly, they found that, within the DCD group, performance on both tasks was related to severity of motor impairments. This suggests that both pathways contribute to the range of severity of the disorder.

Another example comes from studies of Williams syndrome (WS), a genetic disorder associated with severe visuospatial deficits (Chapter 2) which has also been characterized primarily as a deficit of the dorsal stream, with the ventral stream relatively intact. One study used functional imaging to examine whether the ventral stream was truly intact in this population. O'Hearn et al. (2011) compared activation patterns during viewing of human faces, cat faces, houses, and shoes in individuals with WS and mental-age-matched and chronological-age-matched controls. The ventral stream activation of individuals with WS differed from both control groups: they showed the same level of activation to face stimuli as chronological-age-matched typically developing children, but less activation to house stimuli than either mental-age-mathed or chronological-age-matched children. This pattern suggested that ventral stream processing may not be intact in individuals with WS. Together with the study of DCD described above, the results highlight the fact that even if seemingly "intact," the ventral stream may not be functioning in a typical manner in conditions attributed to dorsal stream vulnerability. This can be seen as consistent with the interactive specialization hypothesis, where interactions of brain networks in development shape their trajectory and relation to behavior.

Neural Oscillations and Object Processing

A third general direction for understanding object processing and permanence in the infant brain comes from animal and human research on neural oscillations. A number of electrophysiological studies in animals and scalp-recorded EEG studies in humans have identified sharply timed bursts of high-frequency oscillatory activity that relate to aspects of visual processing and cognition (see Singer & Gray, 1995, for review). For example, Tallon-Baudry and colleagues found a burst of gamma frequency EEG oscillations (around 40 Hz) when adults had to bind together spatially separate features to compose a single object (Tallon-Baudry et al., 1998). This team proposed that such bursts of gamma EEG reflect the computational process of "perceptual binding."

The question of at what age infants are able to bind together separate features to compose a unitary object has been controversial based on behavioral experiments (e.g., Kavsek, 2002). Therefore, Csibra and colleagues, set out to determine when infants will show a gamma burst when viewing a Kanisza figure (a well-known illusory figure; Csibra et al., 2000). They found that a group of 8-month-old infants showed a clear burst of gamma activity over left frontal channels at around the same time after the beginning of stimulus presentation as expected in adults. This burst was not evident to the control, "pacmen," stimulus, which could not be bound into a single object. The team also tested a group of younger infants who showed a general tendency for higher gamma EEG in response to the Kanisza stimulus, but this was not structured as a single burst and seemed to be smeared over a long time period of brain computation.

Given the possible role of gamma EEG in object processing, a further question that can be asked is whether these high-frequency oscillations are sustained when an object is being "kept in mind." This may entail the brain keeping active a representation of an object through oscillatory neural activity (for computationally minded readers, a recurrent network). This issue was addressed by recording EEG while presenting infants with visual arrays in which objects appeared or disappeared behind occluders (Kaufman, Csibra, & Johnson, 2003). A sustained gamma response over temporal lobe channels was observed when infants saw an object pass behind an occluder without reappearing (see Figure 6.3 in the color plate section). Interestingly, results suggest that this response was enhanced when an occluder was removed and the object had unexpectedly disappeared. The finding suggests the possibility that active object representations are enhanced (at least temporarily) when the visual pathways of the brain are faced with conflicting visual input. By this view, sustained gamma is a neural signature of an "active" object representation that needs to be strengthened or weakened depending on the extent to which it conflicts with the current visual input.

General Summary and Conclusions

In adults, there is evidence that the dorsal and ventral visual pathways perform different computations on objects, with the ventral pathway involved in identification and recognition, and the dorsal pathway in actions on objects. Preliminary attempts to relate these two pathways to changes in infant behavior were discussed, and a simple model of the changing degree of integration between the pathways was presented. Functional imaging studies in children show that the pattern of dorsal and ventral stream activity to tools is adult-like by early school age, and that the basic networks may be in place even earlier. Further studies are needed to track development of specialization before this age, and to more directly compare this early-maturing trajectory with the more prolonged one seen for face processing. Bursts of high-frequency neural oscillations have been related to the binding of features to compose objects and to the retention of objects following occlusion. Recent studies have used these neural oscillations to study object processing in infants and may provide a direct marker for active representations of occluded objects in infants.

Key Issues for Discussion

- How can we best characterize the relative development of the dorsal and ventral visual pathways?
- Why might processing of faces show a more prolonged development than that of other categories such as tools?
- How does recent neuroscience evidence inform Piaget's claim that objects out of sight are out of mind for young infants?

7

Perceiving and Acting on the Social World

Cognitive neuroscience studies in adults have revealed a network of structures involved in the perception and processing of social stimuli, including interpreting the thoughts and intentions of other humans. However, considerable debate remains about the developmental origins of this brain network. One basic aspect of the visual social brain is the perception of faces. Evidence from behavioral and neuroimaging studies suggests several factors ensure young infants orient frequently to faces. This early exposure allows faces to capture dedicated regions of cortical tissue, a process of cortical specialization that appears to take months to years. This chapter also considers other aspects of social cognition such as perceiving and acting on information from the eyes, and attributing intentions or goals to other humans (mentalizing). Evidence from two developmental disorders, autism and Williams syndrome, initially appears to provide support for a social module that can be selectively impaired (autism) or spared (Williams syndrome) in the face of other deficits. On closer inspection, however, it becomes apparent that such a clean dissociation is not borne out by the evidence and that social information processing emerges as a result of constraints from interactions with other conspecifics, initial biases toward social stimuli, and the basic architecture of the brain.

The Social Brain

One of the major characteristics of the human brain is its social nature. As adults, we have areas of the brain specialized for processing and integrating information about the appearance, behavior, and intentions of other humans. Sometimes this processing is also extended to other species, such as the family cat, or even to inanimate objects such as our desktop computer. A variety of cortical areas have been implicated in the "social brain." The occipital face area (OFA), superior temporal sulcus (STS), and fusiform face area (FFA) are part of a core system involved in the processing of information about facial identity and emotion. Cortical regions including parts of the prefrontal cortex (mainly orbitofrontal and medial areas), the temporo-parietal junction (TPJ), and the anterior temporal lobes or temporal poles, as well as subcortical structures like the amygdala, play key roles in the mentalistic understanding of others' behavior ("theory of mind" or "mentalizing")

Developmental Cognitive Neuroscience: An Introduction, Fifth Edition. Michelle de Haan, Iroise Dumontheil, and Mark H. Johnson.
© 2023 John Wiley & Sons Ltd. Published 2023 by John Wiley & Sons Ltd.
Companion website: www.wiley.com/go/johnson/devneuro5e

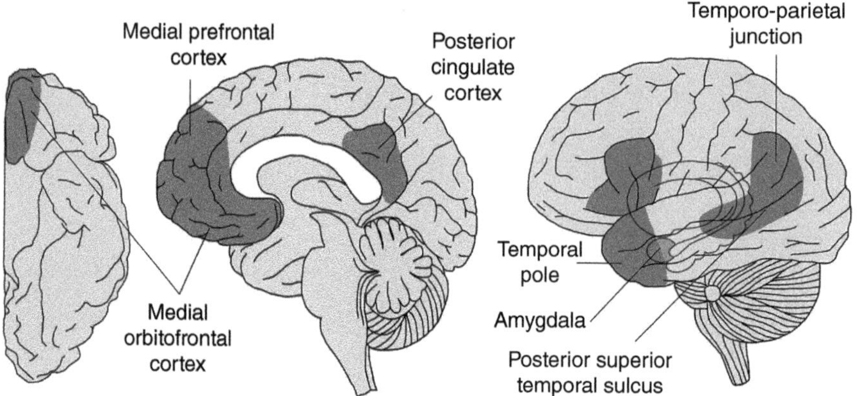

Figure 7.1 Some of the regions involved in the human social brain network. © Iroise Dumontheil.

(see Figure 7.1). Neural activity in these regions may reflect different aspects of mentalizing, as reflected in "key computations" within the adult social brain network (see Molapour et al., 2021). One of the major debates in cognitive neuroscience concerns the origins of the "social brain" in humans, and theoretical arguments abound about the extent to which emerges through experience. The question we will address in this chapter is how this network develops functionality and how particular regions become part of the human social brain.

The three perspectives on human functional brain development discussed in Chapter 1 lead to different expectations about the origins of the social brain. According to the maturational view, through evolution, specific parts of the brain and areas of cortex have become dedicated to processing social information. Plausibly, some of the circuits are present and functioning at birth, while other components of the network become available through maturation later in development. For example, prefrontal areas associated with theory of mind computations may be one of the later components of the social brain to mature. While the maturational timetable may be accelerated or decelerated by experience, the sequence of maturation and the domain-specific wiring patterns are not. According to the skill learning view, at least some parts of the social brain are engaged by social stimuli because these tend to be the visual inputs with which we are most experienced. In other words, we tend to develop a higher level of perceptual expertise for socially relevant visual inputs because of the frequency of inputs and the need to process these quickly and accurately in the social world. By this view we would not expect newborns to have any specific responses to social stimuli, and we should observe parallels between the development of face processing in infancy and the acquisition of perceptual expertise for other stimuli in adults or other ages. The third view, interactive specialization (IS), predicts that the social brain will emerge as a network that becomes increasingly finely tuned to relevant stimuli and events in an activity-dependent manner. We should anticipate interactions between more primitive brain systems, cortical areas, and the environment, to produce the end result of an adult social brain.

In this chapter we initially examine these issues with perhaps the most fundamental visual function of the social brain—the perception and processing of faces. In later sections we go beyond this to examine some of the evidence available on more dynamic aspects of social cognition, such as interpreting eye gaze and the actions of other humans, and the importance of social interactions in adolescence.

Face Recognition

The ability to detect and recognize faces is commonly considered to be a good example of human perceptual abilities, as well as being a basis of our adaptation as social animals. There is a long history of research on the development of face recognition in young infants extending back to the studies of Fantz in the 1960s (e.g., Fantz, 1964). With subsequent developments in neuroimaging, numerous papers have investigated the cortical basis of face processing in adults, including identifying areas that may be specifically dedicated to this purpose (Duchaine & Yovel, 2015; Kanwisher & Yovel, 2006; but see Haxby et al., 2001).

Face recognition skills may be divided into a number of components, including the ability to recognize a face as such, the ability to recognize the face of a particular individual, the ability to identify facial expressions, and the ability to use the face to interpret and predict the behavior of others. In a review of the available literature in the late 1980s, Johnson and Morton (1991a) revealed two apparently contradictory bodies of evidence: while the prevailing view, and most of the evidence, supported the idea that infants gradually learn about the arrangement of features that compose a face over the first few months of life (for reviews see also Maurer, 1985; Nelson & Ludemann, 1989), the results from at least one study indicated that newborn infants, as young as 10 minutes old, will track a face-like pattern further than various "scrambled" face patterns (Goren et al., 1975). Evidence that newborns showed a preferential response to faces was used by some to bolster nativist views of infant cognition (maturational view). In contrast, the evidence for the graded development of face-processing abilities over several months tended to be cited by theorists who believed that such skills need to be learned, and result from experience of the world (skill learning view). To translate these views into the framework introduced in Chapter 1, while some believed that the brain possesses innate representations of faces, others took the view that face representations resulted from the information structure of the environment.

Since the study with newborn infants remained controversial for methodological reasons, it needed to be replicated (Johnson, Dziurawiec, et al., 1991). In the replication, as in the original study, newborn infants (this time around 30 minutes old) were required to turn their head and eyes to keep a moving stimulus in view. This measure contrasts with more standard procedures in which the infant views one or more stimuli in static locations and the length of time spent looking at the stimuli is measured. In the first experiment, three of the four stimuli used in the original Goren et al. (1975) study were used: a schematic face pattern, a symmetric "scrambled" face, and a blank face outline stimulus. While the preferential head-turning-to-follow-the-face pattern was not replicated, the preferential response to the face was successfully replicated using a measure of eye movements (see Figure 7.2).

This experiment confirmed that the brain of the newborn human infant contains some information that facilitates orienting to faces. Many more studies on face preferences in newborns have been published since 1991 (for review see Johnson, Senju & Tomalski, 2015). While there are some differences reached in the conclusions of these

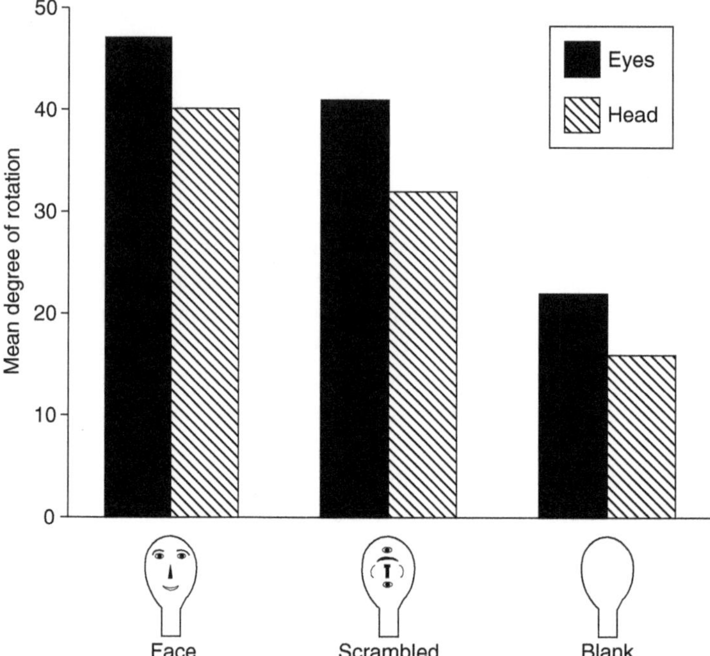

Figure 7.2 Data showing the extent of newborns' head and eye turns in following a schematic face, a scrambled face, and a blank (unpatterned) stimulus.

authors, all of the studies found some evidence for sensitivity to face-like patterns. These have led to several hypotheses about the basis of this newborn preference behavior. Further, these hypotheses may be associated with the three perspectives on functional brain development mentioned earlier.

Further Reading   de Haan (2008); Powell et al. (2018); Johnson et al. (2015).

The *sensory hypothesis* is the hypothesis that the visual preferences of newborns, including face preferences, are determined by low-level psychophysical properties of the stimuli. One common version of this hypothesis is the view that infants show preferential orienting to faces as a consequence of a more general visual bias for "top heavy" patterns that have more elements in the upper than the lower visual field. Faces fit this pattern with two eyes at the top (higher density of elements) than the bottom (lower density of elements with nose and mouth below; Cassia et al., 2004) This hypothesis is consistent with the skill learning perspective since it does not assume any domain-specific bias early in life. Further studies indicate that this top-heavy bias is no longer present by 3–5½ months of age, nor is it present in adults (Chien, 2011). At these ages, infants and adults do prefer upright to inverted faces, suggesting a quick development of response to the usual upright orientation.

A modification of the sensory hypothesis is based on the idea of embodied cognition. This is the view that the mind not only influences the body, but that the features and form of the human body also influence our thinking. Wilkinson et al. (2014) argue that newborns' preference for faces is influenced not only by the sensory information available to newborns, but also by how this relates to the newborns' own sensory receptors: in other words, an important factor is the spatial resonance between the two eyes in the infant's body and the two eyes in the facial stimulus. While this explanation gives an additional possible mechanism that contributes to newborns' preference for looking at faces, it likely is not the only contributing factor as it is not clear how this model would explain, for example, why infants respond differently to upright and inverted faces or to faces adults judge as being more or less attractive (Méary et al., 2014).

Empirical results have led some to the hypothesis that *newborns already have more specific complex representations of faces* per se. This hypothesis is most consistent with the maturational view of functional brain development, since it presumably requires domain-specific circuits to be established prior to experience (Kanwisher & Yovel, 2006). These findings include the preference for attractive faces and data indicating that newborns are sensitive to the presence of eyes in a face (Batki et al., 2001) and prefer to orient toward faces with direct (mutual) eye gaze (Farroni et al., 2002). However, inspection of images of such stimuli through the appropriate spatial frequency filters for the newborn visual system reveals that a mechanism sensitive to low spatial frequency components corresponding to the spatial arrangement of a face (see below) could account for these seemingly more complex preferences.

A third hypothesis argues that the newborn's brain contains a system that biases it to orient to faces, a *face-biasing system*. Johnson and Morton (1991a) referred to such a system as "Conspec." This hypothesis is most consistent with the IS view. In contrast to the previous hypothesis, the functioning of a cortical face module is not assumed. Rather, the bias is presumed to be close to the minimum necessary for picking out faces from a natural environment. In contrast to the sensory hypothesis, the spatial relations between face features are thought to be important, even though the representation underlying this preference may not exactly map on to a face (Simion et al., 2003). In one variant of this hypothesis, Johnson and Morton (1991a) argued that their "Config" stimulus (see Figure 7.3) was the minimal sufficient representation to ensure this preference. In neural network simulations, Bednar and Miikkulainen (2003) found that such a representation could account for the vast majority of documented newborn preferences between schematic and realistic face stimuli (see Figure 7.3).

Another view proposed sees the infant as an active agent seeking out social stimuli, i.e., argues that *newborns have social motivation*. According to this view, social interactions (Powell et al., 2018) are actively sought out by infants, possibly through mediation of the prefrontal cortex (PFC), and this drives the subsequent specialization of cortical face areas. This view emphasizes the findings that infants over the first week of life begin to show preference for dynamic faces that respond contingently. It also notes the multimodal nature of early social interactions, which involve not just the visual input of the face, but also aspects of the voice, social touch, and smells.

In conclusion, it appears that there are potentially a number of factors that may contribute to the newborn and young infant's tendency to orient to faces in the visual environment

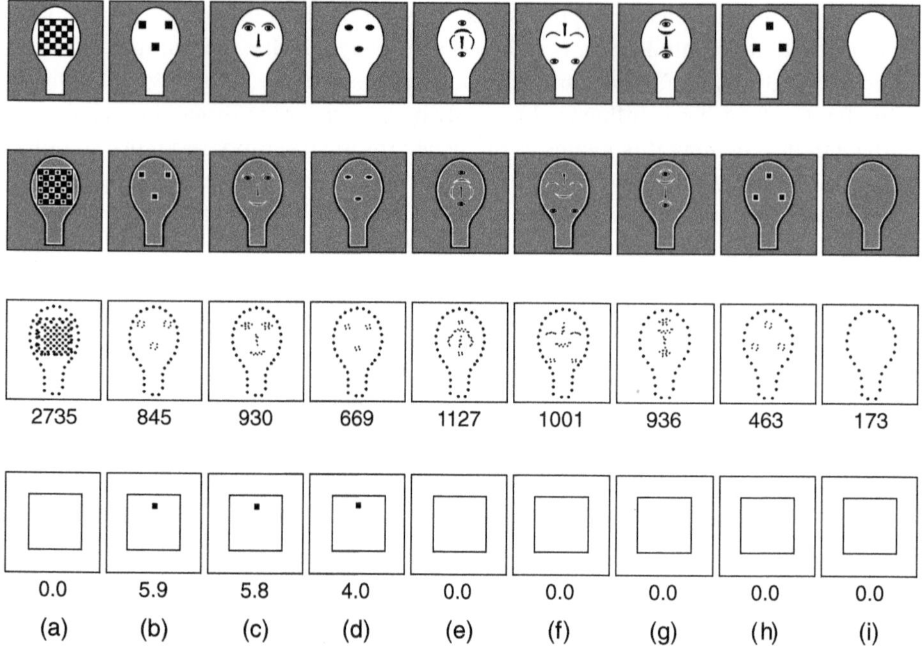

2735   845   930   669   1127   1001   936   463   173

0.0   5.9   5.8   4.0   0.0   0.0   0.0   0.0   0.0

(a)   (b)   (c)   (d)   (e)   (f)   (g)   (h)   (i)

Figure 7.3   A summary of human newborn and model responses to schematic images. The top row represents some of the schematic patterns presented to both newborns and the "retina" of the neural network model. The next two rows illustrate the lateral geniculate nucleus and visual cortex stages of the models processing. The bottom row indicates the output of the model, with the preferred stimuli being b, c, and d. The preferences of the model correspond well to the result obtained with newborn infants.

(see Johnson, Senju & Tomalski, 2015, for review). The available evidence is consistent with the view that subcortical biases contribute to the newborn's preferential orienting to faces in the visual environment, ensuring that these are a frequent input to developing cortical systems which begin to take greater control over behavior during the first year of life. Multimodal social interactions may also play a role.

Surprisingly, many other studies which used more conventional infant testing methods have not found a preference for face patterns over others until two or three months after birth (for review see also Maurer, 1985; Nelson & Ludemann, 1989). For example, Maurer and Barrera (1981) used a sensitive "infant control" testing procedure, in which participants view a series of singly presented static stimuli, and established that while 2-month-olds looked significantly longer at a face-like pattern than at various scrambled face patterns, 1-month-olds had no such preference. Using the same method, Johnson et al. (1992) replicated this result and extended the original findings by including in the stimulus set the "de-focused" face arrangement stimulus used in the newborn studies earlier. The results replicated entirely the previous findings: the face was looked at longer than any of the other stimuli by 10-week-olds, but a group of 5-week-old infants showed no preference. This evidence was consistent with the alternative claim that infants gradually construct representations of faces as a result of repeated exposure to them over the first few months of life (Gibson, 1979). Clearly, these apparently contradictory findings raised a problem for theories of the development of face recognition that involved only one process (either learning or innate face representations). In an attempt to interpret this apparently conflicting behavioral data, Johnson and Morton (1991a,b) turned to evidence from two areas of biology: ethology and brain development.

Brain Development and Face Recognition in Humans

As mentioned in Chapter 5, both neuroanatomical and neurophysiological data indicate that visually guided behavior in the newborn infant is largely (though not exclusively) controlled by subcortical structures such as the superior colliculus and pulvinar, and that it is not until several months of age that cortical circuitry comes to dominate subcortical control over behavior. Consistent with these arguments is the position that visually guided behavior in human infants is based on activity in two or more distinct brain systems. If these systems have distinct developmental time courses, then they may differentially influence behavior in infants of different ages.

Johnson and Morton used an ethological approach to consider brain mechanisms in development of responses to social stimuli: filial imprinting in the domestic chick. Imprinting in chicks was selected because it is well studied both in terms of behavior and in terms of its neural basis. Filial imprinting is the process by which young precocial birds such as chicks recognize and develop an attachment for the first conspicuous object that they see after hatching (for reviews see Bolhuis, 1991; Johnson & Bolhuis, 1991). While imprinting has been reported in the young of a variety of species, including spiny mice, guinea pigs, chicks, and ducklings, only in precocial species (those that are mobile from birth) can we measure it using the conventional measure of preferential approach. Evidence from studies of imprinting in chicks and its neural bases suggested a two-process theory of development of recognition of conspecifics, an orienting system present from birth, and the equivalent of a cortical system that develops specialization for processing visual social stimuli based on this biased input (Johnson et al., 1985). This work in chicks motivated a more detailed study of face recognition in human newborns and application of the two-process theory to human infants.

There is much evidence that the recognition of individual faces in adults involves cortical areas and pathways (Figure 7.4). This evidence comes from three main sources: (a) neuropsychological patients with brain damage who are unable to recognize faces (prosopagnosia); (b) neuroimaging studies of face perception; and (c) single- and multicellular recording studies with non-human primates. To briefly review this literature, prosopagnosia commonly results from damage to the region of cortex that lies between the temporal and occipital (visual) cortex, although the exact neuropathology required is still controversial and may vary between participants. Some cases have suggested that only a right-hemisphere lesion is necessary, while others suggest bilateral lesions are required (see Rossion, 2014, for a review). The deficit resulting from these lesions can be fairly specific. Although prosopagnosic patients have difficulty recognizing individual faces, they sometimes seem to be able to identify other objects. Of course, exceptions to this have been noted, with some patients having difficulty identifying other complex objects. However, not all face processing is entirely abolished in these patients. For example, some

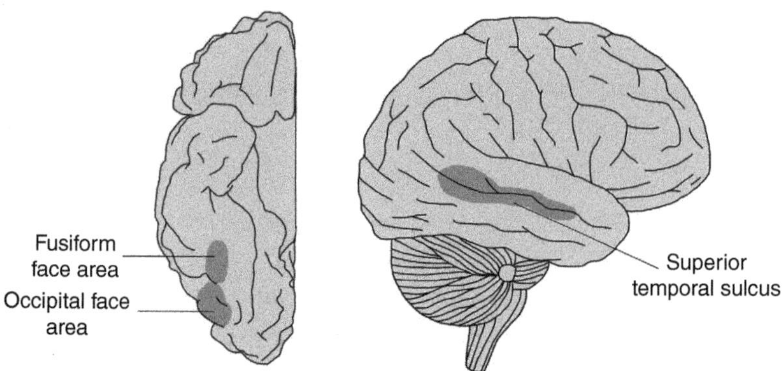

Figure 7.4   Key regions involved in face processing in adults. © Iroise Dumontheil.

prosopagnosic patients seem to have intact facial emotion processing (e.g., Bate & Bennetts, 2015), and many show "covert" recognition of familiar faces as indicated by sensitive measures such as galvanic skin responses (Tranel & Damasio, 1985).

Evidence from neuroimaging studies of human adults has implicated a range of cortical regions within the social brain, including regions of the fusiform gyrus (FG), lateral occipital area, and STS in encoding or detecting facial information. The specificity of responses to faces has been most extensively studied for the FFA, a region which is more activated by faces than by many other comparison stimuli, including houses, textures, and hands (Kanwisher et al., 1997). While the greater activation of the FFA to faces than other objects has led some to propose it is a face module (Kanwisher et al., 1997), others call this view into question. In particular, investigations demonstrating (a) that the distribution of response across the ventral cortex may be more stimulus-specific than the strength of response of a particular region such as FFA (Haxby et al., 2001; Ishai et al., 1999; but see Spiridon & Kanwisher, 2002), and (b) that activation of the FFA increases with increasing expertise in discriminating members of non-face categories (Gauthier et al., 1999), together suggest that the region may play a more general role in object processing. However, the observations do remain that faces more than any other object activate the FFA, and that the distribution of activity over the ventral cortex for faces differs from other objects in that it is more focal and less influenced by attention (Haxby et al., 2001).

> **Further Reading**   de Haan (2008); Duchaine & Yovel (2015).

Thus, there is considerable evidence for specific cortical involvement in face processing among adults. However, there is also evidence for a face preference in newborn infants, whose behavior, as discussed in the previous chapter, is thought to be guided largely by subcortical sensorimotor pathways. Consideration of this evidence led Morton and Johnson to propose a two-process theory of infant face preferences analogous to that in chicks. We argued that the first process consists of a system accessed via the subcortical visuomotor pathway (but possibly also involving some of the deeper, earlier developing, cortical layers) that is responsible for the preferential tracking of faces in newborns. However, the influence of this system over behavior declines (possibly due to inhibition by developing cortical circuits) during the second month of life. The second brain system depends upon a degree of cortical maturity and exposure to faces over the first month or two, and begins to control infant orienting preferences from around 2 or 3 months of age. The newborn preferential visual orienting system biases the input set to developing cortical circuitry (Johnson & Morton, 1990a), likely together with other early multisensory inputs from familiar voices and other multimodal input (e.g., this circuitry is configured in response to a certain range of input before it starts to gain control over the infant's behavior). Once this occurs, the cortical system has acquired sufficient information about the structure of faces to ensure that it continues to acquire further information about them. The proposal is that a specific, early developing brain circuit acts in concert with the species-typical environment to bias the input to later developing brain circuitry. A number of strands of evidence are consistent with this theory. First, let us consider evidence about the neural basis of the face bias in newborns.

Johnson and Morton (1991a) speculated that the face bias (Conspec) was mediated largely, but not necessarily exclusively, by subcortical visuomotor pathways. Owing to the continuing difficulty in successfully using functional imaging with healthy awake newborns, this hypothesis has, as yet, only been indirectly addressed. First, de Schonen and colleagues examined face preferences in infants with perinatal damage to regions of cortex. Even in cases of damage to the visual cortex, the bias to orient to faces remained (Mancini et al., 1998). The second line of evidence used the fact that the nasal and temporal visual fields feed differentially into the cortical and subcortical visual pathways respectively. Specifically, Simion and colleagues predicted that the face orienting bias would be found in the temporal visual field, but not the nasal visual field. This prediction was confirmed (Simion et al., 1998). Control experiments showed equal orienting in both fields to striped, non-face patterns (Simion et al., 1998). A third line of evidence comes from a large number of adult neuropsychological and neuroimaging studies showing evidence for a subcortical "quick and dirty" route for face processing (see Johnson, 2005, for review). Analysis of these studies reveals that the adult subcortical route rapidly processes low spatial frequency "coarse" information about faces, and then modulates activity the face-sensitive cortical areas that process fine detailed information about faces. The visual stimuli which maximally elicit activity in the adult subcortical route are strikingly similar to those to which newborns preferentially orient, strongly suggesting that this route is the basis for newborn behavior (see Figure 7.5).

While Johnson and Morton (1991a,b) identified the superior colliculus as a major visuomotor structure that could be involved in determining Conspec preferences, another candidate structure is the pulvinar. Portions of pulvinar receive input directly from the superior colliculus (as well as from the retina and, at least in adults, striate and extrastriate visual cortex). Additionally, in adults there are reciprocal connections to frontal, temporal, and parietal regions and to the anterior cingulate and amygdala. The advent of new technology suitable for studying the neural correlates of behavior including functional magnetic resonance imaging (fMRI) in newborns may allow further investigation of this issue (Kosakowski et al., 2022).

We now turn to the neurodevelopment of face processing during infancy and childhood. To recap, the three perspectives on human functional brain development discussed earlier make differing predictions about how this specialization arises. By a maturational view we

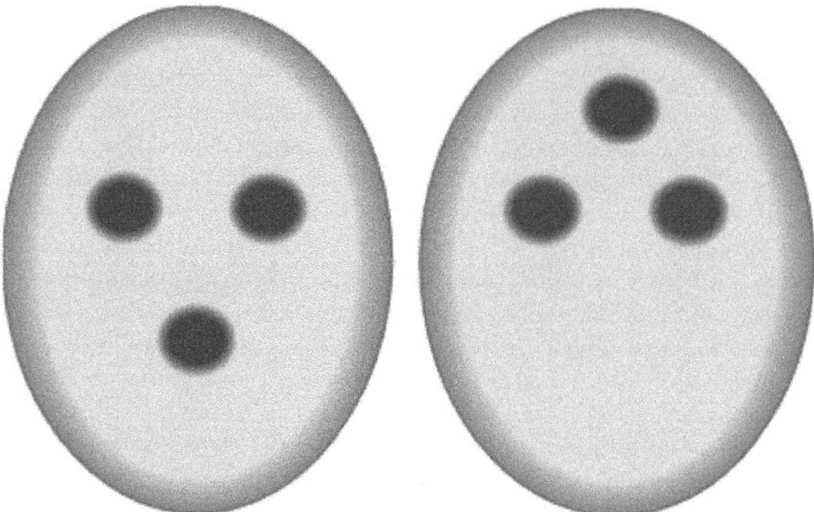

Figure 7.5 Schematic illustration of the stimuli that might be optimal for eliciting a face-related preference in human newborns. These hypothetical representations were created by integrating the results from several experiments with newborns.

expect to see the gradual addition of new brain modules relating to aspects of face-processing ability. From a skill learning perspective, the FFA should become active as the perceptual skill of face processing is acquired. From an IS view, we expect to observe the increasing neural specialization and more restricted localization of face processing with development. With these different predictions in mind, let us review some of the evidence from developmental cognitive neuroscience.

Several labs have examined changes in event-related potentials (ERPs) as adults view faces. In particular, interest has focused on an ERP component termed the "N170" (because it is a negative-going deflection that occurs after around 170 ms), which has been strongly associated with face processing in a number of studies on adults (see de Haan et al., 2003, for review). Specifically, the amplitude and latency of this component vary according to whether or not faces are present in the visual field of the adult volunteer under study. An important aspect of the N170 in adults is that its response is highly selective. For example, the N170 shows a different response to human upright faces than to very closely related stimuli such as inverted human faces and upright monkey faces (de Haan et al., 2002). While the exact underlying neural generators of the N170 are currently still debated, the specificity of response of this component can be taken as an index of the degree of speciali-zation of cortical processing for human upright faces. For this reason de Haan and col-leagues undertook a series of studies on the development of the N170 over the first weeks and months of postnatal life.

The first issue addressed in these developmental ERP studies was when does the face-sensitive N170 emerge? A challenge in this type of developmental research in ERP in gen-eral is that, during development, ERPs thought to reflect similar cognitive processes can be different looking across ages in terms of timing and location. In a series of experiments, a component in the infant ERP that has many of the properties associated with the adult

N170, but that is of a slightly longer latency, was identified (240–290 ms; de Haan et al., 2002; Halit et al., 2003; Halit et al., 2004). In studying the response properties of this N290 component at 3, 6, and 12 months of age we have discovered that (a) the component is present from at least 3 months of age (although its development continues into middle childhood), and (b) the component becomes more specifically tuned to human upright faces with increasing age. To expand on the second point, we found that while 12-month-olds and adults showed different ERP responses to upright and inverted faces, 3- and 6-month-olds did not (de Haan et al., 2002; Halit et al., 2003). Thus, the study of this face-sensitive ERP component is consistent with the idea of increased specialization of cortical processing with age, as well as with some behavioral results (see below). Further studies have confirmed the face sensitivity of the N290 compared to objects in infants from 4½ to 12 months of age, with the component also increasing in amplitude over this time (Conte et al., 2020). Source localization of the N290 in this study showed a prominent source in the middle FG and in surrounding areas, consistent with findings for the adult N170 component.

ERP studies of children 4 years of age and older have found a more stable pattern of specialization for face processing. Kuefner et al. (2010) tested children aged 4–17 and adults and failed to find any evidence of age-dependent changes in size, timing, or scalp distribution of the P1 or N170 that were specific to faces. When they controlled for age-related variations of the P1, the N170 appeared remarkably similar in size and scalp distribution across the ages tested, and there were only small age-related decreases in latencies. At all ages the N170 showed equivalent face-sensitivity: it had the same topography and right-hemisphere dominance, it was absent for meaningless (scrambled) stimuli, and larger and earlier for faces than for cars. This result suggests there is no further specialization for face processing after 4 years of age. However, the IS view argues that the processes of specialization occur first locally and then proceeds to the network level. Thus, with increasing age, greater specialization might be more apparent in studies of functional connectivity across networks, rather than in studies of individual ERP components thought to tap a particular component of the network.

The IS view predicts increases in the degree of both specialization and localization of face-evoked activity in cortex during development. This stands in contrast to a maturational perspective prediction that face processing in the child's brain should be limited to a subset of the areas that can be activated by faces in adults. Several studies have used fMRI to investigate the neurodevelopment of face processing in children (see Haist et al., 2013, for review). Several of these studies have yielded support for the dynamic changes predicted by the IS hypothesis. As far as face processing in general is concerned, fMRI studies have shown that certain regions of the cortex show reliable activation to faces from at least mid-childhood. However, two studies are of particular importance as they specifically address both the localization and the specialization predictions. Scherf and colleagues (Scherf et al., 2007) used naturalistic movies of faces, objects, buildings, and navigation scenes in a passive viewing task with children (5–8 years), adolescents (11–14 years), and adults (see Figure 7.6 in the color plate section). They found that the children exhibited similar patterns of activation of the face-processing areas commonly reported in adults (such as the FFA). However, this activation was not selective for the category of face stimuli; the regions were equally strongly activated by objects and landscapes. Moreover, this lack of fine-tuning of classical face-processing areas stood in contrast to distinct preferential activation patterns for other object categories (occipital object areas and the parahippocampal place area). In a similar study, Golarai et al. (2007) found substantially larger right FFA and left parahippocampal volumes of selective activation to faces compared to objects, places, and scrambled patterns in adults than in children, and Peelen et al. (2009) found that the FFA, but not the fusiform body responsive area, was larger and more selective to faces in adults compared to children. In line with this research, Kamps et al. (2022) used engaging movies as stimuli in 3–12-year-olds and found that, while time courses of activation in face, scene, and object regions already had similarities to adult patterns of activation at age 3, there was also developmental change during childhood with, in particular, increase selectivity for faces with age in the posterior STS. While this increase in functionally defined areas with development may initially appear to contradict predictions of the IS view, it is important to note that the contrasts employed defined the increase as an expansion of the area of category-specific activation of FFA. Overall, fMRI studies suggest that the FFA shows a protracted developmental trajectory extending to at least late adolescence in terms of volume of the FG, the intensity of activation, the spatial location of the FFA within the FG, and the connectivity of the FFA with key areas including the OFA. The developmental changes observed in these studies thus provide strong support for IS as a framework within which to interpret functional brain development (Cohen-Kadosh & Johnson, 2007).

Another area of face-processing research that could potentially contribute to our understanding of the developmental trajectory of face-specialized cortical areas is studies of individuals with developmental prosopagnosia. Developmental prosopagnosia refers to individuals who never develop typical adult face-processing abilities (Duchaine & Nakayama, 2006), and in particular facial identity recognition skills. This condition can occur in the absence of any obvious sensory or intellectual deficit (Avidan et al., 2005; Behrmann & Avidan, 2005; Yovel & Duchaine, 2006). Prosopagnosia that emerges after an acquired brain injury in adulthood generally involves damage to the right ventral occipito-temporal cortex (Rossion, 2014). Yet, while there are cases of developmental prosopagnosia that arise from brain trauma early in life, there is increasing evidence for cases in which typical adult-like face-processing skills fail to develop in the absence of any known brain injury (Duchaine & Nakayama, 2006). Some of these individuals have family members that are also affected, and it is tempting to speculate on a genetic cause. It has been suggested that there is a hereditary subtype of developmental prosopagnosia with a prevalance rate of 2.5% (Grüter et al., 2008). However, since nothing is currently known about the early rearing and social interaction environment of these families, the specificity or directness of this genetic effect remains unknown. Functional MRI studies have generally shown that developmental prosopagnosics activate face-sensitive regions of cortex (for review see Duchaine & Nakayama, 2006). However, the degree of selectivity of this response remains in doubt, and a plausible interpretation of these results is that while adult developmental prosopagnosics show activation in face-sensitive regions, they may also show the lack of specificity in this response reported earlier for typically developing children. Moreover, Avidan et al. (2005) showed that, in four developmental prosopagnosics, face perception recruited additional brain areas (e.g., the inferior frontal gyrus) that were not commonly activated in typical adults. Interestingly, however, inferior frontal gyrus activation has been observed in children in several of the developmental neuroimaging studies (Gathers et al., 2004; Passarotti et al., 2003; Passarotti et al., 2007; Scherf et al., 2007). Thus, the localization and specialization components of the IS account may accommodate data from developmental prosopagnosia. If this account of developmental prosopagnosia is correct then, with sufficient training on faces, the selectivity of the response of cortical areas, and their interconnectivity, should change toward the typical pattern. This was tested in a training study with a developmental prosopagnosic patient that showed that, as behavioral performance improved, the selectivity of cortical processing of faces (as measured by the N170 component) increased (DeGutis et al., 2007). Further, these authors also observed increased functional connectivity between face-selective regions as measured by fMRI, particularly the right OFA and the right FFA.

Converging evidence about the increasing specialization of face processing during development comes from a behavioral study that set out to test the intriguing idea that as processing "narrows" (Nelson, 2003) to human faces, infants will be less able to discriminate non-human faces (Pascalis et al., 2002). Pascalis and colleagues demonstrated that, while 6-month-olds could discriminate between individual monkey faces as well as human faces, 9-month-olds and adults could only discriminate the human faces. These results are particularly compelling since they demonstrate a predicted competence in young infants that is not evident in adults. A similar pattern has been observed for discriminating between own-race and other-race faces. At 6 months of age infants can discriminate individual faces from their own and other races but by 9 months of age this ability to process other-race faces is typically lost (Kelly et al., 2007). One study showed that exposing Caucasian infants to Chinese faces through perceptual training via picture books for a total of one hour between 6 and 9 months allowed the Caucasian infants to maintain the ability to discriminate Chinese faces at 9 months of age (Heron-Delaney et al., 2011). This shows that the development of the processing of face race can be modified by training, highlighting the importance of early experience in shaping the face representation.

Further Reading   Quinn et al. (2019).

The lines of evidence reviewed above are hard to reconcile with a strictly maturational view that posits the addition of new components of processing with development. Rather, the evidence available to date suggests "shrinkage" of the cortical activity associated with face processing during development. While increased specialization and localization are specifically predicted by the IS view, these changes are not necessarily inconsistent with skill learning (Gauthier & Nelson, 2001). Thus, with regard to face perception, the available evidence from newborns allows us to rule out the skill learning hypothesis, while the evidence on the neurodevelopment of face processing over the first months and years of life is consistent with the kinds of dynamic changes in processing expected from the IS, and not the maturational, approach.

Further Reading   de Haan (2008).

Perceiving and Acting on the Eyes

Moving beyond the relatively simple perception of faces, a more complex attribute of the adult social brain is processing information about the eyes of other humans. There are two important aspects of processing information about the eyes. One of these is being able to detect the direction of another person's gaze in order to direct your own attention to the same object or spatial location. Perception of averted gaze can elicit an automatic shift of attention in the same direction in adults (Driver et al., 1999), allowing the establishment of "joint attention" (Butterworth & Jarrett, 1991). Joint attention to objects is thought to be crucial for a number of aspects of cognitive and social development, including word learning. Beginning in the first year of life, activation of the social brain network during coordinated attention may enhance the depth of information processing and encoding; with development, joint attention may become an internalized skill for socially coordinating mental attention to internal representations (Mundy & Jarrold, 2010). A second critical aspect of gaze perception is the detection of direct gaze, enabling mutual gaze with the viewer. Mutual gaze (eye contact) provides the main mode of establishing a communicative context between humans and is believed to be important for normal social development (e.g., Kleinke, 1986; Symons et al., 1998). It is commonly agreed that eye gaze perception is important for caregiver–infant interaction, and that it provides a vital foundation for social development (e.g., Jaffe et al., 1973; Stern, 1977).

With regard to the social brain network, the STS has been identified in several adult imaging studies of eye gaze perception and processing (for review see Adolphs, 2003). However, the connectivity of the network changes with age; children 7–12 years of age exhibit weaker functional connectivity between the OFA and the FG than do adults and show no significant functional connectivity between the OFA and STS (Cohen-Kadosh et al., 2011). As with cortical face processing, above, in adults the response properties of the STS are highly tuned (specialized) in that the region responds only to biological motion (Puce et al., 1998).

Several studies have demonstrated that gaze cues are able to trigger an automatic and rapid shifting of the focus of the adult viewer's visual attention (Driver et al., 1999; Friesen & Kingstone, 1998; Langton & Bruce, 1999). Studies used variants of the spatial cueing paradigm (see Chapter 5), where a central or peripheral cue directs attention to one of the peripheral locations. When the target appears in the location toward which the cue was directed (the congruent position), the participant is faster to look at that target compared to another target at an incongruent position relative to the cue. Human infants start to discriminate and follow adults' direction of attention from the age of 3 or 4 months (Hood et al., 1998; Vecera & Johnson, 1995). In our studies we examined further the visual properties of the eyes that enable infants to follow the direction of the gaze. We tested 4-month-olds using a cueing paradigm adapted from Hood et al. (1998). Each trial began with the stimulus face eyes blinking (to attract attention) before the pupils shift to either the right or the left for a period of 1,500 ms (see Figure 7.7). A target stimulus was then presented either in the same direction as the stimulus face eyes were looking (congruent position) or in a location incongruent with the direction of gaze. By measuring the saccadic reaction time of infants to orient to the target, we demonstrated that the infants were faster to look at the location congruent with the direction of gaze of the face.

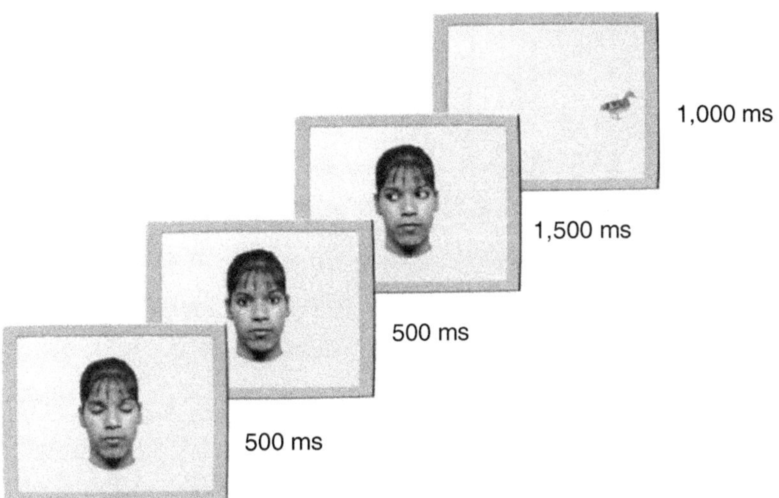

Figure 7.7 Example of the edited video image illustrating the stimulus for experiment 1 in Farroni et al. (2000). In this trial the stimulus target (the duck) appears on the side incongruent with the direction of gaze. Taylor & Francis Group.

In a series of experiments using this basic paradigm, we have established that it is only following a period of mutual gaze with an upright face that cueing effects are observed. In other words, mutual gaze with an upright face may engage mechanisms of attention such that the viewer is more likely to be cued by subsequent motion. In summary, the critical features for eye gaze cueing in infants are (a) lateral motion of elements and (b) a brief preceding period of eye contact with an upright face.

Evidence from functional neuroimaging indicates that a network of cortical and subcortical regions is engaged in eye gaze processing in adults (Senju & Johnson, 2009). This network of structures overlaps with, but does not completely duplicate, the patterns of activation seen in the perception of motion, and the perception of faces in general. While it may be important to activate the whole network for eye gaze processing, one region in particular, the "eye area" of the STS, appears to be critical. The finding that infants are as effectively cued by non-eye motion (Farroni et al., 2000) provides preliminary evidence that their STS may be less finely tuned than in adults.

Following the surprising observation that a period of direct gaze is required before cueing can be effective in infants, several authors have investigated the developmental roots of eye contact detection. It is already known that human newborns have a bias to orient toward face-like stimuli (see earlier), prefer faces with eyes opened (Batki et al., 2001), and tend to imitate certain facial gestures (Meltzoff & Moore, 1977). Preferential attention to faces with direct gaze would provide the most compelling evidence to date that human newborns are born prepared to detect socially relevant information. For this reason, Farroni and colleagues investigated eye gaze detection in humans from birth (Farroni et al., 2002). Human newborn infants were tested by presenting them with a pair of stimuli, one a face with eye gaze directed straight at the newborns and the other with averted gaze (see Figure 7.8). Videotapes of the baby's eye movements throughout the trial were analyzed. The dependent variables were the total fixation time and the number of orienting responses.

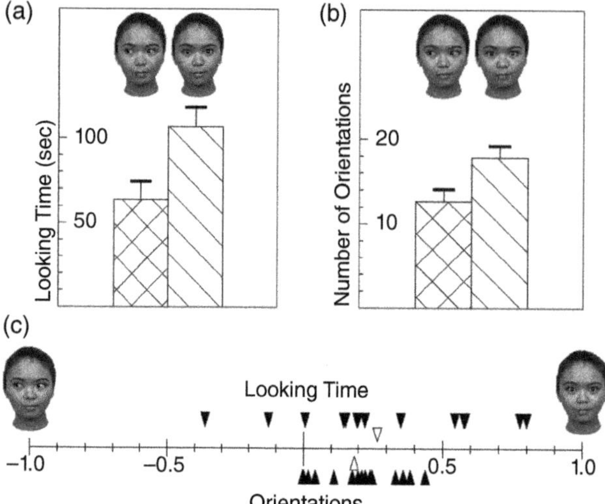

Figure 7.8 Results of the Farroni et al. 2002 preferential looking study with newborns. (a) Mean looking times (and standard error) spent at the two stimulus types. Newborns spent significantly more time looking at the face with direct gaze than looking at the face with averted gaze. (b) Mean number of orientations toward each type of stimulus. (c) Filled triangles indicate reference scores for the direct gaze over the averted gaze for each individual newborn. Open triangles indicate average preference scores. Farroni et al. 2002 / National Academy of Science.

Results showed that the fixation times were significantly longer for the face with the direct gaze. Further, the number of orientations was higher with the straight gaze than with the averted gaze.

In a second experiment, Farroni and colleagues attempted to gain converging evidence for the differential processing of direct gaze in infants by recording ERPs from the scalp as infants viewed faces. Four-month-old babies were studied with the same stimuli as those used in the previous experiment with newborns, and a difference between the two gaze directions was found at the face-sensitive component of the infant ERP discussed earlier (Farroni et al., 2002). The authors' conclusion from the second study was that direct eye contact enhances the perceptual processing of faces in 4-month-old infants.

FNIRS has also been used to investigate brain activation in response to mutual gaze, and gaze shifts. One such study used eye tracking and fNIRS to investigate brain activation in 7-month-olds in a "peek-a-boo" game with infant-directed speech. When the partner presented with direct gaze toward infants during social play, infants fixated the partner's eye region for a longer duration, and hemodynamic activity increased in the dorsomedial PFC. In contrast, brain activity increased in the right dorsolateral PFC whether the partner's eyes were direct or averted in gaze. This suggests a specific role of the medial prefrontal cortex (MPFC) in social interactions involving face-to-face communication with mutual gaze (Urakawa et al., 2015).

The empirical evidence gathered on the development and neural basis of eye gaze processing in infants is consistent with the IS view. Specifically, the same primitive representation of high-contrast elements (Conspec) as underlies the face bias in newborns may also be sufficient to direct them toward faces with direct eye contact. Therefore, the more frequent orienting to the direct gaze in newborns could be mediated by the same mechanism that underlies newborns' tendency to orient to faces in general. A face with direct gaze would better fit the spatial relation of elements in this template than one with gaze averted, suggesting that the functional role of this putative mechanism is more general than previously supposed. This primitive bias ensures a biased input of human faces with direct gaze to the infant over the first days and weeks of life, a bias that helps lay a further foundation stone for the emerging social brain (Senju & Johnson, 2009).

> **Further Reading**   Senju & Johnson (2009); Johnson et al. (2015).

According to the IS view, a network as a whole becomes specialized for a particular function. Therefore, it may be that the eye region of the STS does not develop in isolation, or in a modular fashion, but that its functionality emerges within the context of interacting regions involved either in general face processing or in motion detection. Viewed from this perspective, STS may be a region that integrates motion information with the processing of faces (and other body parts). While STS may be active in infants, it may not yet efficiently integrate motion and face information. In other words, while the 4-month-old has good face processing and general motion perception, it has not yet integrated these two aspects of perception together into adult eye gaze perception. By this account, making eye contact with an upright face fully engages face processing, which then facilitates the orienting of attention by lateral motion. The MPFC, as part of the social brain network, may also play a role (Powell et al., 2012; Richardson et al., 2021). At older ages, eye gaze perception becomes more fully integrated, where even static presentations of averted eyes are sufficient to facilitate gaze.

Understanding and Predicting the Behavior of Others

Beyond face processing and eye gaze detection there are many more complex aspects of the social brain such as the coherent perception of human action, and the appropriate attribution of intentions and goals to other humans. Traditionally, one way in which these issues have been addressed is through behavioral studies with infants, toddlers, and children. However, developmental cognitive neuroscience investigations have also highlighted the developing neural systems underlying these skills. In addition to its sensitivity to eyes, in adults the STS responds to moving biological stimuli (but does not respond as well to non-biological similar moving stimuli, or to static pictures of biological stimuli—Puce et al., 1998). Does this specialization emerge during postnatal development in accord with predictions from the IS view?

Functional MRI studies with children (Mosconi et al., 2005) have shown that the STS can be activated by dynamic social stimuli from mid-childhood. Carter and Pelphrey (2006) used fMRI to test the specificity of response of the STS and other regions in 7- and 10-year-old children while they viewed a variety of biological motion and related stimuli (such as walking robots). They observed that, consistent with predictions from the IS view, the STS became increasingly specific in its response properties to biological motion with increasing age.

Optical imaging (fNIRS) with 5-month-old infants also found patterns of activation consistent with STS activity when they viewed dynamic social stimuli (an actress moving her hands, eyes, and mouth) as compared to dynamic non-social stimuli (moving machinery) (Lloyd-Fox et al., 2009). However, the precise degree of specialization of this response during infancy remains to be determined. In general, while research on the STS is not as advanced as that on the FFA, the evidence currently available suggests that similar processes of emerging functional specialization may occur.

As mentioned above, frontal regions of the brain are also active in response to social stimulus such as eye gaze—in particular, under conditions of joint attention (Carlin & Calder, 2013). An fNIRS study showed that frontal regions are active already in infants. Five-month-olds showed greater left frontal activation when an adult followed their gaze by looking at an object they had looked at, than when the adults did not follow their gaze and instead looked at a different object (Grossmann et al., 2013). The construction of integrated networks shows a more protracted development as mentioned above.

One intriguing discovery in neuroscience is that the perception of others' actions induces sub-threshold motor activity in the adult observer, a phenomenon recorded in both monkeys and humans, and often attributed to a "mirror neuron system" (MNS) (Rizzolatti & Craighero, 2004). A MNS is one in which the same cells and circuits are involved both in generating a motor action, and in processing the visual input that arises from watching someone else performing a similar action. There has been much speculation and controversy about the importance or otherwise of motor mirroring for the perception of action. A promising way to elucidate the function of motor system activation during action observation is to investigate the ontogeny of this phenomenon (Kilner & Blakemore, 2007). As one's own particular motor abilities and skills are a determinant of whether the motor system is recruited during action observation (Calvo-Merino et al., 2005), the limited motor repertoire of infants means that they can shed light on which capacities are modulated by motor system recruitment (van Elk et al., 2008).

One way to observe the functional role of motor activation in the perception of action in infants is to use recordings of sensorimotor alpha EEG activity. However, to make a convincing case for a MNS it is important that this activity be recorded during *both* action and perception in the same participants. Showing changes in one or other task only is not sufficiently compelling evidence. Therefore, Southgate and colleagues (Southgate et al., 2009) used EEG to measure changes in sensorimotor alpha band activity during the observation of predictably occurring repeated actions in nine-month-old infants. Studies with adults have demonstrated that sensorimotor activity is modulated by both the execution and the observation of goal-directed actions (Hari et al., 1998), and likely originates in primary somatosensory cortex (Hari & Salmelin, 1997). Southgate and colleagues found that infants did indeed exhibit sub-threshold motor activity during action observation that matched directly the neural signal occurring during their own similar actions. Interestingly, rather than occurring only in response to an observed action, the motor activation was evident *prior to the onset of the repeated observed action* once infants could anticipate its occurrence. This finding is consistent with previous reports demonstrating that adults activate their motor system when they can predict that someone will perform an action (Kilner et al., 2004). Studies also show that, when infants observe actions, their neural signals show a stronger response to actions that are goal-directed than spatially similar, but not goal-directed, actions (Nystrom et al., 2011). Together, these results fit with proposals that motor activation during action observation may reflect a process of anticipating how an action will unfold (Csibra, 2007).

Beyond the perception of faces, eyes, and the actions of others, the developing child needs to develop neural and cognitive mechanisms for understanding the behavior of others in terms of their intentions, goals, and desires—an ability sometimes referred to as mentalizing or theory of mind. Despite its grand title, theory of mind is a relatively basic and essential function for our understanding of, and interactions with, other people as part of typical social cognition. Specifically, theory of mind refers to the ability most of us have to comprehend another person's thought processes, such as their feelings, beliefs, and knowledge. One type of task that can be used to study theory of mind abilities is the so-called "false-belief" task. For example, the following scenario can be demonstrated with dolls, puppets, or human actors (example of puppet procedure taken from Wimmer & Perner, 1983, with the scenario used by Baron-Cohen et al., 1985).

> Sally has a marble that she puts in a basket. Sally then goes away for a walk. While Sally is away, Anne comes in and moves the marble from the basket to a box. The participant viewing this scenario is then asked "When Sally returns, where will she look for her marble?" If the participant simply tries to predict Sally's response on the basis of their own personal knowledge, then he or she will answer "the box." If, on the other hand, she predicts Sally's response on the basis of Sally's (false) belief, then the participant will correctly predict that Sally will look in the basket for the marble.

Another false-belief scenario involves showing the participant a container that normally holds candy well known to the child and asking the child what is in the container. The participant will reply with the name of the appropriate candy. The box is then opened and the child shown that it contains a pencil, and not candy. The child is then told that a friend will come in a moment and be shown the closed container and asked what is in it. The participant is asked what the friend will say. Once again, only if the child can infer the friend's inevitable false belief will he or she reply with the name of the candy rather than a pencil.

Performance on false-belief tasks shows a consistent developmental pattern. A meta-analysis of 178 studies showed that across countries and various task manipulations the proportion of children correctly responding increases from below chance to above chance during preschool years (Wellman et al., 2001). Successful completion of false-belief tasks is thought to reflect a conceptual change in the understanding of belief, and by extension the mind, in the preschool years. Tomasello (2018) argues that the understanding of false beliefs builds on joint attention and linguistic communication, as well as executive function skills that allow the coordination of different perspectives. Using a standardized paradigm in five different countries (Canada, India, Peru, Samoa, and Thailand) Callaghan et al. (2005) found the children showed an understanding of false beliefs at approximately 5 years of age in all five cultures, suggesting it may be a universal milestone of social cognitive development.

Further Reading   Tomasello (2018).

While children start passing false-belief tasks at around age 5, performance on theory of mind tasks continues to improve during childhood (Richardson et al., 2018). This development of mentalizing, or mindreading, is likely to build on increasing capacity for memory and executive control, practice, and increasing social experience, which would support more effortful mindreading, improved efficiency and automatization, and flexible and appropriate use of mindreading (Apperly, 2021). For example, Richardson et al. (2018) asked 3–5-year-old children to complete an explicit theory of mind task that included a number of false-belief questions and found that children who failed the false-belief questions showed worse response inhibition on the Dimensional Change Card Sorting task than those who showed inconsistent performance or passed the task. Using an implicit task that requires participants to take into account another individual's perspective in a social communicative context, the "Director task," we demonstrated that adolescents were less likely to take into account another person's perspective than adults, showing more of an egocentric bias (Dumontheil et al., 2010). Performance on this task was found to be correlated with performance on a Go/No-Go response inhibition task across development in 9–29-year-olds (Symeonidou et al., 2016). In addition, 11–17-year-olds' performance on this task was found to be more affected by a working memory load than that of adults (Mills et al., 2015). These results highlight how improvements in executive functioning over development (see Chapter 10) can support social cognition.

As in the case of face processing presented earlier, a number of laboratories are studying the neural basis of children's understanding of the mental states of others by using fMRI. There is now a group of fMRI studies with children of various ages using different tasks showing that the MPFC is consistently activated when children are engaged in mentalizing (Blakemore et al., 2007; Kobayashi et al., 2007; Pfeifer et al., 2007). In all of these studies, children/adolescents recruited MPFC more extensively than adults when engaged in mentalizing tasks, even when task performance and possible baseline differences were controlled for (Blakemore, 2008). For example, Wang et al. (2006) employed an irony task to investigate the comprehension of communicative intentions from cartoons in adults and children (9–14 years), and found that children recruited MPFC and left inferior frontal gyrus to a greater extent than adults, whereas adults recruited the FG, extrastriate areas, and the amygdala more strongly than children. Furthermore, in a correlation analysis the authors showed that, within the group of children, there was a positive correlation between age and FG activity and a negative correlation between age and extent of MPFC activity. Similarly, Blakemore et al. (2007) reported that adolescent participants (12–18 years) when thinking about intentions showed more extensive activity in MPFC than adults, whereas adults activated parts of the right STS more than adolescents. Finally, two other developmental mentalizing studies revealed very similar patterns of findings. Pfeifer et al. (2007) examined brain activity during self-knowledge retrieval and found that children (10 years) engaged MPFC to a much great extent than adults. In this study, adults activated the lateral temporal complex (LTC) significantly more than children. Moreover, Kobayashi et al. (2007) presented adults and children (8–11 years) with classical theory of mind tasks and also reported that the children showed greater activity in MPFC than adults, but that adults exhibited great activity in the right amygdala than children. Taken together, these studies show that with age the extent of MPFC activation during mentalizing tasks becomes more focal with development, whereas activity in (task-dependent) posterior (temporal) cortical areas sometimes increases. To assess whether changes in activation in the MPFC and temporal cortex in social cognition tasks during adolescence may be driven by the hormonal changes associated with puberty or by chronological age, Goddings and colleagues (Goddings et al., 2012) scanned females adolescents aged 11.1–13.7 years, an age range when pubertal stage and chronological age can be dissociated. Hormonal assays were collected and the results showed that, in a social emotion task, while MPFC activation decreased with age when controlling for hormone levels, anterior temporal cortex activation increased with hormone levels when controlling for chronological age. These results suggest dissociable effects of pubertal hormones and age on the developing brain. It is important to note, however, that pubertal changes can lead to morphological changes that will affect the social experiences of adolescents, e.g., how others may interact with them. This may in turn affect the maturation of these adolescents' social cognition, beyond direct influence pubertal hormones may have on regions of the social brain.

The developmental patterns observed in social cognition tasks using fMRI generally concord with predictions from the IS, with increasing specialization of brain activation for conditions that involve mentalizing. In a small study 6–10-year-old children heard stories including descriptions of physical facts, characters' appearance and relationships, and their mental states. With increasing age, activation in the right TPJ was found to be increasingly specific to descriptions of characters' mental states rather than their appearance and relationships (Saxe et al., 2009). In a later study a similar pattern was observed bilaterally in the TPJ in children aged 5–11 years; in addition, selectivity of activity in the right TPJ for mental states correlated with children's performance on an independent theory of mind task (Gweon et al., 2012). Further evidence of increasing specialization of activation in the social brain over the course of development was found using the Director task paradigm described above. In this case the MPFC was found to show increasingly greater selective activation in adults than in adolescents (11–16 years old) in trials where participants needed to consider the director's perspective compared to trials when it was not necessary (Dumontheil et al., 2012).

The Atypical Social Brain

Another way to study the emerging social brain is to investigate what happens when it develops differently, as in some developmental disorders like autism and Williams syndrome (see Chapter 2).

Research groups have attempted to locate specific brain deficits in autism. Applying the adult neuropsychology model (see the causal epigenesis view in Chapter 1), it was initially (and incorrectly) assumed that there would be a focal structural deficit somewhere in the brain, and that this "hole" in the brain could be associated with a particular pattern of cognitive deficits. As we will see, although atypical development can result in seemingly quite specific profiles of cognitive differences, the corresponding structural aspects of the brain can be elusive, diffuse, and/or inconsistent across different individuals (see Chapter 2). In the case of autism, structural brain imaging and postmortem neuroanatomical studies have variously implicated the brain stem, cerebellum, limbic system, thalamus, and frontal lobes (for review see Ha et al., 2015). Several studies have reported enlargement of the ventricles, indicating atrophy in adjacent limbic and frontal structures (for review see Pennington & Welsh, 1995). However, like several of the brain atypicalities that have been reported in autism, ventricular enlargement is not specific to autism, since it is also observed in various other conditions, including schizophrenia (Karlsgodt et al., 2008), and is heavily overlapping with the variation between neurotypical individuals. There is an increasing interest in understanding the commonalities and differences between various neurodevelopmental orders in both their brain correlates and their developmental profiles (Patel et al., 2022; Piccardi et al., 2021)

> **Further Reading**   South et al. (2008); Tager-Flusberg (2003).

Another difference observed in several studies on autism is in the cerebellum (e.g., Courchesne et al., 1988). At present it is unclear whether this is a postnatal effect (recall from Chapter 4 that cell migration in this region continues into postnatal life) or whether, as Courchesne et al. (1988) argue, it is caused by atypical cell migration between three and five months of gestation. Some evidence for migration failures in the cortex has also been observed, though these do not appear to be restricted to any particular area (Piven et al., 1990). One problem in determining developmental brain damage is that it is difficult to ascertain which atypicalities are the root cause, and which are subsequent consequences of earlier atypical development. In general, it is likely that structures and areas that develop latest are most likely to be affected by earlier deviations from the normal developmental trajectory. For example, in autism the atypicalities sometimes observed in the cortex, hippocampus, and cerebellum could all possibly be the consequences of a primary difference in the thalamus.

While the brain differences that give rise to autism may be diffuse and variable, the resulting cognitive profile presents a clearer picture. Many social processes observed in early childhood appear to be intact in children with autism; however, they generally later show deficits in aspects of mentalizing or theory of mind. Baron-Cohen, Leslie, Frith and their collaborators have shown in a series of studies that many individuals diagnosed with autism, unlike those with Down's syndrome, may fail false-belief tasks and other theory of mind tasks (for reviews see Baron-Cohen, 1995; Frith, 2003; Happé, 1994), while they are relatively unimpaired (compared to mental-age-matched controls) on a variety of related tasks. Thus, many believe that a difficulty with theory of mind is a central cognitive difference in autism. While there is some agreement on the nature of this cognitive difference, there are a number of detailed variants of this view that are currently being tested.

Other theoreticians have proposed more developmental accounts, arguing that an apparent lack of theory of mind arises from differences in precursors of adults' social cognition skills such as the infants' capacity for imitation (Rogers & Pennington, 1991) or the perception of emotion (Hobson, 1993). Thus, these authors believe that social-cognitive atypicalities from birth, or very shortly thereafter, results in the later impairment of theory of mind.

With the aim of identifying the very earliest precursors of autism, several groups have begun longitudinal studies of babies at risk for a later diagnosis of autism for family genetic reasons. So-called "baby siblings" are baby brothers and sisters of older children already diagnosed with autism and have a higher likelihood of being diagnosed themselves (although it is important to note that this is still a minority). Studies of baby siblings have shown that in many cases the first signs of atypical social interaction skills become evident from 12 to 18 months of age (Zwaigenbaum et al., 2005). While atypical social interaction skills may not emerge until the end of the first year, and diagnosis is usually not possible until 2 or 3 years of age, recent reviews of infant sibling studies have reported a variety of different sensory-motor and brain function differences in early infancy that associate with a later diagnosis of autism (Johnson, Gliga, et al., 2015).

In one example mentioned earlier, an optical imaging (fNIRS) study with 5-month-old infants observed activation of STS while infants viewed dynamic social videos as compared to non-social stimuli (Lloyd-Fox et al., 2009). Later studies built on this work and found that infant siblings as a group tend to show less difference in activation to the social versus the non-social stimulus (Lloyd-Fox et al., 2013) and that this group difference was primarily attributable to those individuals who later went on to a diagnosis (Lloyd-Fox, Blasi, et al., 2017). These results suggest that some early atypicality around STS and nearby areas may be a marker of later atypical social brain function.

It is important to note, however, that some of the early markers that have been identified are not obvious precursors of the social brain. For example, at 14 months, longer latencies to disengage attention from a central stimulus to a lateral one were observed in at-risk infants later diagnosed with autism compared either to other infants at risk or to the low-risk control group (Elsabbagh et al., 2013). Thus, more general atypicalities outside the social brain are also linked to later atypical social processing in autism.

Following on from the infant siblings work, others have studied toddlers that have recently received a preliminary diagnosis of autism. For example, Dawson and colleagues studied effects of familiarity on face and object (toy) processing (Dawson et al., 2002). While a control group of young children showed ERP evidence of recognition of both categories, the potentially autistic sample did not show evidence of recognizing the faces. This suggests some degree of specificity in their neural differences. However, as the authors point out, this difference is probably secondary to an earlier developmental cause, such as failing to orient toward faces at birth.

Given the somewhat mixed structural neuroimaging and neuroanatomical evidence, some research teams have tried to use the pattern of cognitive differences to infer which brain pathways and structures may be implicated in autism. Rather than specific impairments, autism may be associated with atypical patterns of strengths and weaknesses in executive functions tasks that vary across individuals, which may explain the mixed results observed in previous studies (Towgood et al., 2009). At present it is unclear whether these differences are primary to those in theory of mind, independent of theory of mind, or dependent on the same underlying computations (see Pennington & Welsh, 1995). A study in 100 adolescents with autism spectrum disorder found that while theory of mind and executive function skills were correlated, theory of mind, but not executive function, ability was associated with social communication and repetitive behavior symptoms reported by the parents (Jones et al., 2018).

Further Reading South et al. (2008); Elsabbagh & Johnson (2007); Johnson, Gliga, et al. (2015).

In contrast to the cognitive profile seen in autism, at first sight Williams syndrome participants appear to have a typical social brain. Williams syndrome (WS) (also known as infantile hypercalcemia) is a relatively rare disorder of genetic origin (see Chapter 2). Alongside surprising linguistic abilities (see Chapter 9), WS participants perform as well as controls in a face discrimination task (the Benton test; Bellugi et al., 1992), and better than typical adults on the face recognition component of a standard memory test (the Rivermead Behavioral Memory Test; Udwin & Yule, 1991). This pattern of spared abilities suggests approximately the opposite profile of skills to those described earlier for autism, and raised the initial hypothesis that WS individuals have a typical social brain despite difficulties in other domains. As discussed earlier, one prominent account of the cognitive deficit in autism is that people with autism lack a theory of mind. Karmiloff-Smith and colleagues carried out a series of experiments with WS participants to test the hypothesis that there is "a broad cognitive module for representing and processing stimuli relevant to other individuals, including face processing, language, and theory of mind" (Karmiloff-Smith et al., 1995, p. 197). While only about 20% of autistic participants pass theory of mind tasks such as that described earlier, 94% of the WS participants passed these tasks, indicating that theory of mind is typical alongside aspects of language and face processing. From this result, along with some others, it is tempting to portray WS as showing approximately the opposite neurocognitive profile to autism (while acknowledging that both groups may show some general delays), a hypothesis that is enhanced by the differential cerebellar atypicalities observed in autism versus WS (Niego & Benítez-Burraco, 2022).

Further Reading   Karmiloff-Smith (2008); Tager-Flusberg (2003).

However, Karmiloff-Smith et al. (1995) cite evidence from other developmental disorders that is contrary to the simplest view of an impaired or intact prespecified module. For example, in Down's syndrome, a serious deficit in face processing and in the use of morphology in language can co-occur with relatively good performance on theory of mind tasks (Baron-Cohen et al., 1985, 1986). Conversely, in an individual with hydrocephalus with associated myelomeningocele, very competent language output co-exists with serious deficits in face processing and theory of mind (Cromer, 1992; Karmiloff-Smith, 1992). These different patterns of dissociation clearly challenge the notion of a predetermined social module in the brain.

An alternative view consistent with the IS approach (Karmiloff-Smith, 2002) is that some degree of modularization is a result of postnatal development, and not a precursor to it (see also Chapter 13). Specifically, domain-specific biases in the newborn (such as the face preference discussed earlier in this chapter and the speech discrimination abilities discussed in Chapter 9) ensure that cortical circuits are preferentially exposed to socially relevant stimuli like language and faces. With prolonged exposure to such stimuli, plastic cortical circuits develop representations appropriate for processing these inputs, eventually giving rise to an emergent superordinate system for the pragmatics of social interaction in general.

Further Reading   Karmiloff-Smith (2008).

If this general view of the emergence of the social brain is correct, it implies that early sensory or social deprivation may have long-term consequences. Two lines of research have pursued this issue. In the first, Maurer and colleagues have studied people who suffered visual deprivation for varying periods of time following birth due to uni- or bilateral cataracts. These dense cataracts prevent structured visual input until they are reversed by surgery, usually within the first year. By studying aspects of face processing in this clinical population, this research group showed that even after years of normal experience of faces some deficits remained (Le Grand et al., 2001). In other words, visual deprivation over the first months has detectable life-long effects on face processing. Further, by examining cases of unilateral deprivation it has been shown that these effects are more due to right hemisphere (left eye) deprivation. Data such as these present a severe challenge for the "skill learning" approach to human functional brain development, and suggest propensity of the right hemisphere for face processing from the first months. Studies of children with congenital blindness not caused by brain injury show that this more severe and continued visual deprivation can have a serious impact on social development. Infants with severe visual impairment are at risk of serious developmental difficulties, including developmental setback/autistic regression (Dale & Sonksen, 2002), and even those who are developing relatively well by school age still show particular difficulties with social communication and attention (Tadic et al., 2009, 2010). Studies with EEG and ERP further show that infants, toddlers, and older children with congenital visual impairment show altered brain response to social stimuli (e.g., their own name versus another name or social versus non-social sounds; Bathelt et al., 2017; O'Reilly et al., 2017), which is related to their social abilities. This line of study highlights the potential importance of vision for social development; however, it is not clear whether this is a direct impact of vision or whether this occurs via another factor, e.g., a co-occurring disease factor, or via the impact of lack of vision on mother–child interaction, and so on.

While the cataract population can inform us about sensory deprivation, other populations have been studied that suffered from social deprivation. For example, samples of children raised in orphanages (e.g., during the Romanian communist regime) can subsequently have multiple social, cognitive, and sensorimotor problems (for review see Gunnar, 2001). While orphanage rearing can be variable in general care quality, at a minimum, stable long-term relationships with caregivers are missing (Rutter, 1998). The outcome from "good" orphanages can include problems with executive functions (see Chapter 10) and social cognition, while other aspects of sensorimotor, cognitive, and linguistic development can recover well. At the other extreme, in a sample of children raised in Romanian orphanages for at least the first 12 months, 12% exhibited features of autism, although even here these symptoms tended to diminish over time (Rutter et al., 1999).

> **Further Reading**   Shackman et al. (2008).

General Summary and Conclusions

As discussed at the beginning of this chapter, different viewpoints on functional brain development make different predictions about the development of the social brain. Evidence reviewed from the neurodevelopment of face processing appeared inconsistent with a maturational account. The long-term effects of early deprivation and the presence of face and eye gaze biases in the newborn discourage a strictly skill learning perspective. Thus, there is probably neither an innate module for social cognition, nor innate representations relating to these functions that "mature" in the cortex during postnatal life. Rather, complex representations for processing information about other people, their probable thoughts, and likely future actions emerge in the brain as an inevitable result of a combination of factors: initial biases to attend to socially relevant stimuli such as faces and language; the presence of other humans who actively seek interaction with the child; and the basic architecture of the cortex along with its regional biases and patterns of connectivity. Atypicalities in any of these factors could send the infant into a more unusual path in which only some components of typical social cognitive abilities develop.

A challenge for this IS view of the emerging social brain will be to generate hypotheses about the consequences of the network of social processing areas being initially intermixed with other cortical networks. Behavioral studies with infants have demonstrated that, from at least 9 months of age, they attribute goals to appropriately moving objects even when these objects do not physically resemble biological forms (see Csibra, 2003, for review). Indeed, some have suggested that infants and toddlers may over-extend their social cognitive abilities to physical objects that adults would not (Csibra, 2003). Whether this blurring of the boundary between social and non-social perception is a by-product of a social brain that has not yet fully emerged from the rest of the brain remains an interesting topic for future research.

<table>
<tr><td>

Key Issues for Discussion

- Given the complex social cognitive abilities of humans, what is the value of animal models of the developing social brain?
- In what ways might the functions of the components of the social brain network differ in infants from those observed in adults, and how could this be tested empirically?
- What different factors could make the typical development of the social brain take an atypical trajectory?
- What makes a stimulus/situation/event "social" versus "non-social"?

</td></tr>
</table>

8

Learning and Long-Term Memory

Long-term memory is usually divided into different types. One important division is between an explicit form of long-term memory for information we can bring to mind, and an implicit form of long-term memory involving perceptual-motor skills. Each of these also has subtypes: Explicit memory includes semantic memory for facts and episodic memory for events within their spatio-temporal context, and implicit memory includes a range of skills such as conditioning and priming. Initially it was thought that implicit memory was present from birth, while explicit memory emerged later due to the immaturity of the medial temporal lobe (MTL). More recently investigators have found that a form of explicit memory that is dependent on the MTL is also present from birth, with important improvements then occurring at about 8–10 months of age as the MTL matures. While implicit memory performance generally shows little developmental change beyond 3 years of age, explicit memory, particularly the episodic type, continues to develop throughout childhood. There is some debate as to whether this primarily involves an increasing contribution of the prefrontal cortex, or also involves continued development of the hippocampus. The expansion of cortical involvement in the neural basis of explicit memory may be due to an experience-dependent increasing specialization of cortical regions related to the hippocampus. Overall, it seems that several brain systems may be engaged by most memory tasks, a conclusion that echoes those drawn for cognitive domains reviewed in other chapters.

Memory is central to our everyday lives, allowing us to build a knowledge base of facts, rules, and skills and to store details about personal experiences that are central to our sense of who we are as individuals. Our present-day understanding of the cognitive and neural bases of memory was hugely influenced by the study of a single individual, known as HM (Eichenbaum, 2012). He suffered severe memory loss following surgical removal of a large part of both temporal lobes as treatment for medication-resistant epilepsy. After surgery HM was unable to form new long-term memories for facts or events, forgetting such things within minutes. Even so, he remained able to remember things from his life before the surgery and to form new long-term memories under some circumstances, such as when

Developmental Cognitive Neuroscience: An Introduction, Fifth Edition. Michelle de Haan,
Iroise Dumontheil, and Mark H. Johnson.
© 2023 John Wiley & Sons Ltd. Published 2023 by John Wiley & Sons Ltd.
Companion website: www.wiley.com/go/johnson/devneuro5e

learning a motor skill. This advanced our understanding of memory in two ways: (a) by illustrating that memory is not a unitary function and (b) by illustrating that different types of memory rely on different brain networks. In particular, his case highlighted the difference between *explicit memories* (sometimes also called declarative or cognitive memories), which are those memories that we can bring to mind, and *implicit memories* (sometimes also called non-declarative, procedural, habit, or non-cognitive memories), which are memories typically expressed as changes in perceptual or motor performance; only the former type of memory is affected in HM and dependent on the medial temporal lobes (MTL; Cohen & Squire, 1980).

Additional research suggests that there are further divisions within the explicit and implicit forms of memory. Explicit memory includes semantic memory for facts that are not associated with a particular context (e.g., A car has wheels), and episodic memory for events associated with the spatial-temporal context in which we experienced them (e.g., I parked my car under a tree this morning). Explicit memory relies on an MTL (see Figure 8.1) cortical circuit, involving the hippocampus, the surrounding cortex (perirhinal, entorhinal, and parahippocampal cortices) and diencephalon, as well as regions of the prefrontal cortex. Within this system, the hippocampus is believed by many to be particularly involved in episodic memory and less involved in semantic memory (Mishkin et al., 1997; though see Squire et al., 2004 for a different view). Implicit memory involves a range of different skills including motor learning, conditioning, visual discrimination learning, and priming, mediated by different though sometimes overlapping brain circuits. For example, some forms of implicit motor learning and conditioning have been linked to the basal ganglia, cerebellum, and other motor structures, while perceptual priming, which is the greater ease with which a perceptual stimulus is processed if it has been encountered before, has been linked to sensory cortex (Batterink et al., 2019).

The multi-system view of memory has been a dominant framework for those investigating the cognitive neuroscience of memory development. Researchers have taken advantage of the wealth of data from animal and human adult cognitive neuroscience studies on the neural bases of different learning and memory systems in the brain (for reviews see Bachevalier, 2008; Squire et al., 2004) to design tasks and develop theories of the neural

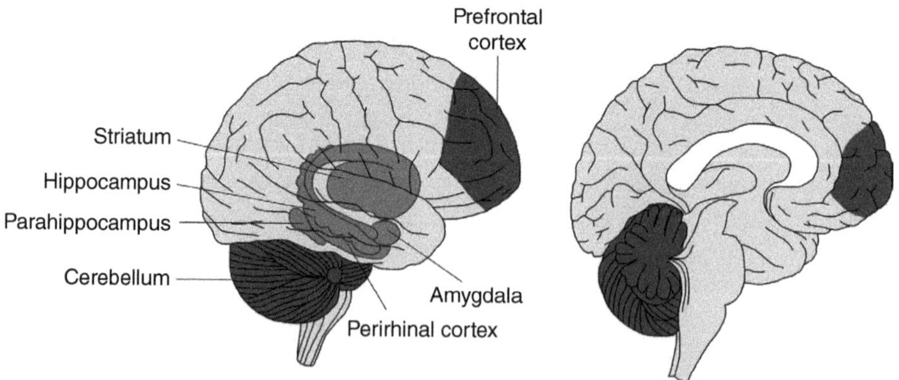

Figure 8.1 Brain areas involved in memory. Figure provided by Iroise Dumontheil.

bases of memory development. Nevertheless, this developmental cognitive neuroscience approach to memory faces a number of challenges:

- Behavioral development in a memory task may be due to the emergence of function in non-memory-related brain systems. For example, improvements in motor or verbal skills may allow better expression of the contents of memory even if these themselves have not changed. Further, at the molecular and cellular level, development and learning often share common mechanisms, making the distinction between them unclear.
- As discussed above, behavioral neuroscience and neuropsychological investigations have revealed that there are multiple brain systems supporting different types of memory. Even in the adult these various systems have proved difficult to disentangle and there is still some debate as to whether it is really appropriate to apply this same framework to the developing system.
- Some forms of "explicit" memory in human adults involve conscious awareness of information, something that is difficult to ascertain in preverbal infants.

Despite these difficulties, we will see in this chapter that considerable progress has been made by adopting a cognitive neuroscience approach to the development of memory systems. As a starting point, most researchers have turned to studies of the neural basis of various memory systems in adults and animals. They have then attempted to develop marker tasks for the functioning of these systems, often speculating that limits in memory abilities in young infants and children may be due to the differential development of neural systems supporting particular types of memory function. Similarly, differential development and involvement of neural systems has been highlighted for improvements in memory in later childhood and adolescence, with a particular focus on the emerging influence of the prefrontal cortex and its interaction with the MTL.

Development of Explicit Memory

Initial cognitive neuroscience hypotheses about infant memory development proposed that infants initially rely on implicit memory, with explicit memory emerging later (Bachevalier & Mishkin, 1984; Schacter & Moscovitch, 1984). This idea emerged from observations including the similarities between the amnesic syndrome, in which medial temporal system damage in adults results in deficits in explicit recognition memory without deficits in learning implicit stimulus–response links, and the profile of memory abilities in infants.

Further Reading   Bachevalier (2008); Bachevalier & Vargha-Khadem (2005).

To test the hypothesis that the putative explicit "cognitive memory" and implicit "habit" systems show a different ontogenetic timetable, Bachevalier and Mishkin (1984) tested infant monkeys aged 3, 6, and 12 months on two types of tasks. These tasks were built around the view that explicit memories can be established quickly after a single learning experience, whereas implicit memories are typically built up slowly with repeated exposures. The first task was a visual recognition task that involved the infant monkey learning to reach for the novel object of a pair following the earlier presentation of the familiar one (delayed non-match to sample). This task involved just a single presentation of the to-be-learned object and was thought to require a "cognitive memory" system as it could only be successfully performed by bringing to mind what had been observed in the single learning episode. In the second task, a visual discrimination habit task, the infant monkey was sequentially exposed to the same 20 pairs of objects every day. Every day the same object of each pair was baited with a food reward, even though their relative positions were varied. The monkeys had to learn to displace the correct object in each pair. This task was thought to involve the "habit" memory system as it could be successfully performed by learning the association between an item and reward across the repeated trials.

Infant monkeys failed to learn the "cognitive memory" task until they were approximately 4 months old, and did not reach adult levels of proficiency even by the end of the first year. In contrast, infant monkeys of 3 or 4 months of age were able to learn the visual "habit" as easily as adults. Bachevalier and Mishkin suggested that this dissociation in memory abilities in infant monkeys is due to the prolonged postnatal development of the MTL system delaying the ability of recognition and association ("cognitive") memory relative to sensorimotor habit formation. This explanation could be extended to human infants since they are also unable to acquire the delayed non-match to sample task in early infancy (until around 15 months), and still have not reached adult levels of performance at 6 years old (Overman et al., 1992).

A key line of evidence to challenge this view came from the visual paired comparison task. This task is a bit like the delayed non-match to sample task, as participants are first familiarized to a stimulus, and then their memories are tested by presenting the familiar stimulus alongside a novel one. However, in the visual paired comparison task, longer looking (instead of reaching) to the novel compared to the familiar stimulus provides evidence of recognition memory. Between 15 and 30 days of age, normal infant rhesus monkeys develop a strong preference for looking toward the novel stimulus, indicating recognition of the familiar one, and this preference becomes stronger over the next months (Bachevalier et al., 1993; Zeamer et al., 2010). This preference is not found in monkeys that received MTL lesions including the hippocampus in early infancy. These surprising results suggest that medial temporal structures do make a significant contribution to visual recognition memory even at this very early age. Further evidence in infant monkeys with neonatal lesions in the MTL (in this case, the perirhinal cortex) show that these lesions reduce, but do not eliminate, novelty preference; in contrast, adult monkeys receiving similar lesions showed an absence of novelty preference (Zeamer et al., 2015). These findings confirm the role of the MTL in this early form of memory, but also suggest that the neural substrates of memory are more broadly distributed in infants than in adults. Together, the findings also suggest that factors other than MTL immaturity might contribute to the more prolonged development of memory performance on the delayed non-match to sample task. For example, slower development of the ability to learn the rule ("Choose the novel object") might be responsible.

Human infants also show evidence of recognition memory after a delay in visual paired comparison tasks: 3–4-day-olds look longer at a novel face than a familiar one even when a 2-minute delay is imposed (Pascalis & de Schonen, 1994), and 3-month-olds can do the

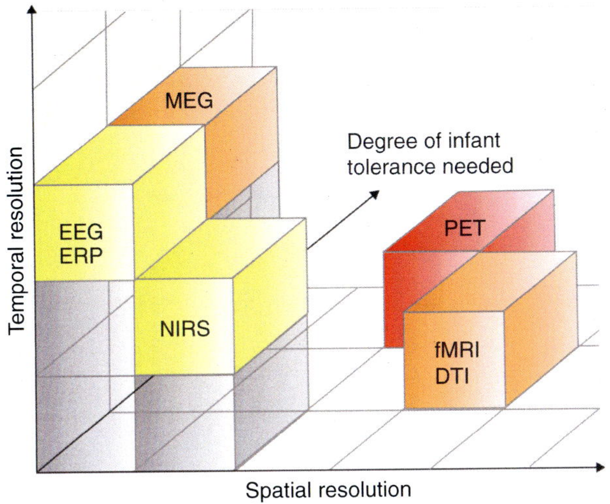

Figure 2.1 An illustration of the relative strengths and weaknesses of different functional brain imaging methods used with infants and children. Lloyd-Fox et al., 2010. Reproduced with permission of Elsevier.

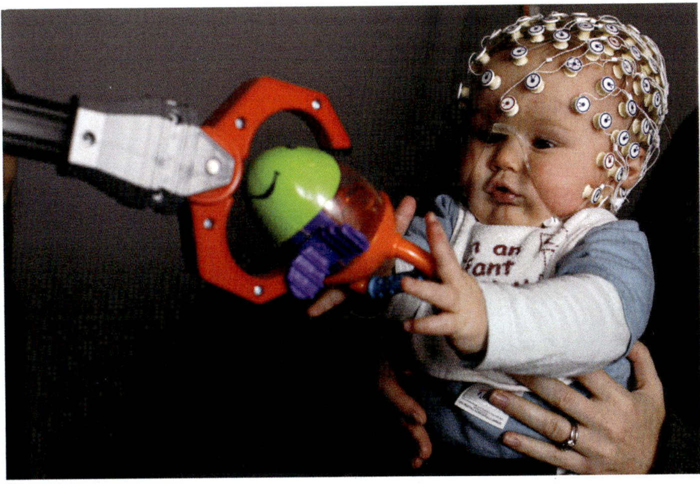

Figure 2.2 An infant wearing a high-density ERP/EEG system (EGI Geodesic Sensor Net) during a study on the "mirror neuron system". The sensor net consists of damp sponge contacts that rest gently on the scalp. © Michael Crabtree. Reproduced with permission.

Developmental Cognitive Neuroscience: An Introduction, Fifth Edition. Michelle de Haan, Iroise Dumontheil, and Mark H. Johnson.
© 2023 John Wiley & Sons Ltd. Published 2023 by John Wiley & Sons Ltd.
Companion website: www.wiley.com/go/johnson/devneuro5e

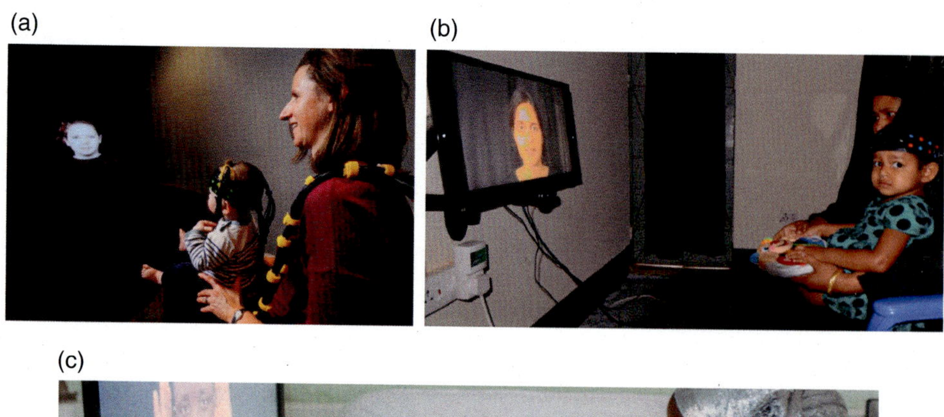

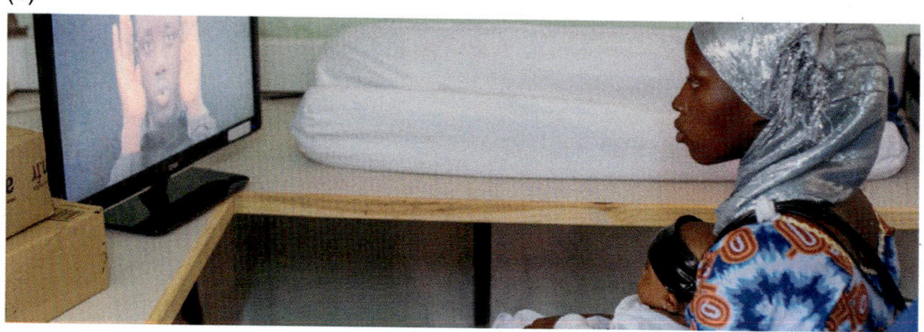

Figure 2.3 Infants in (a) the UK, (b) Bangladesh, and (c) The Gambia engaged in functional near-infrared spectroscopy studies. Light emitters and detectors are incorporated into the head caps. From Blasi et al. 2019 with permission.

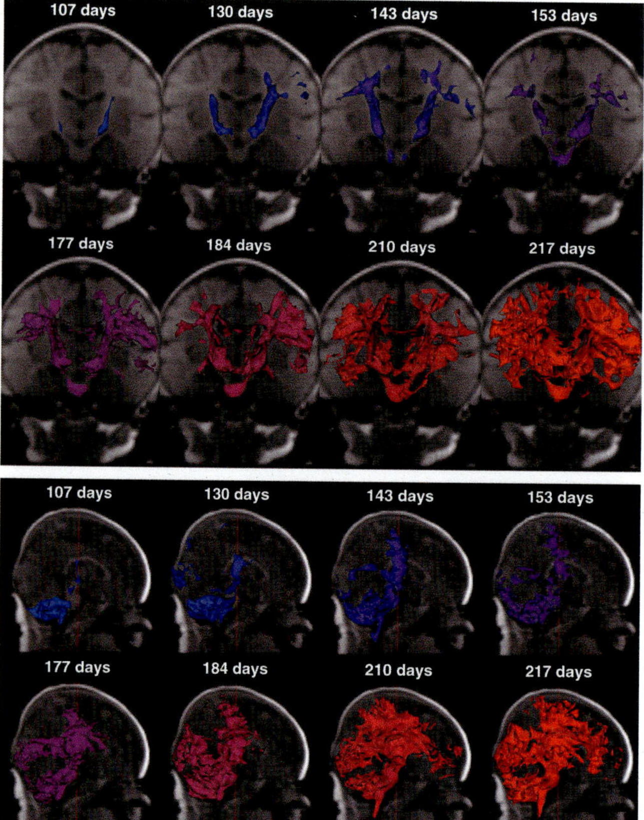

Figure 2.4 The expansion of myelinated fibers over early postnatal development as revealed by a new structural MRI technique. © Dr Sean Deoni. Reproduced with permission.

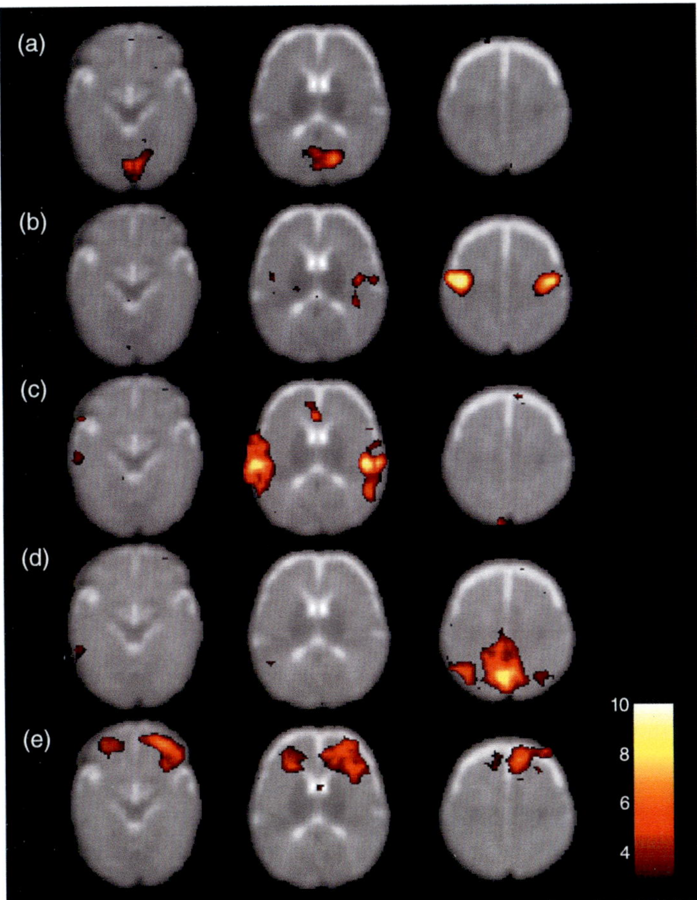

Figure 4.7   Resting state networks in a single representative infant. Rows A to E each show one resting state network at three axial sections. Fransson et al., 2007. Copyright © (2007) National Academy of Sciences, USA.

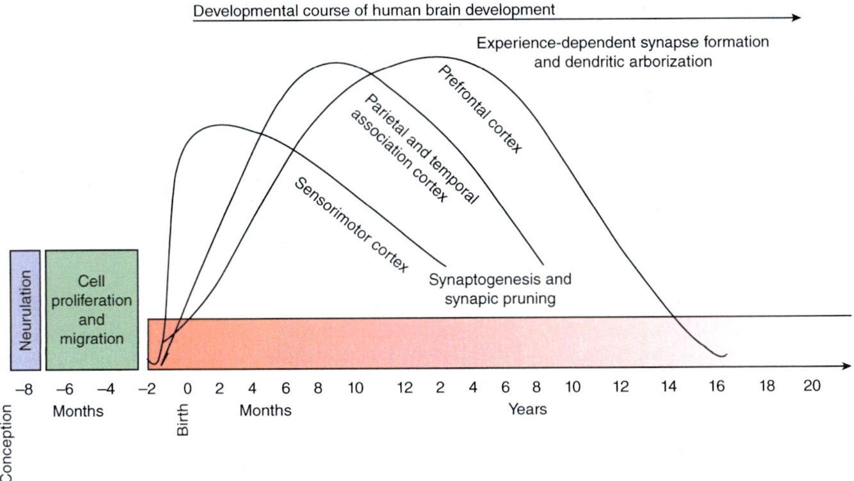

Figure 4.8   Figure illustrating the approximate timeline for some of the most important changes in human brain development, including the characteristic rise and fall of synaptic density.

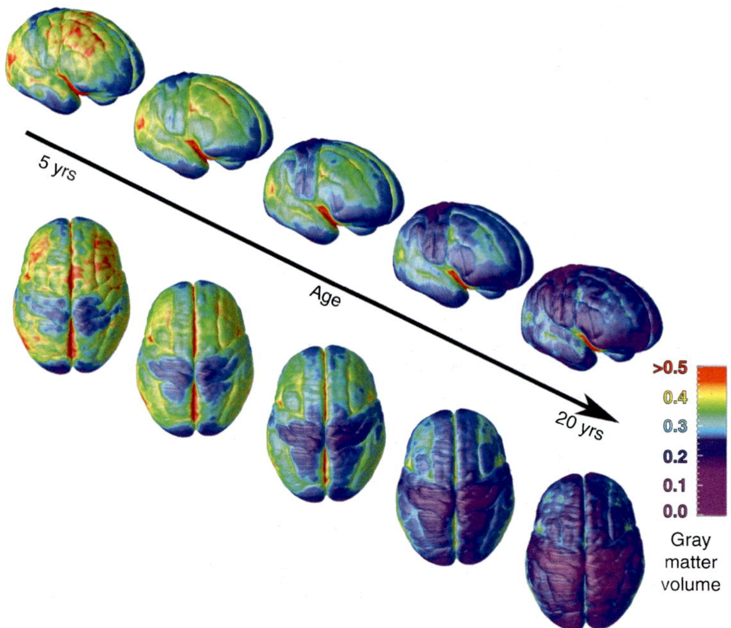

Figure 4.10 A color-coded map of changes in cortical gray matter with development. The maps illustrate regional variations in decreases in gray matter density between the ages of 5 and 20 years. Toga et al., 2006. Reproduced with permission of Elsevier.

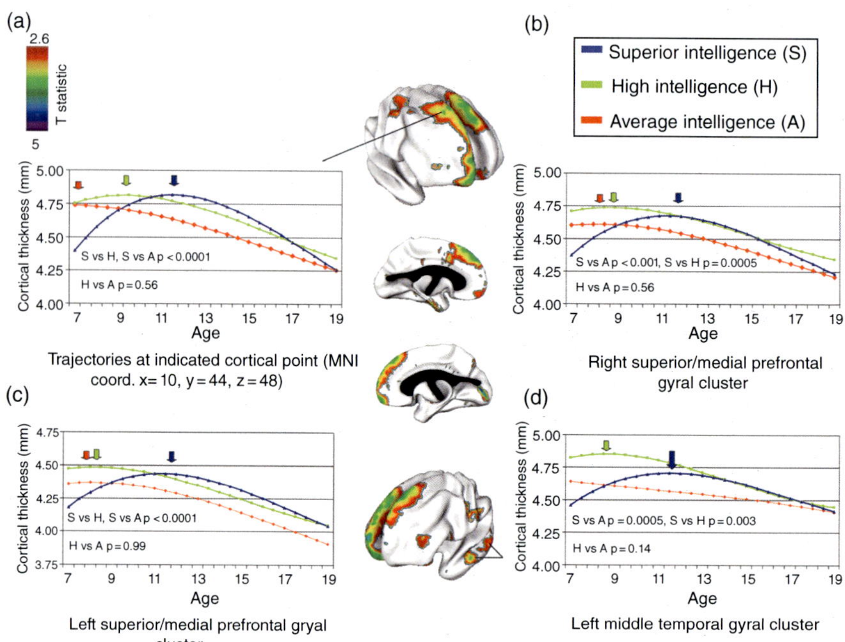

Figure 4.11 The brain maps (center panel) show prominent clusters where "superior" and "average" intelligence groups differ significantly in the trajectories of cortical development. The graphs show the developmental trajectories for these regions. The age of peak cortical thickness is arrowed for each of the three groups in each region. Reprinted by permission from Macmillan Publishers Ltd: *Nature, 440,* 676–679, copyright (2006).

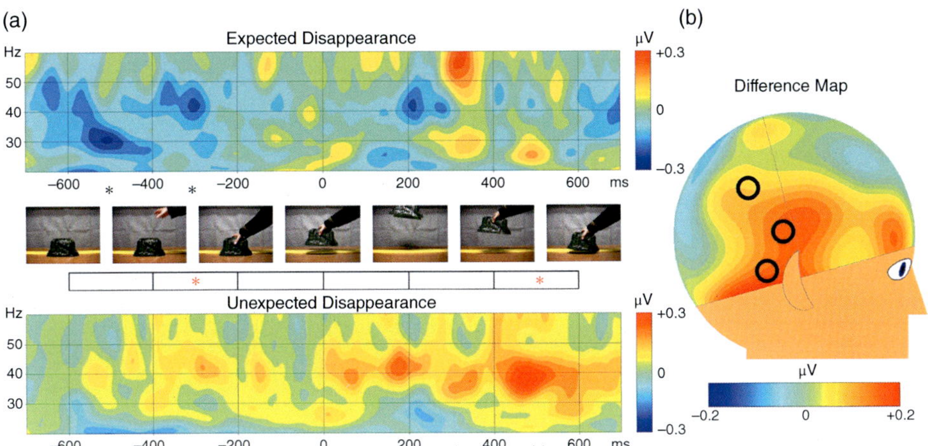

Figure 6.3 Gamma-band EEG activity recorded from infants in the Kaufman, Csibra, and Johnson (2003a) experiment. (a) Time-frequency analysis of the average EEG at three electrodes over the right temporal cortex (around T4) during the phase in which the tunnel was lifted showed higher activations when the object should have been below the tunnel. Black asterisks below the maps indicate a significant difference from baseline; red asterisks indicate a significant difference between conditions in the average gamma activity in 200 ms-long bins. (b) A topographical map of the between-condition difference of gamma-band (20–60 Hz) activity during the occlusion-related peak gamma activity (from −400 to −200 ms) revealed a right-temporal focus. Circles signify right-temporal electrode sites. Reproduced with permission of the Royal Society and the authors.

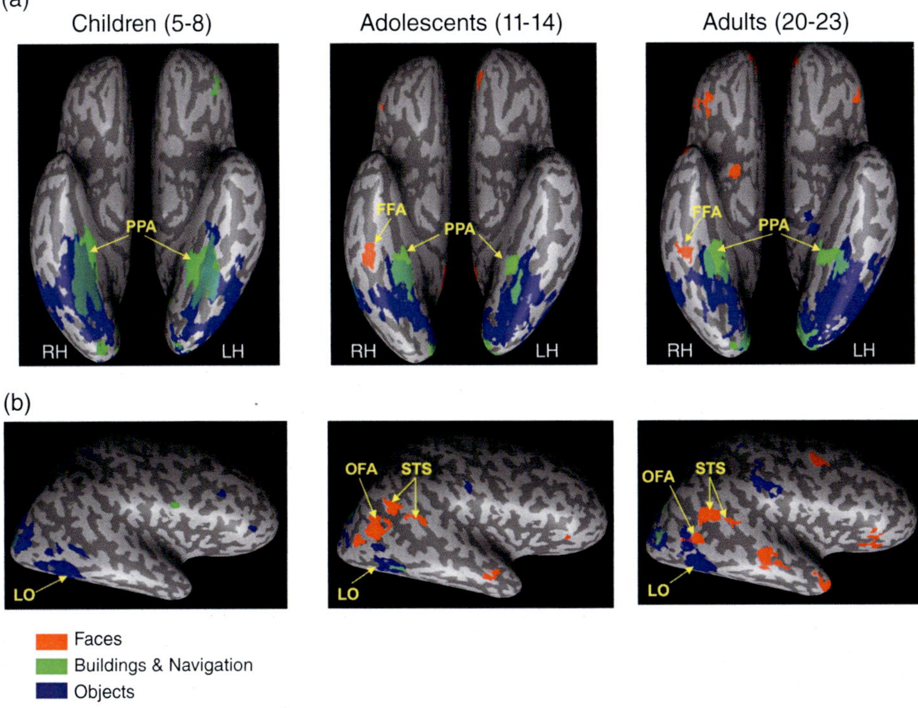

Figure 7.6 Differential activation for each stimulus category mapped onto an inflated brain: (a) ventral view and (b) a lateral view of the right hemisphere for all three age groups. In contrast to older groups, young children showed no face-selective activation in face-related areas. However, objects and buildings or navigation yielded similar patterns of selective activation at all ages. Abbreviations: FFA, fusiform face area; LO, lateral occipital object area; OFA, occipital face area; PPA, parahippocampal place area; STS, superior temporal sulcus. Scherf et al., 2007. Reproduced with permission of John Wiley and Sons.

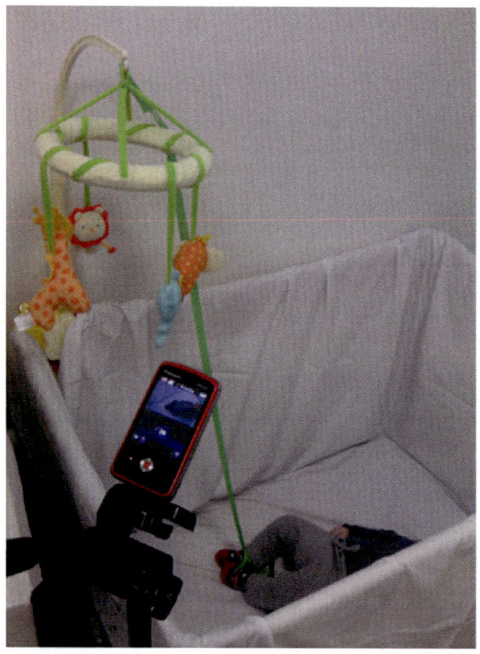

Figure 8.2 (a) Example of a 2-step imitation task—the actions are to put down the ramp, and roll the car down. (b) Example of the mobile conjugate reinforcement task, showing the baby with the ribbon attached to their leg, the mobile they can control, and the phone camera recording the session. Provided by Kayleigh Day with permission.

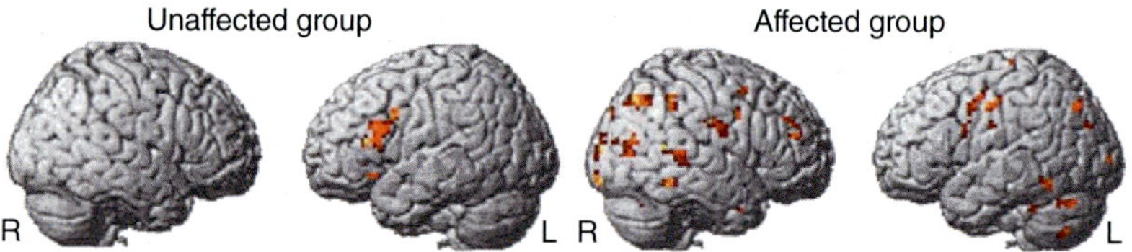

Figure 9.3 Covert language task: group average fMRI activation in the unaffected and affected members of the KE family. Activated regions are projected onto the surface rendering of a typical 3D individual brain, displayed at a statistical threshold of $p < .05$, corrected for multiple comparisons. L, left hemisphere; R, right hemisphere. Liégeois et al., 2003. Reproduced with permission of Nature Publishing Group.

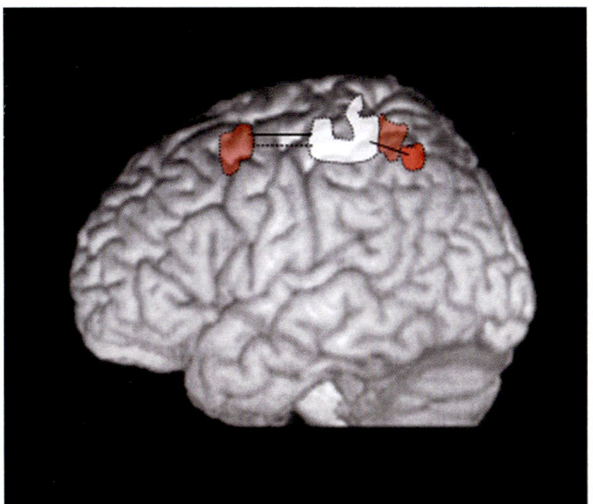

Figure 10.2 A summary of the superior frontal-intraparietal network involved in the development of visuo-spatial working memory. Regions in red show a correlation between brain activity and development of working memory capacity, and regions in white show a correlation between white matter maturation and development. Klingberg, 2006. Reproduced with permission of the AAAS.

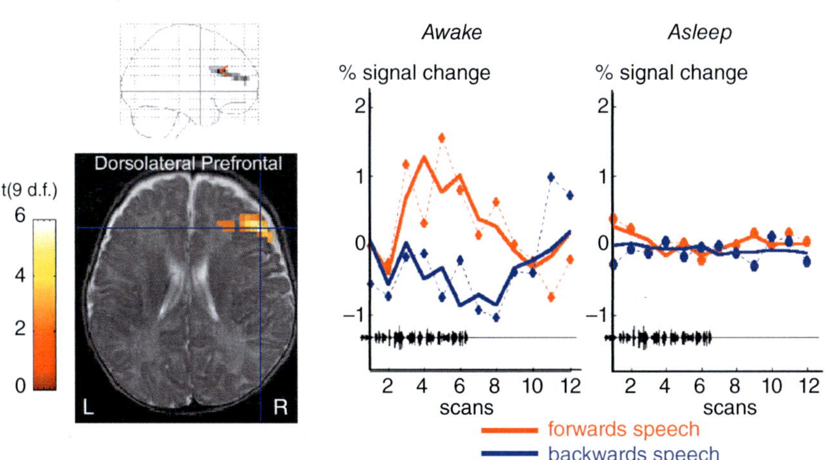

Figure 10.4 Interaction between wakefulness and the linguistic nature of the stimuli. This comparison isolated a right dorsolateral prefrontal region that showed greater activation by forward speech than by backward speech in awake infants, but not in sleeping infants. Dehaene-Lambertz et al., 2002. Reproduced with permission of Elsevier.

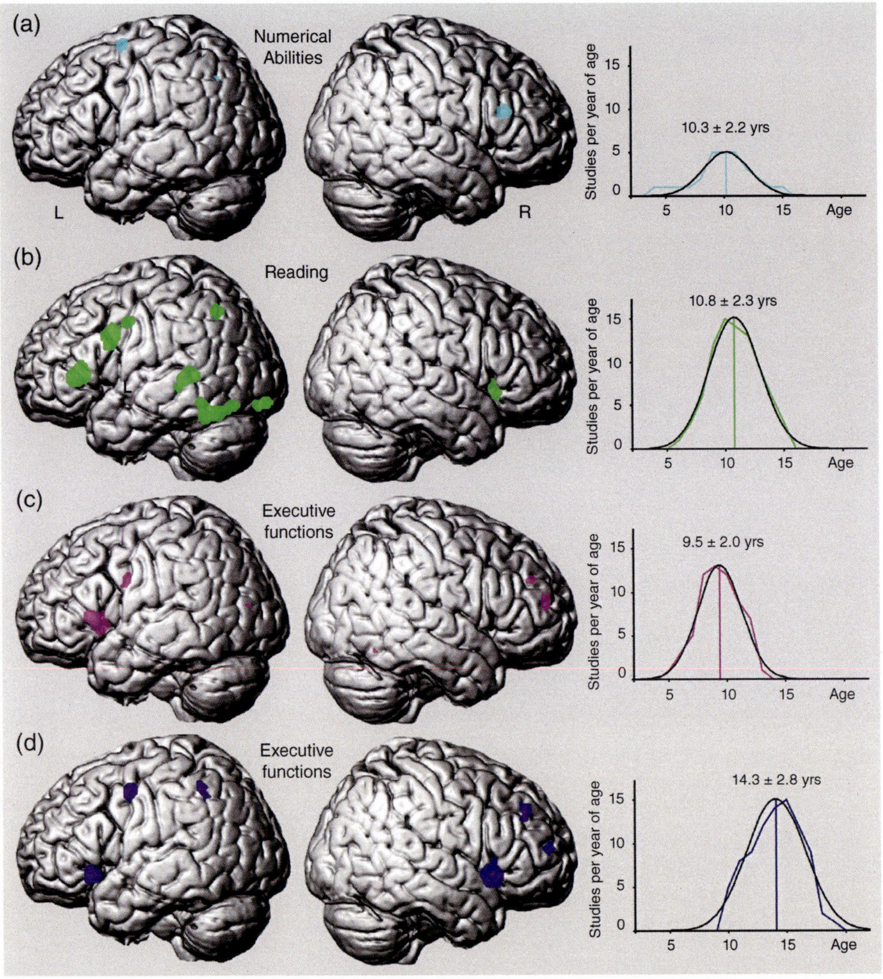

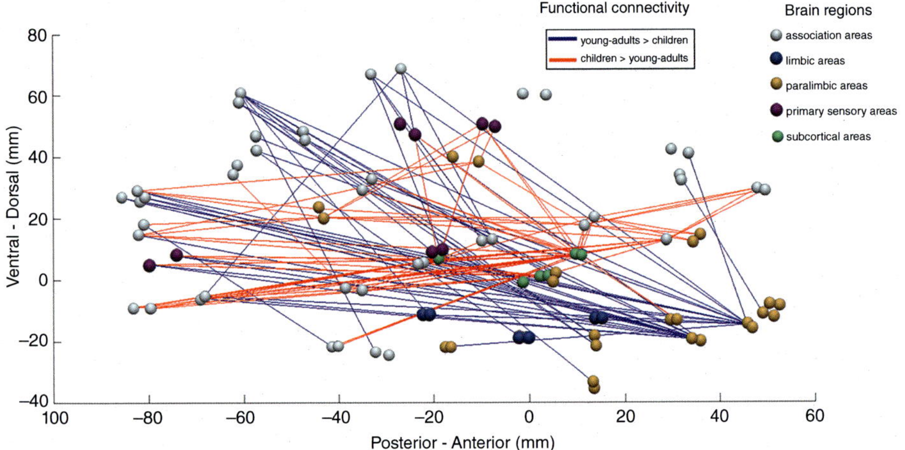

Figure 11.3 A summary over 52 fMRI studies of healthy children and adolescents showing (a) the locations of brain areas involved in numerical abilities (left), (b) reading (left), and (c, d) executive function (left). On the right, the distributions show the number of studies per year of age.

Figure 13.3 Developmental changes in inter-regional functional connectivity. A graphical representation of developmental changes in functional connectivity along the posterior-anterior and ventral-dorsal axes of the brain highlighting higher sub-cortical connectivity and lower paralimbic connectivity in children compared to young adults. Supekar et al., 2009. Reproduced with permission of the authors and of Public Library of Science, Biology.

same over a 24-hour delay (Pascalis et al., 1998). Evidence suggests that, as for monkeys, humans' performance on the task requires an intact MTL. Compared to adults without brain injury, adults who sustained selective, bilateral injury to the hippocampus as infants show reduced looking at novel patterns following delays of 30 s (Munoz et al., 2011). This is not because they cannot perform the task at all, as they show a normal preference for looking at novel patterns if there are only a few seconds between study and test. The relation between the hippocampus and performance on the visual paired comparison task also holds for young human children: one study showed that the degree of novelty preference following a 15-minute delay is related to the structural integrity of the hippocampus in toddlers who experienced prolonged febrile seizures, a condition known to compromise the hippocampus (Martinos et al., 2012). These findings are consistent with a study that examined functional magnetic resonance imaging (fMRI) activation in sleeping toddlers while they were exposed to a song they were familiar with and a new song. Results showed stronger activation in the hippocampus to the familiar compared to the novel song; moreover, this effect was stronger for toddlers who remembered where they had heard the song and which toy character was present at the time (Prabhakar et al., 2018). Together, these results suggest that the MTL memory circuit is functioning to some extent in human infants and toddlers.

To account for these results, Nelson (1995; Nelson & Webb, 2003) proposed a different, but still maturational, view of early memory development. In this view, an immature form of explicit memory, called "pre-explicit memory," is present from birth, is reliant mainly on the hippocampus, and mediates novelty preferences in the visual paired comparison task. At around 8–10 months of age, developments within the hippocampus, the surrounding cortex, and their connections enables a more mature form of explicit memory to emerge that supports a wider range of memory performance. Thus, Nelson believes that early-maturing components of the MTL circuit allow an immature form of explicit memory to function in the first postnatal months, but that an important advance in explicit memory abilities occurs around 8–10 months due to further maturation of the MTL circuit, probably involving the dentate gyrus of the hippocampus. Neuroanatomical studies with monkeys support the idea of earlier- and later-developing circuits involving the hippocampus, with some areas of the hippocampus, specifically the cornu ammonis 1 area (CA1), subiculum, and their associated subcortical connections, developing earlier, and other regions, such as the cornu ammonis 3 area (CA3), dentate gyrus, and their associated cortical projects, developing later (Lavenex et al., 2007).

Further Reading   Johnson et al. (2020).

Another memory task, the deferred imitation task, provides evidence for a striking improvement in memory skills around 8–10 months. In this task, participants are first given a set of toys to play with for a baseline assessment of spontaneous actions. Then, the toys are used to model a specific sequence of target actions to produce an interesting result. The example sequence shown in Figure 8.2a in the color plate section involves putting down a ramp and then letting the car roll down it. Immediately or after a delay, the toys are re-presented to the participant and the number of target actions produced is noted.

Increases in the number of target actions produced in the correct order from baseline to after modeling are taken as evidence of memory. Patients with damage to the MTL show impaired performance in this task (McDonough et al., 1995), even if the injury was sustained in infancy and restricted to the hippocampus (Adlam et al., 2005). Carver, Bauer, and colleagues have found that only about half of typical 9-month-old infants are able to remember any actions in the correct order after a delay (Carver & Bauer, 1999), and, even then, they cannot retain this information longer than about 4 weeks (Carver & Bauer, 2001; Bauer et al., 2006). By 10 months there is a substantial improvement: most 10-month-olds are able to recall some information (Bauer et al., 2006) and they can still do so up to 6 months later (Carver & Bauer, 2001). This sudden improvement in memory skill is attributed to the maturation occurring around this time in the MTL.

> **Further Reading** Bauer (2008); Lynch et al. (2019).

However, this view is not without its opponents. Most notably, Rovee-Collier (Rovee-Collier & Cuevas, 2009) has argued that: (a) infant memory is unitary and does not consist of multiple systems that develop at different rates, and (b) memory development is a continuous process, with no transition occurring at 9–10 months. In this view, it is argued that the basic memory process does not change, but what infants and adults select to encode for learning does. Evidence in favor of this view comes from studies using the mobile conjugate reinforcement task. In this task, there is a learning phase where a ribbon is used to connect the infant's ankle to a mobile so that the mobile moves when the infant kicks (see Figure 8.2b in the color plate section). During baseline and memory tests, the ribbon is disconnected. Changes in the kicking response are then used as a way of measuring what the infant remembers (e.g., characteristics of the mobile or the context in which it is seen): more kicking than in baseline indicates they recognize the mobile. Using this task, and a similar task modified for older infants, Rovee-Collier has found a linear increase in how long information is remembered over the first 18 months of postnatal life, with no indication of a sudden improvement at 9–10 months as predicted in the multi-system view (Hartshorn et al., 1998). Proponents of the multi-system view, in turn, refute this evidence by arguing that the mobile conjugate reinforcement task is a motor skill and thus an implicit memory task reliant on the cerebellum and subcortical structures (Nelson, 1995). Seen this way, the continuous development of performance of the task is unsurprising and not damaging to the multi-system view, as it would expect implicit memory functions to develop early and continuously. Only one study has attempted to relate performance on the mobile task to hippocampal measures. This study found that stronger functional connectivity between the hippocampus and the dorsal anterior cingulate cortex at birth related to better performance on the mobile task at 4 months of age (Scheinost et al., 2020). Since the neural correlates of infants' memory in the mobile conjugate reinforcement task have not been directly studied, the question of whether the task relies on MTL-based memory systems remains unresolved.

> **Further Reading** Rovee-Collier & Giles (2010).

Explicit memory continues to develop beyond the infant and toddler period. In particular, episodic memory appears to show a protracted course of development, attributed to continued development of the MTL circuit and its connections with the frontal lobes. Since episodic memory is defined as memory associated with the spatio-temporal context in which information is encoded, it is commonly studied by investigating how well people can remember details about the "source" of their memories. Drummey and Newcombe (2002) tested children of 4, 6, and 8 years of age in a source memory paradigm adapted from those previously used with adults. In this task the children were first presented with 10 facts (on a variety of topics) by either an experimenter or a puppet. After a delay of one week the children were asked questions on the facts ("item memory"), and also asked to identify the source of the fact (experimenter, puppet, teacher, or parent; "source memory"). Children showed a steady improvement with age in their ability to remember the facts, but showed an abrupt improvement between 4 and 6 years in their ability to monitor the source of those facts. In particular, the 4-year-olds made many errors in identifying the source of the facts compared to the older children. However, while 4-year-olds are less skilled than older children at recalling the sources of their memories, they are still able to do so at a level greater than chance (Sluzenski et al., 2004).

Investigators have disagreed about the brain bases of these improvements in episodic memory throughout middle childhood. Some argue that improvements are mainly due to strategic aspects related to the increased cognitive control allowed by maturation of fronto-parietal circuits. In this view, the MTL is seen as maturing early, and thus contributing little to changes in memory skill during middle childhood and adolescence (Richmond & Nelson, 2007). This view is supported by structural magnetic resonance imaging (MRI) studies showing little change in overall hippocampal volume over middle childhood and fMRI studies showing greater involvement of prefrontal cortex with age (Chiu et al., 2006; Menon et al., 2005; but see counter arguments below).

There is additional evidence that improvements in source memory are probably related to developments in the frontal cortex and its connections with the MTL. First, performance on source memory has been found to be correlated with performance on both "frontal lobe" tasks (Drummey & Newcombe, 2002) and "MTL tasks" (such as cued recall; Sluzenski et al., 2004), pointing to the importance of both regions, and possibly their interconnections, to the development to episodic memory.

Others accept the contribution of the fronto-parietal circuit, but argue that further development in the hippocampus continues to make additional contributions to gains in memory skills over middle childhood (de Haan et al., 2006; Ghetti & Bunge, 2012). These authors point out that, while there is little change in overall hippocampal volume over middle childhood, there are regional changes, with the left anterior hippocampus increasing in volume, and other regions decreasing (Gogtay et al., 2006). The view that the hippocampus continues to develop over childhood and adolescence is also supported by longitudinal studies of hippocampal shape maturation (Lynch et al., 2019). These developmental changes may parallel functional development of the hippocampus. This is highlighted in one study using a source memory task in which children were trained on a set of objects surrounded by colored frames (Ghetti et al., 2010). They were then tested for recognition of the objects (item memory), and for their memory of the color of the frame that had surrounded the objects (source memory). In 8-year-olds, activity in the left hippocampus differentiated between correctly remembered old objects and new objects, but was not affected by whether the child correctly remembered the source. By contrast, 14-year-olds and adults showed differential hippocampal activity dependent on correct recall of both the item itself and the source. Together, these results support the idea that continued development of the hippocampus during middle childhood contributes to improvements in episodic memory over this time.

The role of the hippocampus in episodic memory development is also demonstrated by a unique group of patients who suffered bilateral damage mainly restricted to the hippocampus when they were infants or children (Vargha-Khadem et al., 1997). A report of three cases with such injury, Jon, Kate, and Beth, showed that they all had difficulty remembering information after a delay in spite of normal performance on tests of immediate memory and of memory span (see Figure 8.3). Even though they were unable to retain much information for more than a minute or so, all three children also performed in the normal range on tests of academic achievement (with the exception of spelling) and on the subtests of intelligence quotient (IQ) that tap general knowledge. In other words, the children appeared to have a selective impairment in episodic memory, with relatively intact semantic memory and working memory. This pattern of memory impairment occurred together with selective, bilateral reduction of the volume of the hippocampus to a level approximately 40–60% of normal size, with brain imaging also showing that even the little remaining hippocampal tissue was abnormal. Further study of these patients showed that the extent of their memory impairments correlate with the degree of hippocampal volume reduction, such that those with greater memory impairments show greater volume reductions (Patai et al., 2015).

From a maturational viewpoint, this outcome following injury is easy to explain. The hippocampus normally mediates episodic memory, and thus if it is injured a child will fail to develop episodic memory. However, an alternate explanation is that memory outcome following pediatric hippocampal injury is the result of a reorganization of brain memory systems. Studies comparing infant and adult monkeys suggest that connections of the temporal lobe are more widespread in infants and become more refined with age, providing a possible neuroanatomical basis for reorganization (Webster et al., 1995). From an interactive specialization viewpoint, the loss of the hippocampus might mean that these extra connections are retained and, as a consequence, brain memory circuits are left more widespread and less specialized than normal. In support of this view, the patient Jon shows more widespread activation and atypical connectivity than controls when he is using his residual episodic memory ability (Maguire et al., 2001).

Further Reading de Haan et al. (2006); Vargha-Khadem & Cacucci (2021).

While episodic memory skills tend to reach adult levels by the end of childhood, other aspects of explicit memory seem to show a more prolonged development. Specifically, there is evidence of an increase in strategy use and metamemory skills during adolescence, whereby participants increasingly organize the information to be remembered to facilitate recall and increasingly reflect on the content and accuracy of their memories (Ghetti & Fandakova, 2020). The sort-recall task is a task that can be used to assess the development

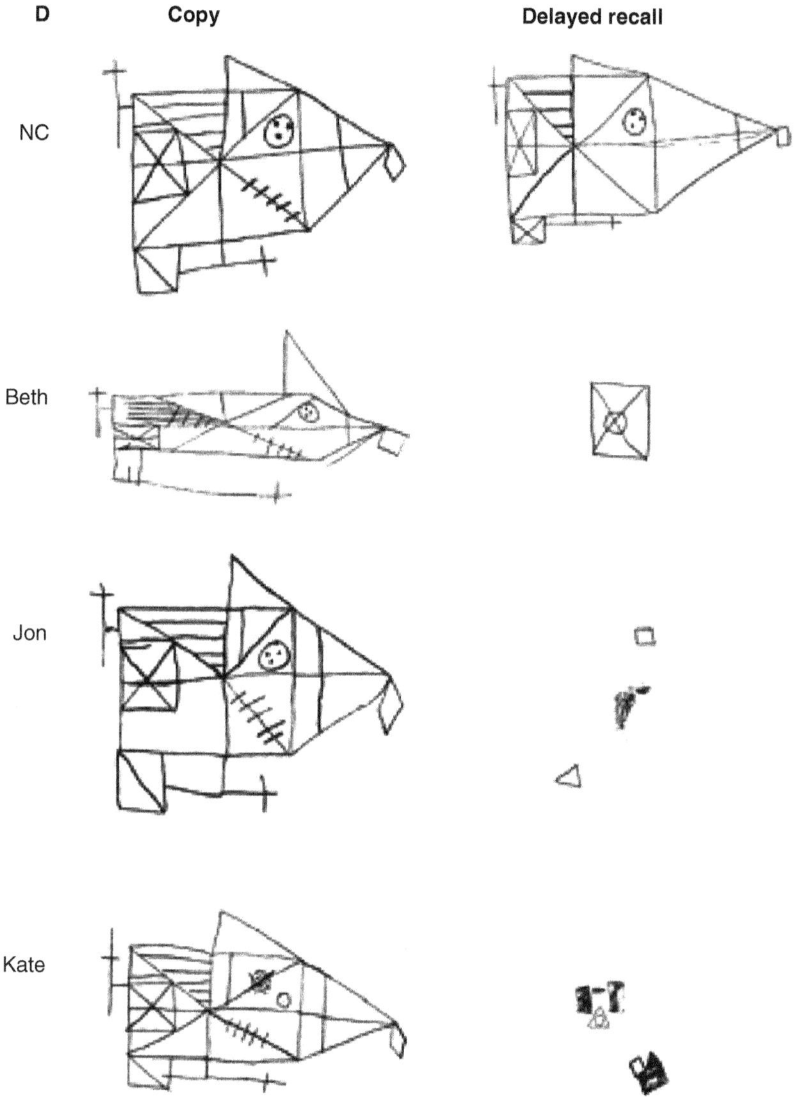

Figure 8.3 Memory performance for Beth, Jon, and Kate on the Rey-Osterrieh Complex Figure. Left column shows their normal copying of the figure, while the right column shows how much less they were able to recall after a 40-minute delay compared to control participants.

of organizational strategies. Participants are shown a set of stimuli that can be categorized semantically (e.g., animals, food, furniture) and are told they have to remember as many stimuli as possible and to do whatever may facilitate learning and remembering the items. Strategic behavior during encoding is reflected by sorting, and during recall by clustering. Studies shows that recall, and sorting and clustering, increase between childhood and adolescence, and that recall performance correlates with the use of strategy (Schneider et al., 2002). As for episodic memory, this prolonged development of strategic processing and metamemory is thought to reflect greater interactions between the prefrontal cortex and hippocampus, but likely also involve the maturation of other networks supporting uncertainty monitoring and subjective experience (Ghetti & Fandakova, 2020). It has also been proposed that greater interactions between neural circuits supporting episodic memory and those supporting executive function may allow adolescents to integrate prior experiences to support goal-oriented behavior (Murty et al., 2016).

Further Reading   Ghetti & Fandakova (2020).

Other aspects of cognition and the environment may influence episodic memory over the life span. There is cross-cultural evidence of a "reminiscence bump" during late adolescence and early adulthood, where individuals asked to recall (either through free or cued recall) autobiographical knowledge from across their life span recall a greater number of memories of their adolescence and early adulthood than of their early childhood (with the phenomenon of childhood amnesia aged 0–5 years old) and mid-adulthood (Conway et al., 2005). Various theoretical accounts have been proposed for this bump, suggesting that the novelty of events occurring in the second and third decades of life lead to increasing remembering, or that brain maturation leads to improved memory skills, but there are also cultural accounts (Munawar et al., 2018). Interestingly, while the timecourse of the reminiscence bump is similar across cultures, the content of autobiographical memories recalled varies; for example, memories from a Chinese sample were more often events with a group or social orientation than were memories from a US sample, which were more often of events oriented toward the individual (Conway et al., 2005).

While episodic memory allows us to remember the past, another aspect of explicit memory is prospective memory, which allows us to plan for the future. Prospective memory enables the execution of an intended action at an appropriate moment, for example remembering to buy a birthday card before going to a party. A distinction is made between event-based prospective memory, which refers to instances in which the action needs to be performed in response to a cue (event), for example buying a card when walking past the shop, and time-based prospective memory, when the intended action has to be carried out at a certain time or after a specific delay, for example taking a cake out of the oven after 40 mins. Some prospective memory capacity is evident by the age of 4 years. In one study, children were asked to draw a picture and name pictures in a stack of cards, but, when the card was an animal, they had to put it away in a box (prospective memory). Most 4-year-olds were able to recall the prospective memory instruction at the end of the experiment and 25–50% of them remembered to perform the prospective memory intention on at least one of four occasions. Performance improved between 4 and 7 years of age and was found to be poor when the ongoing task needed to be interrupted to carry out the prospective memory intention (Kvavilashvili et al., 2001), which suggests that young children have not developed strategic monitoring processes or do not have the attentional resources to deploy them during task performance. Both event-based and time-based prospective memory skills continue to improve during preschool years and middle childhood, with more mixed evidence of development during adolescence; children become more able to use external reminders to cue their remembering and to interrupt ongoing task performance when necessary and become increasingly proficient at using time-checking strategies (Dumontheil, 2014; Magis-Weinberg et al., 2020). Prospective memory recruits a range of neural systems that support intention maintenance, monitoring, and attention capture by the cue, and also relies on retrospective memory for intention retrieval; however, there is very little work done looking at the neural correlates of prospective memory development (Dumontheil, 2014; Magis-Weinberg et al., 2020). Animal models of prospective memory have recently been developed and may lead to new insight about how brain development supports improvements in prospective memory skills (Crystal & Wilson, 2015).

To summarize, the maturational viewpoint has dominated research on the brain bases of explicit memory development. In the current view, components of the MTL circuit, particularly the hippocampus, are functioning to some extent soon after birth to mediate the beginning form of explicit memory. Subsequent development within the MTL and its connections with the frontal cortex mediate further improvements in explicit memory. This framework is based both on (a) the use of tasks whose brain bases is known in monkeys or human adults, with the assumption that the development of human infants' or children's performance reflects maturation of these brain regions and (b) (especially in older children) direct information about the pattern of brain activity associated with memory abilities. Further studies using neuroimaging methods will help provide more detail on the functional roles of the MTL and prefrontal cortex, subregions within them, and connections between them, and may ultimately force an update of these ideas.

Implicit Memory

While there has been great debate about the development of explicit memory, there has been much more agreement about the development of implicit memory. Most investigators agree that implicit memory is in place within the first months of postnatal life, and that there is little if any further development in performance beyond 3 years of age. This has been held true even though different forms of implicit memory, such as perceptual priming and motor sequence learning, rely on different neural substrates.

Several investigators have contrasted the development of implicit and explicit memory in children within the same task and shown that implicit memory develops earlier than explicit memory. For example, Drummey and Newcombe (1995) tested 3-year-olds' priming and recognition memory for pictures of animals from a children's book they had read three months earlier. While the children showed no evidence of explicit recognition of the pictures, they did show evidence of implicit priming by identifying blurred images of the animals more quickly (perceptual facilitation) than control children who had not previously read the book. A second experiment showed that explicit memory improved between 3 and 5 years whereas priming did not.

Studies using a different type of implicit memory task, the serial reaction time task, have reached similar conclusions. In this task, participants must learn the association between a set of response keys and a corresponding set of locations on a display screen, so that they press the correct button when a location is cued. The participant does not know that the locations are actually cued in a repeating sequence. The classic result is that participants become faster at pressing the buttons as they perform the task, even though they are not explicitly aware that there is a pattern. Studies with children show that these changes in reaction time are observed at 4–6 years and do not change with increasing age (e.g., Meulemans et al., 1998; Thomas & Nelson, 2001). However, there is an increase with age in children's ability to explicitly report on the sequence (Thomas & Nelson, 2001).

A neuroimaging study does suggest that the neural correlates of implicit memory in the serial reaction time task may change with age. Thomas et al. (2004) found a similar network of cortical and subcortical structures, including the striatum, activated in adults and 7–11-year-old children (Thomas et al., 2004). At both ages, activity in the right striatum correlated with implicit memory performance. However, there were also differences in activation between adults and children: children showed greater subcortical activation while adults showed greater cortical activation. This could be related to differences in explicit awareness.

While the studies described above suggest that implicit memory abilities are largely in place by 3 years of age, they do not provide insight into how this accomplishment is achieved. Some tasks used to assess implicit memory in children are not suitable for use with infants and toddlers. However, some attempts have been made to bridge this gap by using tasks that are suitable over a wider age span or by drawing analogies between tasks used in older children and those suitable for infants. Another challenge is that few studies with human infants have used tasks whose brain bases have been investigated. In several of the explicit memory tasks outlined above, infants have been tested in behavioral tasks that are close analogues of the monkey and adult tasks, which allowed inferences to be made about the neural bases of infants' performance. In contrast, with implicit forms of memory there has been little attempt to develop marker tasks for infants that closely correspond to tasks used with animals, neuropsychological patients, or functional brain imaging.

One exception is the conditioned eyeblink, a paradigm in which a harmless puff of air blown on to the eye, or a gentle tap on the forehead, causes the infant (or animal) to blink. In a conditioning procedure, the forehead tap, or air puff can come to be predicted by an unrelated sensory stimulus such as an auditory tone. Studies with animals point to the importance of the cerebellum for this type of implicit memory. The early development of the cerebellum may be responsible for the fact that at between 10 and 30 days the human infant starts to show the conditioned eyeblink response (see Lipsitt, 1990; Dziurawiec, 1996).

In other cases, tasks can be adapted or analogies between tasks drawn. For example, one study has attempted to adapt the serial reaction time task for use with infants by using eye movement latencies, rather than button press latencies, as the dependent variable. Results showed that 9-month-olds demonstrate the classic pattern of longer reaction times on random blocks than on the sequenced blocks that preceded or followed (Koch et al., 2020). While explicit knowledge of the sequence cannot be tested at this age, the results do confirm the early appearance of adult-like patterns of implicit memory performance. The neural bases underlying infants' performance has yet to be determined. It is reasonable to speculate that the striatum may be involved as it seems likely from the evidence reviewed in Chapter 4 that this structure is already functional by around birth.

Priming has also been investigated in infants using event-related potentials. In adults, repeating an image after a series of intervening items leads to a priming effect about 200–600 ms after stimulus onset (Henson et al., 2004; Webb & Nelson, 2001). Webb and Nelson found that 6-month-olds showed a priming effect in a similar time window, though the shape of the component influenced by priming, and its location on the scalp, differed compared to adults. Importantly, they found no priming effects for the infant late slow wave, which has been linked to recognition memory (Nelson, 1995) and which is generated by the temporal lobes (Reynolds & Richards, 2005). The absence of modulation of the slow wave in this priming task is consistent with the idea that infants were not using the MTL circuit in this task, though it is not clear from this study which brain circuits were mediating the priming effect.

General Summary and Conclusions

While there is some capacity for explicit memory in the newborn, this expands over the first year of life. Currently, the predominant views emphasize the importance of maturation of components of the MTL circuit and its connections with the frontal lobe for the development of memory skills continuing into adolescence. However, it is important to consider the limitations of this view as it tends to impose the adult organization of memory systems onto infants and young children, which may not be appropriate. Future studies which take fuller advantage of neuroimaging techniques to allow a better, more direct characterization of brain memory systems in developing human infants and children will help to evaluate the validity of this viewpoint and also better explore how the interactive specialization framework applies. For example, at the neural level memory development may involve regions of the temporal cortex becoming coordinated with the hippocampus through a process of input-dependent increasing specialization.

It is clear that several types of implicit memory probably dependent on subcortical structures such as the cerebellum and basal ganglia are present from birth or shortly thereafter. Other types of implicit memory such as perceptual priming may rely on the related primary sensory cortices. There are currently few marker tasks that allow the direct comparison of cognitive neuroscience data to developmental findings, but this seems a promising area for future research. Advances in the application of imaging methods such as functional near-infrared spectroscopy (fNIRS) and fMRI to infants and young children in innovative paradigms can also play a role.

From the discussion above it is evident that most memory tasks likely engage multiple memory systems, in a similar way to the partially independent brain pathways that are engaged in eye movement control and attention shifts. A lack of maturity in one or other pathway may be masked in some tasks due to compensatory activity in other pathways. Possibly it is the extent of integration between different memory pathways that is the most significant change with postnatal development. If this is the case, it is not until we have a more integrative account of the relations between different brain memory pathways that we will be able to make sense of the developmental data.

<table>
<tr><td>

Key Issues for Discussion

- How does evidence of prefrontal cortex involvement in infant cognition challenge the view that this region's involvement in explicit memory begins later in childhood?
- How might information from neuroimaging studies help to understand the phenomenon of infantile amnesia—the fact that as adults we have very few memories from the first years of life?
- Research shows that the hippocampus is one of the few brain regions that can continue to produce new neurons postnatally. What implications might this have for children with early MTL injury?

</td></tr>
</table>

9

Language

This chapter starts with the question of whether language is "biologically special." This issue refers to the extent to which the human infant's brain is predisposed to learn about language. Cognitive neuroscience approaches to this question are outlined. First, studies from early brain damage indicate that different regions of cortex are capable of supporting nearly normal language acquisition. Second, neuroimaging studies with congenitally deaf participants show that regions which normally support oral language can also support other functions. While these two lines of evidence indicate that there are no innate language representations in the cortex, there is strong functional neuroimaging evidence that the left temporal lobe is biased to process speech input from shortly after birth.

An increasing number of studies have examined how experience with language shapes the development of brain language systems. Studies of perceiving speech sounds of the native language are consistent with the idea that the brain activity related to language becomes more specific and focalized as processing of language becomes more efficient and automatic. Studies examining experience with language have increasingly focused on understanding the influence of exposure to more than one language during infancy, and documenting consequences for brain language systems and outcomes.

The effects of genetic mutations in the FOXP2 gene are associated with difficulties in speech and language; in other developmental disorders, such as Williams syndrome, language can be relatively proficient as compared to other domains of cognition. While these apparent dissociations have sometimes been characterized as evidence for an "innate language module," they are also consistent with an interactive specialization view.

Introduction

Language is a system for communicating with other people using sounds, symbols, and words to express a meaning, idea, or thought. A network of regions of the adult cortex has been typically associated with language functions, and this network is illustrated along

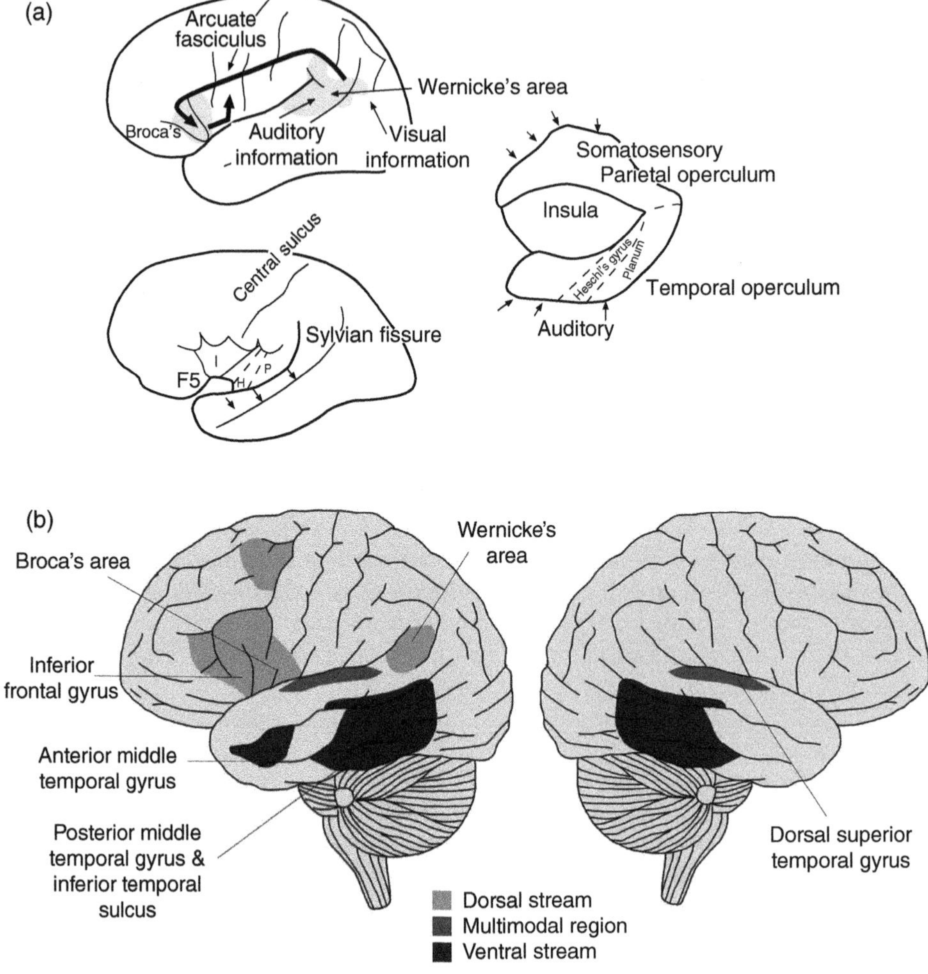

Figure 9.1 (a) Some of the key structures involved in language processing. Top left: Schematic view of information flow from posterior sensory areas to frontal response areas through the inferior temporofrontal loop. The shaded regions show where brain damage causes fluent (Wernicke's) and nonfluent (Broca's) aphasia. These regions are conceptual rather than anatomical. Bottom left: The Sylvian fissure has been pulled out and down in the direction of the arrows to reveal the insula (I) and the auditory cortex (H, P) on the superior surface of the temporal lobe. The region of the frontal operculum indicated as F5 contains mirror neurons in the monkey. It is thought that these neurons play a crucial role in imitation learning. Right: Enlarged view of Heschl's gyrus and planum temporal. (b) Dorsal and ventral language streams. Figure provided by Iroise Dumontheil.

with the putative functions of each region in Figure 9.1a, b. Language includes both primary functions, such as perception and processing of incoming speech and production of meaningful speech output, and secondary functions, such as reading and writing. Primary language functions normally seem to emerge without much effort as children develop, whereas mastery of secondary language functions requires extensive practice. This chapter will focus mainly on primary language functions, and secondary language functions will be discussed in Chapter 11. Within primary language functions, it is important to distinguish between speech perception, which involves the complex sensory processing required to discriminate between and analyze the sound patterns of an incoming speech stream, and language processing, which refers to a variety of skills including understanding word meaning and applying of rules for grammar. The majority of theorists believe that there is a developmental relation between the two skills, with early speech perception abilities providing the foundation for later construction of successful language processing (e.g., Kuhl, 2000; Werker & Tees, 1999).

A challenge for developmental cognitive neuroscientists studying language acquisition is that it is not possible to use animal models to directly study an aspect of cognition unique to humans. This is not to say that animal studies have not made any contribution to our understanding of language acquisition. There is work on the auditory aspects of language such as speech perception in various species. For example, song birds share developmental characteristics with humans in terms of having critical periods for acquisition and in the basic brain networks (see e.g., Aamodt et al., 2020). There is also work on the genes involved in language, such as FOXP2, in mice and other species (see e.g., Fischer & Scharff, 2009). However, the bulk of our understanding about the brain bases of language development has come from studies of humans, including studies of functional neuroimaging during speech and language processing, studies of infants with focal brain damage, studies of congenital atypicalities, and correlations between phases of language acquisition and neuroanatomical development. An implicit or explicit motivating question for many researchers in this field has been the extent to which language is "biologically special." This issue refers to the extent to which the human child is predisposed to learn language. To translate this question into the framework presented in Chapter 1, does a region of the cortex have innate representations for language, or are language representations the emergent result of a variety of constraints, including the basic architecture of the cortex and inter-regional connectivity?

Further Reading   Dehaene-Lambertz (2017).

Two cognitive neuroscience approaches to this question have been pursued. The first of these has been to investigate whether there are particular parts of the cortex critical for language processing or acquisition. If particular parts of the cortex are critical for language, some assume that this region(s) contains innate language-specific representations. In contrast, if several regions of cortex can support language acquisition, this suggests that the representations involved will emerge given the combination of language input and basic cortical architecture. The second cognitive neuroscience approach has been to attempt to identify neural correlates of language-related processing abilities very early in life. We will begin by reviewing evidence relevant to the first of these approaches.

Are Some Parts of Cortex Critical for Language Acquisition?

Language acquisition has become a focal point for studies designed to investigate the extent to which particular cortical areas are prespecified to support specific functions. Two complementary directions have been taken, with one set of studies examining the extent to which language functions can be supported by other regions of the cortex, and another group of studies concerned with whether other functions can "occupy" regions that normally support language. The first of these lines of research has asked whether children suffering from perinatal lesions to "language areas" can still acquire language. The second line of research has involved testing congenitally deaf children to see what functions occupy regions of cortex that are normally (spoken) language areas, and also studying the consequences of brain damage on the ability to produce sign language in a previously fluent signer.

If particular cortical regions are uniquely prespecified to support language, then it is reasonable to assume that early damage to such regions will impair the acquisition of language. This implicit hypothesis has motivated a large body of research, the conclusions of which still remain somewhat controversial. Language is left-lateralized in the majority of typical adults (about 95% of right handers and 75% of left handers; Knecht et al., 2000) and largely remains left-lateralized regardless of whether it is spoken or signed (MacSweeney et al., 2008). In an influential book, Lenneberg (1967) argued persuasively that if localized left-hemisphere damage occurred early in life, it had little or no effect on subsequent language acquisition. This contrasted both with the effect of similar lesions in adults or older children, and with several congenital atypicalities in which language is delayed or never emerges. Lenneberg's view lost adherents in the 1970s, as evidence accumulated from studies of children with hemispherectomies suggesting that left-hemisphere removal always leads to selective subtle deficits in language, especially for syntactic and phonological tasks (e.g., Dennis & Whitaker, 1976). Similar results were also reported for children with early focal brain injury due to strokes (Vargha-Khadem et al., 1994). These findings were compatible with studies of normal infants indicating a left-hemisphere bias at birth in processing speech and other complex sounds (see below) and led some researchers to the conclusion that functional asymmetries for language in the human brain are established at birth, and cannot be reversed. However, some of the secondary sources that summarized the work on hemispherectomies and/or early focal injury failed to note that the deficits shown by these children are far more subtle than the frank aphasias displayed by adults with homologous forms of brain damage (see Bishop, 1983, for a critique of the Dennis & Whitaker, 1976, study). Indeed, most of the children with focal left-hemisphere injury who have been studied to date fall within the normal range, attend normal age-appropriate schools (Stiles et al., 2002), and do far better than adults with equivalent damage.

While most studies to date have involved assessing language competence in children who acquired perinatal lesions many years beforehand, Bates, Stiles, and their colleagues carried out *prospective* studies of language and spatial cognition in children who have suffered a single unilateral injury to either the right or the left hemisphere, prenatally or before 6 months, confirmed by at least one radiological technique (Bates & Roe, 2001; Stiles & Thal, 1993; Stiles et al., 2002). Children in these studies were identified prior to the onset of measurable language skills and were examined longitudinally. This team studied more than twenty cases between 8 and 31 months of age. Regardless of the lesion site, infants with focal brain lesions were delayed, suggesting, not surprisingly, that there is a general non-specific cost to early brain injury. However, the prospective study of these infants also produced a number of rather surprising results. Based on the adult aphasia literature, delays in word comprehension would be expected to be most severe in children with damage to left posterior sites. In contrast to this, it appears that comprehension deficits are actually more common in the right-hemisphere infant group. Such a deficit is rarely reported in adults with right-hemisphere damage. This is consistent with another study of 1½–11-year-olds showing that language deficits with left-hemisphere injury were worse if it was sustained after 2 years of age. Together, these results suggest that the regions responsible for first-language learning in young children are not necessarily the regions responsible for language use and maintenance in older children and adults.

Further Reading   Martin et al. (2022); Stiles et al. (2002).

Another complicating factor revealed by prospective studies of infants with focal lesions came from a study of language production by Reilly et al. (1998). This study used the same population with pre- or perinatal lesions just discussed. Reilly and colleagues looked at many different aspects of lexical, grammatical, and discourse structure in a storytelling task, in children with focal lesions and typically developing controls between 3 and 8 years of age. In this study (like many others within this age range), there were no significant differences between the left- and right-hemisphere groups on any of the language measures. However, the infants with focal lesions as a whole performed worse than typically developing controls on several measures of morphology, syntax, and narrative structure. Each of these disadvantages appeared to resolve over time in the children with focal lesions group, but each time the child moved on to the next level of development (in language acquisition), differences between the infants with focal lesions and typically developing infants reappeared. Thus, functional recovery does not appear to be a one-off event, but rather may reoccur at several critical points during acquisition.

More recently, investigators have used functional neuroimaging to understand how brain language systems reorganize following injury early in life. In other words, if the brain regions that normally subserve language are injured, what brain regions then take on language function? Several studies have now looked at this question by studying brain activation during language processing in children with injury to Broca's area due to stroke or epilepsy (Guzzetta et al., 2008; Liegeois et al., 2004). One pattern of reorganization that has been reported is for activation to be observed in the right hemisphere in areas homologous to those that would normally be activated in the left hemisphere. For example, Guzzetta and colleagues studied five children all of whom had damage to Broca's area in the left hemisphere due to perinatal stroke. Brain activation was measured while children carried out a task involving generating a rhyme in response to a cue word. All five children showed activation in the right hemispheres in areas that were homologous to those activated in the left hemisphere in a healthy control group. Another pattern of reorganization that has been reported is for activation to be observed in the left hemisphere in the tissue surrounding Broca's area. For example, Leigeois and colleagues (2004) found that in four out of five of their cases with injury in or near Broca's area, language activation occurred not in the right hemisphere but in the left-hemisphere cortex surrounding the lesioned area.

What factors determine whether reorganization takes place within the damaged hemisphere or in the intact hemisphere? A recent study examined the potential role of asymmetry the *planum temporale*, a region within Wernicke's area that is typically larger on the left than the right side of the brain. In a group of patients with left-sided focal epilepsy, Pahs et al. (2013) measured the length of the planum temporale in each hemisphere and determined which hemisphere was dominant for language using several metrics. They found that the length of the planum temporale in the right hemisphere was the main predictor of language lateralization, accounting for 48% of the variance. A longer planum temporale in the right hemisphere was related to atypical (right or bilateral) language dominance, suggesting that the size of the right, contra-lesional planum temporale may reflect a "reserve capacity" for interhemispheric language reorganization in patients within left, focal lesions in language regions. A second study highlights another potential factor that may influence language outcomes: connections between the hemispheres. Diffusion tractography was used to examine intrahemispheric and interhemispheric connections in adolescents who had been born preterm, as preterm birth is also associated with language difficulties, occurring in 38% of participants in this study (Northam et al., 2012). Analysis of intrahemispheric pathways included the arcuate fasciculus (dorsal language pathway) and uncinate fasciculus/extreme capsule (ventral language pathway). Analysis of interhemispheric pathways (in particular, connections between the temporal lobes) included the two major commissural bundles: the corpus callosum and anterior commissure. Language impairment was present in 38% of adolescents born preterm. Volume reduction was most striking in the posterior corpus callosum (splenium), with only those interhemispheric fibers connecting the temporal lobes, and not those connecting the occipital or parietal lobes, being reduced. Language impairment was only detectable if the anterior commissure (a second temporal lobe interhemispheric pathway) was also small, with a combined measure of interhemispheric connectivity explaining 57% of the variance in language abilities. Together, these studies provide some clues as to the factors that might influence reorganization of language, though prospective studies would be needed to determine whether these neuroanatomical features (length of right planum temporale or amount of interhemispheric connection) are causes or consequences of the pattern of reorganization ultimately observed.

A final interesting point to note from such studies is that they have also indicated that activity may be more widely distributed in patients compared to healthy controls. For example, Tillema and colleagues (Tillema et al., 2008) found that children with left-hemisphere stroke not only showed language-related activation in the right-hemisphere homologues but also showed more bilateral activation in the anterior cingulate cortex and more activation in primary visual cortex. Overall, these studies demonstrate that brain regions other than the traditional left-hemisphere "language areas" can support language function and suggest that, following left-hemisphere injury, language processing may be less focally represented in the brain. This pattern is consistent with an interactive specialization (IS) view, where the ultimate brain correlates of language will be determined by interaction among brain areas during development so that alternate patterns can emerge following injury. However, as the IS model suggests, brain representation of cognitive processes may be more distributed and less focal following brain injury.

To summarize, the effects of early focal lesions on language acquisition are complex. However, in general the evidence supports the following conclusions:

1) Most children with early left-hemisphere damage go on to acquire language abilities within the normal range (although often at the lower end).
2) In general, there is no significant difference in language outcomes when direct comparisons are made between children with early left-versus right-hemisphere damage.

3) Different regions of the cortex may be involved in the actual acquisition of language from those that are important for language use in adults.
4) Functional compensation may have to reoccur at several points in language acquisition, and is not a one-off event.
5) Functional compensation may involve a within-hemisphere or across-hemisphere neural reorganization.

Evidence from visual and spatial tasks in deaf participants suggests that visual processing in the cortex is different in individuals reared in the absence of auditory input. Specifically, the congenitally deaf seem more sensitive to events in the peripheral visual field than are hearing participants. Event-related potentials (ERPs) recorded over classical auditory regions, such as portions of the temporal lobe, are two or three times larger for deaf than for hearing participants following peripheral visual field stimulation. This pattern of findings was confirmed in another study showing that congenitally deaf, but not hearing, adults showed a greater change in functional magnetic resonance imaging (fMRI) signal in primary auditory cortex in response to visual stimuli in the periphery compared to the center of vision (Scott et al., 2014). Thus, the lack of auditory input has resulted in a normally auditory area becoming at least partially allocated to visual functions. Importantly, the whole-brain analysis in the fMRI study just mentioned also showed differences between deaf and hearing individuals in other brain regions, including visual brain regions involved in processing motion and multisensory/supramodal regions like the posterior parietal cortex and anterior cingulate cortex. These results are consistent with the idea that networks in the brain adapted to altered experience, rather than single regions altering their function in isolation.

Further Reading   Neville Bavelier (2002).

Thus, the evidence described so far indicates that regions of the cortex are not prespecified for language processing, and that other functions may be able to "capture" areas that normally end up processing language. However, it is still possible that small variations on the basic architecture of the cortex, or inter-regional connectivity patterns, may be sufficient to "attract" language processing to certain regions during normal functional brain development.

Some evidence for this view comes from studies of deaf signers with acquired focal brain lesions. After reviewing evidence that sign language has most of the same formal properties as spoken languages, Bellugi et al. (1989) found that deaf signers with left-hemisphere lesions acquired in adulthood were aphasic for sign language, while showing intact performance on several visuospatial tasks. Adult patients with recently acquired right-hemisphere lesions, on the other hand, showed the reciprocal pattern of performance. In this latter group, signing was fluent even to the extent of being able to sign fluently about the contents of their room while showing gross spatial distortion in their description of its contents! This, and other evidence, led Bellugi and colleagues to propose that it is the computations required for linguistic processing, rather than its modality, that determine cortical localization.

This conclusion received partial support from an fMRI study in which hearing and deaf participants were scanned while reading sentences in English or viewing an American Sign

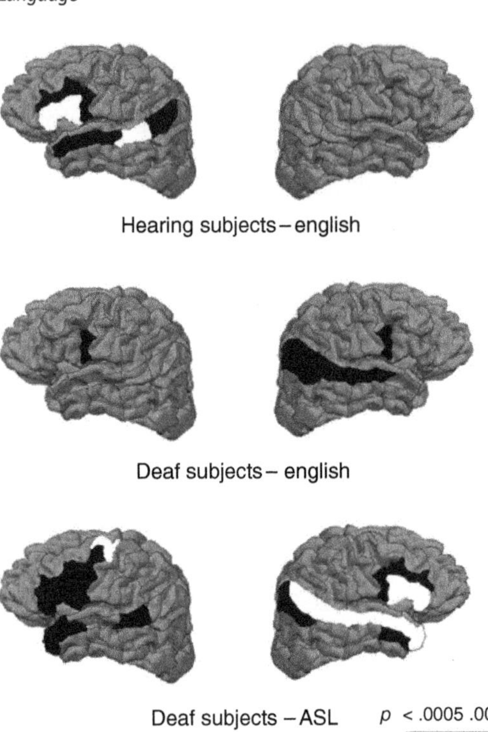

Figure 9.2 Cortical areas showing increases in blood oxygenation on fMRI when normal hearing adults read English sentences (top), when congenitally deaf native signers read English sentences (middle), and when congenitally deaf native signers view sentences in their native sign language (American Sign Language).

Language (ASL) signer perform the same sentence (Neville et al., 1998; see Neville & Bavelier, 2002). The top of Figure 9.2 shows patterns of cortical activation when hearing adults read English. Consistent with previous research, there was robust activation within some classical left-hemisphere language areas, such as Broca's area. No such activation was observed in the right hemisphere. When deaf people viewed an ASL signer performing the same sentences (bottom of figure), they showed activation of most of the left-hemisphere regions identified for the hearing participants. Since ASL is not sound-based, but does have all of the other characteristics of language, including a complex grammar (Klima & Bellugi, 1979), these results suggest that some of the neural systems that mediate language can do so regardless of the modality and structure of the language acquired. This pattern of findings is supported by other work with deaf individuals who acquired sign language early (native) or late (after 15 years) showing that left posterior superior temporal cortex activation was greater in early compared to late signers in response to signed sentences (Twomey et al., 2020). This latter study is in support of the idea of the importance of early language experience, regardless of modality, for the emergence of typical functional networks for language. Having said this, there were also some clear differences between the hearing and deaf activations, with the deaf group activating some similar regions in the right hemisphere. One interpretation of the right-hemisphere activation is that it is evoked by the biological motion inherent in sign, but not spoken, language.

Further Reading   MacSweeney et al. (2008).

Neural Basis of Speech Processing in Infants

The second general approach to investigating the extent to which language is "biologically special" involves attempting to identify speech-relevant processes in the brains of very young infants. A further question then concerns how specific to speech these processes are. The logic here is that if there are specific neural correlates of speech processing observable very early in life, this may indicate the capacity for language-relevant neural processing prior to significant experience.

One example of this approach concerns the ability to discriminate speech-relevant sounds such as phonemes. Behavioral experiments have revealed that young infants show enhanced (categorical) discrimination at phonetic boundaries used in speech such as /ba/, /pa/. That is, a graded phonetic transition from /ba/ to /pa/ is perceived as a sudden categorical shift by infants. These observations initially caused excitement as evidence for a speech-perception-specific detection mechanism in humans. However, over the past decades it has become clear that other species, such as chinchillas, show similar acoustical discrimination abilities, indicating that this ability may merely reflect general characteristics of the mammalian auditory processing system, and not an initial spoken-language-specific mechanism (see Werker & Vouloumanos, 2001). However, Werker and Polka (1993) have reported that while young human infants can discriminate between a very wide range of phonetic constructs, including those not found in the native language (e.g., Japanese infants, but not Japanese adults, can discriminate between "r" and "l" sounds), this ability becomes restricted to the phonetic constructs of the native language around 10 months of age.

Further Reading Aslin et al. (2008); Friederici (2008).

If brain correlates of this process could be identified, it may be possible to study the mechanisms underlying this speech-specific selective decrease of sensitivity. As mentioned in Chapter 2, ERPs offer a good opportunity to study the neural correlates of cognition in normal infants in a non-invasive manner. The excellent temporal resolution of ERPs can be complemented by some spatial resolution through the use of high-density ERPs (HD-ERP). When components of ERP differ in both latency (following the event) and spatial distribution, we may be confident that different neural circuitry is being activated. An example of this approach comes from an HD-ERP study of phonetic discrimination in 3-month-old infants.

Dehaene-Lambertz and Dehaene (1994) presented their participants with trials in which a series of four identical syllables (the standard) was followed by a fifth that was either identical or phonetically different (deviant). They time-locked the ERP to the onset of the syllable and observed two peaks with different scalp locations. The first peak occurred around 220 ms after stimulus onset and did not habituate to repeated presentations (except after the first presentation) or dishabituate to the novel syllable. Thus, the generators of this peak, probably primary and secondary auditory areas in the temporal lobe, did not appear to be sensitive to the subtle acoustical differences that encoded phonetic information.

The second peak reached its maximum around 390 ms after stimulus onset and again did not habituate to repetitions of the same syllable, except after the first presentation. However, when the deviant syllable was introduced, the peak recovered to at least its original level.

Thus, the neural generators of the second peak, also in the temporal lobe but in a distinct and more posterior location, are sensitive to phonetic information. Further studies have found that such mismatch responses are also found for vowel discriminations even in newborns (Cheour-Luhtanen et al., 1995).

Researchers have also used methods with better spatial resolution to investigate early correlates of speech perception. Dehaene-Lambertz et al. (2002) measured brain activation with fMRI in awake and sleeping healthy 3-month-olds while they listened to forward and backward speech in their native tongue (French). The authors assumed that forward speech would elicit stronger activation than backward speech in areas related to the segmental and suprasegmental processing of language, while both stimuli would activate mechanisms for processing fast temporal auditory transitions (see Werker & Vouloumanos, 2001, for a discussion of the appropriate control stimuli for human speech). Compared to silence, both forward and backward speech activated widespread areas of the left temporal lobe, which was greater than the equivalent activation on the right for some areas (planum temporale). This activation pattern is consistent with the ERP experiment described above. However, forward speech activated some areas that backward speech did not, including the angular gyrus and mesial parietal lobe (precuneus) in the left hemisphere. The authors suggest that these findings demonstrate an early functional asymmetry between the two hemispheres. However, they acknowledge that their results cannot discriminate between an early bias for speech perception or a greater responsivity of the left temporal lobe for processing auditory stimuli with rapid temporal changes (both of which could start during the final trimester in uterus). Further, in awake infants only, the right dorsolateral prefrontal cortex showed greater activation to forward speech.

The general conclusions from the above studies are reinforced by converging results from functional near-infrared spectroscopy (fNIRS; Chapter 2). In this experiment, Mehler and colleagues (Peña et al., 2003) played normal infant-directed speech or the same utterances played in reverse while they measured changes in the concentration of total hemoglobin over the right and left hemisphere. They observed that left temporal areas showed significantly more activation when infants were exposed to normal speech than to backward speech or silence, leading them to conclude that neonates are born with a left-hemisphere bias for speech processing.

Taken together, the experiments on speech perception just described provide perhaps the best example in developmental cognitive neuroscience of the power of converging methodology. It is encouraging that similar conclusions can be drawn from experiments using behavioral, ERP, fMRI, and fNIRS methods.

Influence of Experience on Brain Language Processing

Whether or not there are biologically "special" areas for processing language in the brain, language inputs must play some role in shaping the brain's response to language. Children grow up hearing a specific language(s) in their environment and develop expertise for that language.

Studies of infants' processing of speech sounds (see above) have demonstrated not only that infants can discriminate these sounds early in life, but that the type of language they hear in their environment subsequently shapes how infants perceive speech sounds. More specifically, behavioral experiments show that infants younger than about 6 months can discriminate speech sounds that are present in their native language, as well as those that are not. However, by about 10–11 months, their ability to discriminate non-native contrasts diminishes. ERP studies have shown similar results. For example, Rivera-Gaxiola and colleagues (Rivera-Gaxiola et al., 2005) found that American infants showed discrimination of both English (native) and Spanish (non-native) consonant contrasts at 7 months in the N250-550 response, whereas 11-month-olds did so only for native contrasts. Moreover, they found that there was an increase in response to native contrasts only from 7 to 11 months of age. If infants' changing response to native contrasts depends on their experience hearing the language, an interesting question is whether earlier exposure to speech would accelerate this process. Peña et al. (2012) addressed this question by using ERPs to study the response to a change of phoneme at a native and non-native phonetic boundary in full-term and preterm human infants. Although babies can hear sounds from outside the womb before they are born, the womb acts like a high-frequency filter so the phonemes heard inside the womb would be highly degraded. By contrast, preterms, who of course leave the womb earlier, might be able to benefit from earlier and richer exposure to speech in the environment outside the womb. The results of the study showed that preterms did not, in fact, appear to benefit from their earlier exposure, and the change in response to native and non-native speech depended on maturational age rather than duration of experience. These results do not cast the importance of speech exposure into doubt, but do suggest that the impact of that exposure depends on the state of the brain at the time of the experience.

These early discriminative capabilities are related to later language abilities. For example, one ERP study showed infants who at 7½ months of age showed a larger ERP to deviant than standard native speech sounds produced a larger number of words at 18 and 24 months, had faster vocabulary growth from 14 to 30 months, and produced more complex sentences at 24 months than infants who did not show this response (Kuhl et al., 2008). Supporting this work, a study using magnetoencephalography (MEG; a technique that uses magnetic fields to measure brain activity) to assess the mismatch response found that this not only predicted grammar abilities in the native language at 6 years of age but also was useful for classifying those children who had speech and language disorders at this age (Zhao et al., 2021).

> **Further Reading**   Dehaene-Lambertz (2017).

When studying the impact of the early environment for language development, one factor to consider is the number of languages to which an infant is exposed. Some infants are exposed to two (bilingual) or even more (multilingual) languages as they grow up. The language input in bilingual/multilingual environments is more varied, and infants will have less exposure to each individual language compared to the input to the single language in a monolingual environment. Key questions for researchers include: What impact does this more varied exposure have on the development of brain language systems, particularly over the period of "narrowing in" to the native language? And are later language outcomes influenced? Understanding the developmental impact of multilingualism is also of interest as there is literature suggesting that multilingualism can benefit higher cognitive functions as early as the first year of life (Kovács & Mehler, 2009), although there is some debate about the nature and developmental persistence of any such advantages (e.g., Bialystok, 2017).

There are as yet few studies directly addressing the brain correlates of exposure to multiple languages over the key period up to 10–11 months (discussed above). One ERP study demonstrates the flexibility in infants to respond to more than one language over the flexible period in infancy. Conboy and Kuhl (2011) gave Spanish exposure to English speaking monolinguals from 9 months. Using ERPs, they found that infants showed no discrimination of a Spanish phoneme contrast before exposure to the language at 9 months, but did so at 11 months. They concluded that rapid learning is still possible even as the flexible period is coming to a close. Other ERP studies have also shown that the difference in response between different speech sounds (native, non-native) in infants is influenced by bilingualism, but these studies also highlight the complexity of factors influencing the response, including age, gender, and attention (e.g., Shafer et al., 2011). Overall, studies conclude that bilingual infants show perceptual narrowing to the native languages they hear, though they may retain more sensitivity to non-native contrasts and reach a less "narrow" end state than monolinguals

---

Further Reading   Byers-Heinlein et al. (2013).

A further question is how infants might cope with different modalities of language (signed or spoken, as discussed above) together with number of unimodal languages. Using NIRS, Mercure et al. (2020) studied bilingual and multilingual hearing infants (4 to 8 months of age) and bimodal, bilingual infants exposed to English and British Sign Language (BSL). They found that all infants activated areas in the language network – the inferior frontal and posterior temporal areas. Among other results, they also found that multivariate pattern analysis, a method which allows the consideration of subtle differences in activation across brain voxels between conditions, was able to classify whether activation within the left hemisphere was elicited by spoken or signed language in monolinguals but was not able to differentiate the two in unimodal or bimodal bilinguals. These findings are in general consistent with the idea that, in infancy, bilingual input may affect the specialization process.

To summarize, language input from spoken words has an important impact on the neural bases of language. Generally speaking, the findings discussed above are consistent with an IS view that the brain activity underlying language functions becomes more focal with experience, as language skills become more efficient and automated.

Neural Correlates of Typical and Atypical Language Acquisition

Bates et al. (1992) reviewed and identified a number of correlations between neuroanatomical developments in the human cerebral cortex and "landmarks" in language acquisition. Such correlations cannot be used to establish firm links between neural events and cognitive ones. However, the approach is useful for identifying specific hypotheses that can then be tested with more direct methodologies.

Consistent with the findings reviewed so far in this chapter, Bates et al. (1992) suggested that the evidence for some neuroanatomical differences between the right and left hemispheres at birth may set up computational differences that "bias" language toward the left hemisphere. It is likely that these differences between the hemispheres are due to differential timing of development, rather than a genetic encoding of a specific architecture. Bates et al. (1992) identify a peak transition in both behavioral and neural development at around eight to nine months after birth. They discuss the establishment of long-range connections (especially from the frontal cortex) and the onset of an adult distribution of metabolic activity (Chugani, 1994) at this age, and suggest that these neural developments enable the onset of a number of language-related skills such as word comprehension and the inhibition of non-native speech contrasts.

Around 16–24 months there is a "burst" in vocabulary and grammar, and Bates and colleagues argue that this correlates with a steep increase in synaptic density in many relevant cortical regions. The increased synaptic density, they speculate, enables a larger capacity for storage and information processing. At around 4 years of age most normal children have acquired the basic morphological and syntactic structures of their native language, and thus have reached the end of the "grammar burst." Bates and colleagues point out that this "stabilization" of language coincides with the decline in overall brain metabolism and synaptic density (although, as mentioned in Chapter 4, the peaks of these two measures may not coincide exactly).

Developmental disorders involving language can also be informative about the neurodevelopmental basis of these abilities. Some authors have argued that developmental disorders in which language is relatively deficient, despite a lack of problems in other domains, constitute evidence for an innate language module. Conversely, cases where language is supposedly intact, despite impairment in other domains, could provide powerful converging evidence. An example of the former kind that has attracted much scientific and media interest in recent years concerns three generations of the KE family and the forkhead box protein P2 (FOXP2) gene. As reviewed in Chapter 3, while initial reports suggested that about half of the members of this family had an inherited specific impairment in grammar (Gopnik, 1990), claims for a "grammar gene" have been diluted by the finding that the deficits in the affected family members are much broader than grammar alone. Vargha-Khadem and colleagues have conducted extensive work on the family and found deficits in coordinating complex mouth movements (orofacial dyspraxia), timing of rhythmic motor movements, and various aspects of language outside grammar (e.g., lexical decision), with IQ scores 18–19 points below unaffected members of the same family (Vargha-Khadem et al., 1995). Moreover, there is not a clear association of mutations of the FOXP2 gene with developmental language disorders. For example, one study of 270 4-year-old children with poor language functioning found that none had a mutation of the FOXP2 gene (Meaburn et al., 2002).

Nevertheless, the relatively simple single-gene basis of the condition, and its association with language, make it worthy of further study (see Marcus & Fisher, 2003, for further discussion). One such study has involved fMRI during verb generation and repetition tasks (Liégeois et al., 2003). This study demonstrated that affected members of the family showed significant under-activation mainly in two regions: Broca's area and the putamen. However, as can be seen in Figure 9.3 in the color plate section, the affected members also showed greater activation than did controls in a variety of other cortical regions, suggesting widespread adjustment of cortical networks in response to the aberrant gene. This conclusion is consistent with a case study of an 8-year-old boy with a FOXP2 mutation who was not related to the KE family, who showed widespread neural differences from controls beyond the language network on structural MRI imaging (Liégeois et al., 2016).

Further Reading   Liégeois et al. (2014).

While the KE family may not have turned out to have the specific deficits initially hypothesized, similar claims about the specificity of language deficits have been made for so-called "specific language impairment" (SLI). While there is much controversy about the behavioral specificity or otherwise of this condition (see Bishop, 1997), most neuroanatomical studies have tended to focus on the particular regions thought to be important for language as discussed above. Thus, it is difficult to know whether or not regions outside the adult classical language regions are atypical as well. However, a recent study set out to examine the whole brain of SLI participants using MRI (Herbert et al., 2003). Compared to controls, these children showed a substantial increase (more than 10%) in cerebral white matter throughout the brain. This atypical pattern was no more severe in the classical language areas than anywhere else in the brain, suggesting to the authors that the population they studied had a "generalized systems impairment" that differentially affected language.

A case where language is supposedly spared relative to other domains of impairment is Williams syndrome (WS). In WS (see Chapters 2 and 7), language can be strikingly proficient, despite severe deficits in other domains. Several studies have demonstrated that linguistic and face-processing skills are surprisingly good in WS participants, despite low IQs (typically in the 50s–60s) and serious difficulties on visuospatial, number, motor, planning, and problem-solving tasks (see Karmiloff-Smith, 2008, for review). The linguistic ability of these participants is not just superficial (such as repetition of previously heard sentences), but is generative in the sense that they can create new sentences with appropriate grammatical structure. The depth of this linguistic knowledge is currently the subject of debate. In certain cases, the dissociation between linguistic ability and other abilities can be very marked. For example, the following is the spontaneous speech of a 21-year-old WS participant (studied by Annette Karmiloff-Smith and colleagues) with performance equivalent to a mental age of 5½ years as assessed on a standardized spatial cognition task (Ravens Matrices):

[talking about her job at the hospital where she helps with the catering staff]

Well sometimes I do and sometimes I don't. They just tell you how many cups you put on the tray and the saucers and how many (um) spoons and sugars and creams you put on the tray and how many sandwiches they want for how many people.

[*Experimenter*: But do you count them yourself, the cups?]

No, they count them and I just put them on the trays. So, I put them in with clingfilm on top so they don't get wet.

In this example, while this participant is unable to count relatively small numbers (the number of cups on a tray), she is able to express herself grammatically and has a fairly wide vocabulary. While it seems initially attractive to account for such cases of isolated proficient language in terms of innate representations for language, the evidence on cortical development in Chapter 4 should make us wary of jumping to this conclusion. A better way to view such cases is that some deviation from the normal trajectory of brain development results in a slightly maladaptive architecture in some structures (Karmiloff-Smith, 2008). This slightly deviant aspect of neural architecture has much more effect on some domains of processing than others. Once early postnatal development deviates from the normal trajectory, then it is likely that this deviation will be amplified by atypical interactions with the environment.

General Summary and Conclusions

The degree to which the acquisition of language is "biologically special" has been hotly contested on the basis of linguistic and behavioral data. While the addition of cognitive neuroscience data has not resolved this debate, it has illuminated some of the constraints and biases that ensure that most adults acquire a specialized brain network for the production and processing of language. For example, converging evidence from several cognitive neuroscience methods indicates that, from birth or even earlier, regions of the left temporal lobe are engaged by, and process, auditory inputs that require rapid temporal processing. While these regions may be the ideal "home" for speech processing, if they are damaged early in life then other parts of cortex can also support this function, though perhaps not quite as efficiently. In accord with the IS view, it is likely that processing speech within this region from early in life influences its synaptic and dendritic microstructure, resulting in it becoming increasingly specialized and different from other regions. Some degree of plasticity may remain during development, resulting in a reconfiguration of the microcircuitry of the region following specific training. Interestingly, recent studies have also focused more attention to the white matter tracts connecting brain language regions. For example, the dorsal pathway projecting from Broca's area to the superior temporal region appears to be especially important for higher-language functions and is weaker in non-human primates compared to humans, as well as in human children compared to adults (reviewed in Frederici, 2009). Better understanding of such pathways may also provide new insights into the brain bases of reorganization, since the IS view indicates that interactions among brain regions are important in this process.

Developmental disorders in which language functions are supposedly selectively impaired, or remain "intact" in the face of other deficits, have been used as examples in support of claims for an "innate language module." Closer examination of such cases reveals that they are better viewed as deviations from the typical trajectory of brain development that have more effect on some domains of processing than others. Once early postnatal development deviates from the typical trajectory, then it is likely that this deviation will be amplified by atypical interactions with the environment.

Language acquisition and speech perception have been some of the most active areas of developmental cognitive neuroscience. The use of converging methodologies, and frequent comparisons between typical and atypical trajectories of development, make it the general domain most likely to see major breakthroughs over the next decade.

Key Issues for Discussion

- What methods and approaches are likely to be most successful for revealing the genetic contribution to language acquisition?
- To what extent are there parallels between the emergence of functional specialization of the fusiform face area (Chapter 7) and those areas that support speech and language processing?
- How does speech perception contribute to language acquisition and its disorders?

10

Prefrontal Cortex, Executive Functions, and Decision-Making

The prefrontal cortex shows a prolonged course of development compared to most other cortical regions. This chapter outlines two approaches to understanding the role of this part of cortex in cognitive development. In the first of these, the maturational approach links structural development of the prefrontal cortex to the emergence of specific cognitive skills. One example is work associating maturation of the dorsolateral prefrontal cortex with developments in working memory. Perseverative errors of young infants on object permanence tasks have been linked with maturation of the dorsolateral prefrontal cortex by means of data from human infants, infant monkeys, and adult prefrontal-cortex-lesioned monkeys. Further, the development of working memory and inhibitory control, two core executive functions, has been linked with maturation of lateral prefrontal cortex by means of data from structural and functional neuroimaging. This maturational account requires modification in the light of recent evidence suggesting a broader neural network is involved in executive functions, and highlighting the role of experience in shaping this network. Another example of the maturational approach has involved associating the differential rate of development of prefrontal cortex compared to other cortical and subcortical regions with self-regulation difficulties, increases in risky decision-making, and mental illness observed in adolescence. The mismatch between the slowly maturing prefrontal regulation systems and the earlier-maturing emotion-processing systems is said to account for adolescents' increased vulnerability for risky decision-making and mental illness. This view may need modification in light of evidence indicating that the dichotomy between prefrontal control systems and subcortical systems is an over-simplification.

A second view of the role of the prefrontal cortex in cognitive development is that it is needed for skill learning and plays an important and early role in brain organization. Studies of electroencephalogram coherence suggest a role for prefrontal regions in the cyclical reorganization of cortical representations during cognitive development. Some of the problems with this approach are outlined. It is concluded that the specificity of prefrontal cortex functioning probably arises from a combination of initial neurochemical and connectivity biases and the relatively protracted plasticity of the region.

Developmental Cognitive Neuroscience: An Introduction, Fifth Edition. Michelle de Haan, Iroise Dumontheil, and Mark H. Johnson.
© 2023 John Wiley & Sons Ltd. Published 2023 by John Wiley & Sons Ltd.
Companion website: www.wiley.com/go/johnson/devneuro5e

Introduction

As discussed in Chapter 4, the prefrontal cortex shows a prolonged period of postnatal development, with detectable changes in gray matter volume, white matter volume, and cortical thickness throughout childhood and into the teenage years (Brain Development Cooperative Group, 2012; Giedd et al., 1999; Gogtay et al., 2004; Shaw et al., 2008). For this reason, considerable research has focused on studying how the development of the prefrontal cortex may support the development of a range of cognitive functions during childhood and adolescence.

Two major unresolved issues concerning frontal cortex development are as follows:

- Are the specialized computations performed by the frontal cortex due to a unique neuroanatomy/neurochemistry or to other factors?
- How are we to align evidence for prefrontal cortex functioning within the first six months of life with evidence for the continuing neuroanatomical development of the region until the teenage years?

These issues will be revisited at the end of the chapter. First, the chapter will describe two alternative views that have been taken of the relation between prefrontal cortex structural development and increases in cognitive ability in childhood. The first of these is that structural development in the prefrontal cortex (PFC) occurs at a particular age, allowing certain increases in cognitive ability. This view is in line with the maturational perspective outlined in Chapter 1. Two examples of this view will be presented, one highlighting the role of the PFC in the development of object permanence and working memory, and the other examining the role of the PFC in the increased vulnerability to risky behavior and mental illness observed in adolescence. A different view is that the PFC is consistently involved in the acquisition of new skills and knowledge from very early in life, and may also play a role in organizing other parts of cortex (e.g., Thatcher, 1992). By this second view, PFC regions play a fundamental role in cognitive transitions primarily because of the region's involvement in the acquisition of *any* new skill or knowledge. A corollary of this hypothesis is that prefrontal involvement in a particular task or situation may decrease with experience of the task. This perspective on PFC development is inspired by the skill learning model of functional brain development outlined in Chapter 1. With regard to PFC development, many of the predictions from this view overlap with those from the interactive specialization (IS) view. Consequently, these two perspectives will only be differentiated at the end of the chapter.

Prefrontal Cortex and Object Permanence

One of the most comprehensive attempts to relate a cognitive change to underlying brain development has concerned the emergence of object permanence in infants. In particular, Diamond, Goldman-Rakic, and colleagues (Goldman-Rakic, 1987; Diamond & Goldman-Rakic, 1989; Diamond, 1991) have argued that the maturation of PFC during the last half of the human infant's first year of postnatal life accounts for both Piaget's (1954) observations about object permanence and a variety of other transitions in related tasks which involve working memory and the inhibition of prepotent responses (e.g., Diamond et al., 1994).

The region of the frontal lobe anterior to the primary motor and premotor cortex, commonly called the prefrontal cortex, accounts for almost one-third of the total cortical surface in humans (Brodmann, 1909, 1912) and is considered by most investigators to be the locus of control for many abilities central to higher-level cognition. Extensive clinical (Milner, 1982) and experimental observations of the effects of injury to this region have also supported the notion that the PFC supports important aspects of cognition (for reviews see Owen, 1997; Duncan, 2001). While there are no universally accepted theories of frontal cortex functioning, some particular forms of cognitive processing that have been consistently linked to the PFC in adults pertain to the planning or carrying out of sequences of action, the maintenance of information "on-line" during short temporal delays (working memory), and the ability to inhibit a set of responses that are appropriate in one context but not another (inhibitory control) (Diamond, 2013).

It was Piaget who first observed that infants younger than around 7 months fail to accurately retrieve a hidden object after a short delay period if the object's location is changed from one where it was previously and successfully retrieved. In particular, infants of this age make a particular preservative error. That is, they often reach to the hiding location where the object was found on the immediately preceding trial. This characteristic pattern of error, called "A not B," was cited by Piaget (1954) as evidence for the failure of infants to understand that objects retain their existence or permanence when moved from view. Between 7½ and 9 months, infants begin to succeed in the task at successively longer delays of 1 to 5 seconds (Diamond, 1985, 2001). However, their performance is unreliable: infants continue to make the A not B error up to about 12 months if the delay between hiding and retrieval is incremented as the infant's age increases (Diamond, 1985).

> **Further Reading**   Diamond (2013); Hendry et al. (2021).

Diamond and Goldman-Rakic (1989) tested monkeys in a version of Piaget's object permanence task. In the object permanence task, participants are shown an object hidden at location A and are permitted to retrieve it. After a predetermined number of successful retrievals at location A (usually three), the object is then hidden at location B. Infant monkeys failed to retrieve the hidden object at location B when the delay between hiding and retrieval was 2 seconds or more. Diamond and Goldman-Rakic (1989) found that animals with damage to the parietal cortex, a brain region closely associated with spatial processing, did not show this performance deficit; nor did animals with lesions to the hippocampal formation, a region known to be crucial for other memory-related tasks (Diamond et al., 1989). However, damage to the dorsolateral prefrontal cortex (DLPFC) severely impaired the adult monkeys' performance when a delay was imposed between hiding and search, indicating that this region plays a central role in delayed response tasks that require the maintenance of spatial information over temporal delays.

Developmental evidence that links maturation in frontal cortical regions to the emergence of working memory abilities comes from studies (Diamond & Goldman-Rakic, 1986, 1989) showing that infant monkeys initially fail the A not B task in ways similar to monkeys with PFC damage. However, between the ages of 1½ and 4 months, the infant monkeys show a similar progression as that seen in human infants between 7½ and 12 months, with both becoming able to succeed in the task and to withstand longer and longer delays. Lesions to the DLPFC in infant monkeys who have acquired the ability to succeed in the task abolish this skill, confirming the importance of this region in the task. In humans, non-invasive imaging methods have also been used to link maturation of the PFC to development of infants' performance in the A not B task. In one series of studies, increases in frontal electroencephalogram (EEG) responses in human infants have been shown to correlate with the ability to respond successfully over longer delays in delayed response tasks (Bell, 1992a, 1992b; Bell & Fox, 1992; Fox & Bell, 1990) as well as how quickly they acquire proficiency in the task over 6–12 months of age (MacNeill et al., 2018). Another study using optical imaging (near-infrared spectroscopy) showed a correlation between behavioral demonstration of object permanence and blood oxygenation in the PFC (Baird et al., 2002).

In Diamond's (1991) view, the emergence of the ability to demonstrate knowledge about an object's permanence results from the maturation of the DLPFC between the ages of 5 and 12 months. Diamond proposed that this region of the cortex is important when a participant has to both retain information over spatial delays and inhibit prepotent (previously reinforced) responses. Let's briefly review the evidence for this position. Infants younger than 7½ months fail to retrieve hidden objects when any delay is imposed between hiding and retrieval. In this sense, human infants behave like the adult monkeys with DLPFC lesions and infant monkeys, as discussed previously. Human infants of this age make similar retrieval errors in both the A not B task, in which the side of hiding is switched after several repeated hidings on one side, and in an object retrieval task when the side of hiding is varied randomly (Diamond & Doar, 1989). This suggests that Piaget's observations about object permanence reflect the state of development of one or more underlying neural mechanisms common to performance on both the object permanence and retrieval tasks. Since successful performance on these tasks requires both memory for the most recently hidden location and the inhibition of an incorrect reach to the last rewarded location, the underlying neural mechanism must subserve both computations. Extensive neurobiological evidence implicates the PFC in spatial working memory performance, and lesions of this region frequently reveal patterns of behavior in which the inhibition of inappropriate responses is impaired. Consequently, Diamond and colleagues have presented a strong case for the importance of DLPFC development.

Prefrontal Cortex and Executive Functions Development During Adolescence

The maturational approach to PFC development has also been extended to later childhood and adolescence. The results from a variety of behavioral tasks designed to tap into advanced PFC functions have demonstrated that adult levels of performance are not reached until adolescence or later (see Fabiani & Wee, 2001; Lee et al., 2013; Luciana, 2003; Luna et al., 2015). For example, participants from 3 to 25 years of age were tested on the CANTAB (Cambridge Neuropsychological Testing Automated Battery), a well-established and well-validated battery of tests previously used on adult human and animal populations

(Fray et al., 1996). This battery assesses several measures, including working memory skills, self-guided visual search, and planning. Importantly for developmental studies, the battery is administered with touch-screen computer technology and does not require any verbal or complex manual responses. Using the CANTAB, Luciana and Nelson (1998, 2000, Luciana, 2003) found that while measures that depend on posterior brain regions (such as recognition memory) were stable by 8 years, measures of planning and working memory had not yet reached adult levels by age 12. While rates of improvement in working memory skills slow down, longitudinal studies show that after a linear increase into early adolescence, working memory continues to improve in mid- to late adolescence (Dumontheil et al., 2011; Lee et al., 2013; Tamnes et al., 2013). Measures of inhibitory control can also show that adult performance is not reached until mid-adolescence; however, developmental trajectories seem more sensitive to the type of task used (Lee et al., 2013; Luna et al., 2004).

Further Reading Olson & Luciana (2008).

While such behavioral measures are useful as marker tasks, it is even better to use functional imaging while children perform tasks likely to engage PFC regions. Studies in adults identify a set of fronto-parietal regions as being consistently recruited during cognitive control or executive function tasks (Figure 10.1), which include lateral aspects of the PFC, extending into the anterior insula, as well as the anterior cingulate cortex, pre-supplementary motor area, and lateral parietal cortex.

Several functional magnetic resonance imaging (MRI) studies by Klingberg and colleagues (summarized in Klingberg, 2006) document that the DLPFC (in particular, the superior frontal sulcus) is involved in working memory in both children and adults, but also show that it is activated as part of a network also involving the intraparietal cortex (see Figure 10.2 in the color plate section). These studies show that stronger activation of the fronto-parietal network is related to greater working memory capacity, and that activation

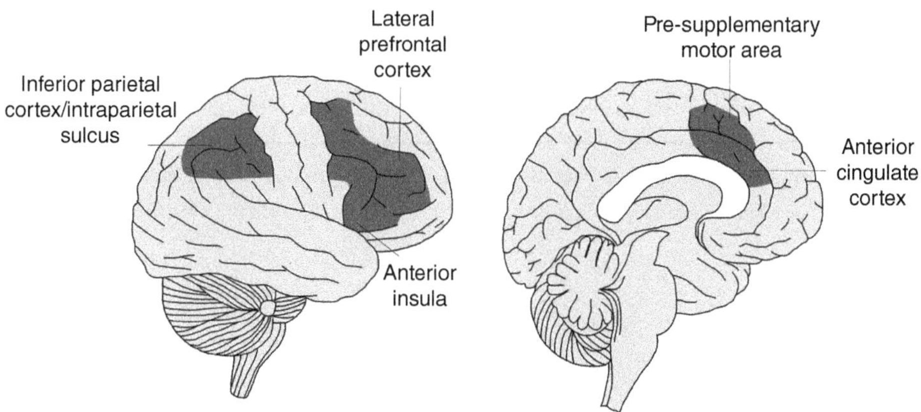

Figure 10.1 Key regions of the cognitive control network. © Iroise Dumontheil.

of the network also increases with age independent of performance (Klingberg et al., 2002; see also Andre et al., 2015 for a meta-analysis). A longitudinal structural MRI study found that across the ages of 8 to 19 years, longitudinal improvements in working memory correlated with a decrease in cortical volume in a range of frontal, parietal, and occipital regions not necessarily overlapping with those regions showing greater activation during working memory tasks in functional MRI studies, and therefore implicated a broader network (Tamnes et al., 2013). Importantly, these associations do not simply reflect the fact that gray matter volume reduction and improvement in cognitive ability are concurrent during development, as analyses controlled for changes in general cognitive ability. More specific findings have been observed when focusing on white matter tracts. Indeed, the development of white matter tracts connecting the frontal and parietal regions seems to play a role in this process: maturation of these tracts relates to working memory (but not reading) performance, and to the degree of cortical activation in the frontal and parietal gray matter (Olesen et al., 2003; Figure 10.2). At a cellular level, computational modeling indicates that stronger synaptic connectivity between the prefrontal and parietal regions, and not faster transmission of neural signals or stronger connections within each region, can by itself account for observed changes in brain activity associated with the development of working memory in childhood (Edin et al., 2007).

> **Further Reading**   Klingberg (2006, 2008).

A similar importance of the maturation of specific white matter tracts has been observed for the development of inhibitory control. One study reported that fronto-striatal tracts' radial diffusivity increased with age and correlated with performance on a no-go task, while age effects only were observed for cortico-spinal tracts (Liston et al., 2006). Functionally the evidence is somewhat mixed, with the observation of both decreases and increases in PFC activation during inhibitory control tasks (e.g., Tamm et al., 2002; see Crone & Dahl, 2012, for review). It may be that, overall, the development of inhibitory control tends to be associated with gradually more specific and localized brain activation. For example, in an event-related potential (ERP) study, while 6–7-year-olds, 9–10-year-olds, and adults all showed a frontal source close to the anterior cingulate cortex for the no-go N2 ERP component, 6–7-year-olds recruited an additional occipito-temporal source, while 9–10-year-olds recruited a parietal generator (Jonkman et al., 2007). Functional connectivity analyses of an anti-saccade task in 8–27-year-olds suggest that increased top-down control from PFC regions to down-stream cortical and subcortical regions may support the development of response inhibition (Hwang et al., 2010).

> **Further Reading**   Crone & Steinbeis (2017).

While this body of evidence is convincing, it also suggests that the PFC maturation hypothesis is not the whole story, and that some modification or elaboration of the original account will be required. A further line of evidence that can potentially challenge the maturational view comes from studies showing how experience can influence working memory capacity and prefrontal activation. Traditionally, working memory capacity was thought to be a fixed, unchangeable trait once we become adults. More recent studies show greater flexibility in working memory capacity. For example, one study showed that after 5 weeks of training on a working memory task, adults showed both an increased working memory capacity and increased working memory-related brain activity in the fronto-parietal network (Olesen et al., 2004). Meta-analyses have indeed shown that working memory can be trained through repeated practice at gradually increasing load (e.g., Melby-Lervåg et al., 2016). These findings have potential implications for understanding the development of working memory. The maturational viewpoint would argue that development of the fronto-parietal network enables improvements in working memory performance. However, the training study suggests an alternative viewpoint: experience using working memory and/or the increased demands on working memory as children develop might drive changes in brain networks underlying this ability, a view more consistent with the IS and skill learning perspectives. However, there is also evidence that there may be genetic constraints that limit working memory capacity development and susceptibility to training (see Chapter 11).

Longitudinal studies of the neural correlates of working memory have also provided evidence that maturation of fronto-parietal areas alone may not fully account for improvements in working memory over childhood. These studies show that, while current working memory performance is related to measures of fronto-parietal activation, parietal cortex thickness, and fronto-parietal and fronto-striatal white matter integrity, future working memory performance is predicted by fronto-striatal white matter and activity in the basal ganglia and thalamus (Darki & Klinberg, 2014; Ullman et al., 2014). This is consistent with the possibility that subcortical activity plays a role in driving the specialization of the fronto-parietal network for working memory. This is similar to the ideas discussed in Chapter 7, describing the role of subcortical regions in driving the specialization of ventral cortex for face processing.

Another line of evidence suggests that the neural bases of infant working memory may be more broadly based than initially thought. For example, EEG studies have reported that in 6-month-old infants there is a burst of high-frequency EEG activity over right temporal regions during the period of maintenance of the representation of the object, after it disappears, and before it reappears (Kaufman, Csibra, & Johnson, 2003, 2005; see Chapter 6). Another EEG study found that, in infancy, working memory was associated with changes in EEG power from baseline to task across all brain regions, whereas by 4 years of age working memory was associated with changes in EEG over frontal regions only (Bell & Wolfe, 2007). A recent study of the development of working memory in infants born preterm showed that neonatal hippocampal volumes correlated with working memory ability at 2 years of age but cortical volumes, even from the dorsolateral prefrontal region, did not (Beauchamp et al., 2008). Studies with adolescents also show that, while the DLPFC is consistently active in working memory tasks throughout adolescence, the contribution of the hippocampus diminishes from early to late adolescence (Finn et al., 2010). Findings such as these hint the neural network underlying working memory may be different, possibly more diffuse, in infants, and may become increasingly focused on the fronto-parietal network with development (Kaldy & Sigala, 2004), a pattern consistent with the IS view. By middle childhood, it seems the fronto-parietal and cingulo-operculum networks supporting executive functions (Figure 10.1) may be in place, and that further development is driven by quantitative, rather than qualitative, changes in brain activation (Engelhardt et al., 2019).

Social Decision-Making and Self-Regulation During Adolescence

Another domain in which the PFC is believed to play an important role is in decision-making. Following structural MRI research findings of prolonged structural development of the brain during adolescence, investigators have been particularly interested in understanding how maturation of the prefrontal cortex contributes to decision-making in adolescence. Adolescence is a period of clear gains in intellect and cognitive skills, but it also marks a period of intense emotions and impulsive behavior that seem not to reflect these cognitive improvements, as well as a time when there is a clear increase in incidence of antisocial and risky behaviors and mental illnesses including depression, anxiety and mood disorders, eating disorders, and substance abuse (Duell et al., 2018; Rapee et al., 2019).

Several investigators have argued that this profile of behavior can be explained by the differential rate of maturation of the PFC compared to other brain regions. One specific model of this general idea is the social information processing network (SIPN) model (Nelson et al., 2005). In this model, social behavior involves three interacting "nodes": (a) a detection node which detects social information, involving mainly occipito-temporal cortex; (b) an affective node which processes the emotional significance of a social stimulus and influences the behavioral and emotional responses to it, involving subcortical structures and orbitofrontal cortex (Figure 10.3), and (c) a cognitive regulatory node which is important for goal-directed behavior, impulse control, and theory of mind, involving much of the PFC. According to this model, the three nodes mature at different rates, with the detection node maturing very early in life, the affective node maturing in early

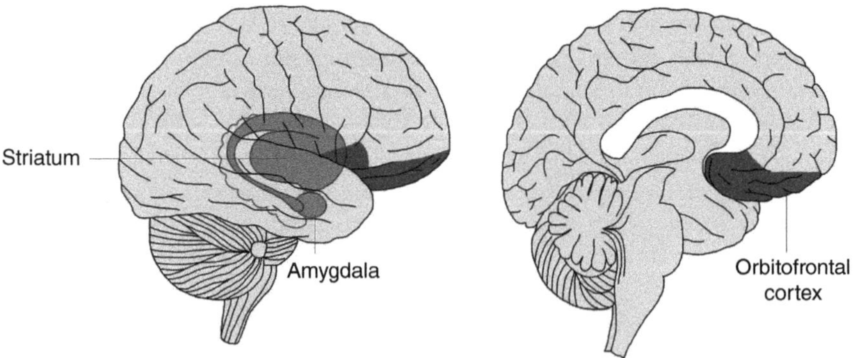

Figure 10.3 Key regions supporting the processing of emotions and rewards. © Iroise Dumontheil.

adolescence, and the cognitive regulatory node maturing much more slowly than the other two nodes. This mismatch in maturation among the nodes creates a vulnerability in adolescence, when strong emotional responses to social stimuli are not yet properly moderated by the as-yet immature cognitive regulatory system. This vulnerability then accounts for the rise in risky decision-making.

In order to study decision-making more systematically, Bechara and colleagues developed a task called the Iowa Gambling Task (Bechara et al., 1994). This task is meant to tap into the type of decision-making we are often faced with in everyday life, when we must choose between options that have different short-term and long-term advantages. In the task, participants are asked to choose cards one at a time from four decks, two of which can result in immediate big rewards (gains) and two of which result in immediate smaller gains. The twist in the game is in how losses are delivered: the two decks that result in high gains are also accompanied by large losses and thus are not advantageous in the long term as they will result in a net loss. By contrast, the decks that result in low immediate gains are accompanied by small losses and so more advantageous as they result in a net gain in the long term. Participants are not told about the differences between the decks and must discover the differences themselves as they choose cards and receive the resulting gains and losses. Healthy adults come to learn that it is best to choose from the advantageous decks during the course of the task, whereas patients with PFC damage, in particular the ventromedial PFC, keep selecting from the disadvantageous decks that give large short-term gains but result in a loss over the long term. Thus, they seem to be influenced more by the immediate reward and to fail to consider the longer-term consequences of their actions and regulate their choices accordingly (Bechara et al., 1997; however, see Chiu et al., 2018, for discussion of limitations of the task). This description is quite similar to that often applied to adolescents, who are often said to have difficulty in seeing the future consequences of their actions.

Crone and Westenberg (2009) have carried out studies of children and adolescents using a similar task, and found that there is a slow course of development, with children not consistently learning to make advantageous choices until they are 16–18 years old. Brain imaging studies suggest that this behavioral pattern relates to an adolescent imbalance between subcortical reward-processing systems and frontal control systems: adolescents are more driven by reward systems, leading to suboptimal choices in social decision-making tasks, and improvements in performance between adolescence and adulthood may be linked to the slow maturation of the ventromedial PFC until early adulthood (Crone & Westenberg, 2009). However, a review suggests the pattern of choices made by children and adolescents on the Iowa Gambling Task or child-adapted versions does not in fact align with what is observed in patients with ventromedial PFC lesions. Children are found to prefer options associated with infrequent losses, regardless of whether these packs of cards are advantageous or disadvantageous in the long term, a frequency bias that is found to decrease with age and is not observed in patients with ventromedial PFC lesions (Cassotti et al., 2014). Healthy adults appear to be able to inhibit a tendency to shift response after a loss, which allows them to better integrate information about which choices may be advantageous in the long term (Cassotti et al., 2014).

Further Reading   Cassotti et al. (2014).

Another strand of research on decision-making has focused on risk taking and its sensitivity to the social context. Along similar lines to the SIPN model, researchers proposed that during adolescence there is a "mismatch" between more rapidly developing subcortical regions processing emotions and rewards (Figure 10.3) and the slower developing PFC supporting planning, cognitive control, and mentalizing (Figure 7.1 and Figure 10.1). This leads to behavior and choices being driven by the socio-affective context rather than by a consideration of risk and rewards and long-term consequences of actions (Casey et al., 2008; Steinberg, 2008).

What is the evidence for this potential mismatch? A study analyzed longitudinal structural data from 33 participants aged 7–30 years. Results showed indeed that a majority of these participants showed relatively earlier maturation of amygdala and/or nucleus accumbens gray matter volumes compared to the PFC (Mills et al., 2014). However, subcortical regions cannot be considered unitarily as their timecourses of structural development vary and are differentially susceptible to gender and puberty (Goddings et al., 2014).

Functional evidence of unique patterns of activation during adolescence have focused on emotional and reward reactivity. Activity in the amygdala in response to emotional faces has been found to be greater in adolescents than in children and adults (Hare et al., 2008). One study was designed to assess risk taking in the lab. Participants were asked to play a car-driving game where they had to complete a circuit as quickly as possible. When they reached a crossroad the traffic lights could be amber, in which case participants had to decide whether to wait for the light to turn green again, and lose time, or take a risk and cross. In the latter case the light may turn red and the participant would get into a crash, slowing them down further. The measure of risk taking was the number of crashes participants had. The results showed that male adolescent participants (aged 13–16 years) had a similar number of crashes to youths (18–22) and adults (24 and older) when they completed the task alone. However, in a second condition they carried out the task while being watched by two friends, who were not allowed to speak. In this social context, adolescents more than doubled the number of crashes they had (Gardner & Steinberg, 2005). This study highlights the sensitivity of adolescents to the social context and peer influence.

In a follow-up study, a similar design was used while participants were in a functional MRI scanner. Adults (24–29 years) were found to show greater lateral PFC activation than adolescents whether in the peer or alone condition, while adolescents (14–18 years old) showed increases in activation in the ventral striatum and orbitofrontal cortex while they were being watched, compared to when they were not watched, a pattern not observed in adults (Chein et al., 2011). These results were interpreted as showing that the social context increases risk taking in adolescents by increasing the value of potential rewards associated with risky decisions (see Blakemore & Mills, 2014, for a discussion of the balance between physical and social risk during adolescence). Indeed, another study has demonstrated that adolescents showed greater activation in part of the ventral striatum when receiving a reward when they were watched by a friend, compared to when they were alone (Smith et al., 2015). Animal studies concur with the observation that social context may affect behavior in adolescence. For example, a study showed that juvenile (the equivalent to adolescent) mice spent more time drinking alcohol when they were in the presence of litter mates than alone, a pattern that was not observed in adult mice (Logue et al., 2014).

Further Reading   Steinberg (2008); Steinberg et al. (2018).

While these findings are promising, the maturational view may need modification. One reason is that there is evidence showing that rewards are also processed cortically, and that control systems also rely on subcortical regions (Phillips et al., 2003). Thus, the processes involved in decision-making are more broadly based and developments at the behavioral level may not only be attributable to maturation of a specific component. A case study also demonstrated that problems with emotion regulation and attention engagement begin in the first months of life following perinatal prefrontal injury (Anderson et al., 2007), indicating that prefrontal systems may be already involved in regulation of behavior from early in postnatal life.

There is also evidence that areas of the social brain involved in decision-making, including regions of the PFC, undergo both structural changes and functional reorganization during the second decade of life, possibly reflecting a sensitive period for adapting to one's social environment (Blakemore & Mills, 2014). The amygdala may be a key structure that mediates reorganization of brain systems during adolescence. In one view, there is a shift in the attribution of relevance to stimuli such as peers' faces and proximity of parents that is mediated by the amygdala, possibly due to the action of hormones. As a result, amygdala inputs to existing stable neural networks, such as those involving the PFC, are increased or decreased. This destabilizes the functional interactions among regions within these networks and allows for a critical restructuring of the networks' functional organization. This process of network reorganization enables processing of qualitatively new kinds of social information and the emergence of novel behaviors that support mastery of adolescent-specific developmental tasks (Scherf et al., 2013).

Crone and Dahl (2012) integrated a range of cognitive and neuroimaging research and proposed that rather than focusing on a potential mismatch of brain systems maturation during adolescence, excessive risk taking, and negative impacts of peer influence, adolescence could be viewed as a time of flexible adjustments of cognitive capacities and skills to adapt to changing socio-affective goals. Maturing social brain (Figure 7.1) and cognitive control (Figure 10.1) networks supporting social cognitive skills and executive functions interact with puberty-driven maturation of subcortical systems processing emotions and rewards (Figure 10.3) over the course adolescence and eventually lead to mature social competence and positive long-term goals, or can lead to excessive orienting toward rewards and addiction, social withdrawal or anxiety, or diminished goals seen in depression. Crone and van Duijvenvoorde (2021) suggest that individual differences in reward drive may be a key factor underlying adolescents' sensitivity to their environment and their specific patterns of risk taking.

> **Further Reading**   Crone & Dahl (2012); Crone & Van Duijvenvoorde (2021).

Prefrontal Cortex, Skill Learning, and Interactive Specialization

An alternative approach to understanding the role of the PFC in cognitive development has been advanced by several authors who have suggested that the region plays a critical role in the *acquisition* of new information and tasks. From this perspective, the challenge to the infant brain in, for example, learning to reach for an object is equivalent in some respects to that of the adult's brain when facing complex motor skills like learning to drive a car. Three concomitants of this general view are that (a) the cortical regions crucial for a particular task will change with the stage of acquisition, (b) the PFC plays a role in organizing information within or allocating it to other regions of cortex, and (c) that development involves the establishment of hierarchical control structures, with the frontal cortex maintaining the currently highest level of control. Three recent lines of evidence indicating the importance of prefrontal cortical activation early in infancy have given further credence to this view: (a) functional MRI and positron emission tomography (PET) studies, (b) psychophysiological evidence, and (c) the long-term effects of perinatal damage to the prefrontal cortex.

The limited number of functional MRI and PET studies that have been conducted with infants have often surprisingly revealed functional activation in the PFC, even when this would not be predicted from adult studies. For example, in a functional MRI study of speech perception in 3-month-olds, Dehaene-Lambertz et al. (2002) observed a right DLPFC activation that discriminated (forward) speech in awake, but not sleeping, infants (see Figure 10.4 in the color plate section and Chapter 9). Similar activation of DLPFC was found in response to faces at the same age (Tzourio-Mazoyer et al., 2002). While this is evidence for activation of at least some of the prefrontal cortex in the first few months, it remains possible that this activation is passive as it may not play any role in directing the behavior of the infant. Two other lines of evidence, however, suggest that this is not the case.

Developmental ERP studies have often recorded activity changes over frontal leads in infants, and some recent experiments suggest that this activity has important consequences for behavioral output. These experiments involve examining patterns of activation that precede the onset of a saccade (Chapter 5). In one example, Csibra et al. (1998, 2001) observed that pre-saccadic potentials that are usually recorded over more posterior scalp sites in adults are observed in frontal channels in 6-month-old infants. Since these potentials are time-locked to the onset of an action, it is reasonable to infer that they are the consequence of computations necessary for the planning or execution of the action.

Further evidence for the developmental importance of the prefrontal cortex from early infancy comes from studies of the long-term and widespread effects of perinatal damage to the area. In several of the previous chapters we have seen how selective perinatal damage to the relevant regions activated in adults often only has, at worst, mild effects on infants, with subsequently almost complete recovery of function. In contrast, perinatal damage to frontal cortical regions often results in long-term difficulties that can become increasingly apparent with development. This generalization from several domains suggests that the PFC plays an important structuring or enabling role from very early in postnatal development. One example of this theoretical approach to functional PFC development has been advanced by Thatcher (1992) and Case (1992) (for another, related approach see Stuss, 1992).

Thatcher (1992) and colleagues have analyzed EEG data from a large number of participants at various ages between 2 months and 18 years. During recording, the participants sat quietly and, as far as can be ascertained, were not responding to any stimuli. From 16 leads placed evenly across the scalp, the spontaneous EEG rhythms of the brain were recorded before being subjected to a complex analysis that ascertained the extent to which the recordings from each lead "cohered" (roughly speaking, correlated in their activity) with other leads. From the large amount of raw data that such an analysis generates, the major factors (i.e., which leads cohere) could be adduced for each age group, and the age "peaks" of rate of increase in coherence for these leads could be computed. Thatcher then added two working assumptions to these data, namely that scalp electrode locations provide reasonable indications of underlying cortical activity, and that the extent of correlation between the activity of regions (leads) reflects the strength of neural connections between them. The result was a hypothesis about recurring cycles of cortical reorganization orchestrated by the frontal cortex.

Figure 10.5 illustrates the hypothesis put forward by Thatcher on the basis of a complex analysis of a subset of his data. The lines in between points indicate the leads that show the greatest rate of increase in cohesion at that age. Thatcher's claim is that there are cycles of cortical reorganization which begin with microcycles involving reorganization of short-range connections in the left hemisphere. Following this, longer-range frontal connections become important, at first only on the left side, but then bilaterally. The cycle is then completed by reorganization of short-range connections on the right-hand side. According to Thatcher, a complete cycle of this kind takes approximately four years, though subcycles and microcycles provide the basis for shorter-term periods of stability and transition. By virtue of the longer-range connections related to the frontal cortex, Thatcher suggests that this part of cortex plays an integral role in the reorganization of the cortex as a whole.

Case (1992) attempted to use these EEG findings to relate brain reorganizations to cognitive change by arguing (a) that cognitive change has a similarly "recursive" character, and (b) that many limitations on cognitive performance are due to functions such as working memory commonly associated with the frontal cortex. One example used to illustrate that neural and cognitive changes may both be manifestations of a common underlying process is the rate of change in EEG coherence between a particular pair of leads (frontal and parietal) and the rate of growth in working memory span during the same ages (see Figure 10.6). While the change in the rate of growth in the two variables may seem compelling at first

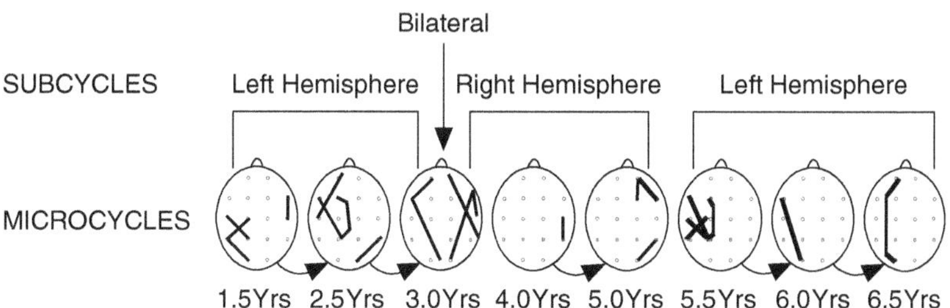

Figure 10.5 Summary of the sequence and anatomical distribution of the coherence patterns reported by Thatcher (1992). Lines connecting electrode locations indicate a measure of strong coherence. A microcycle is a developmental sequence that involves a lateral-medial rotation that cycles from the left hemisphere to bilateral to right hemisphere in approximately 4 years. Note the hypothesized involvement of the frontal cortex in the "bilateral" subcycle. Thatcher 1992 / with permission of Elsevier.

(a)

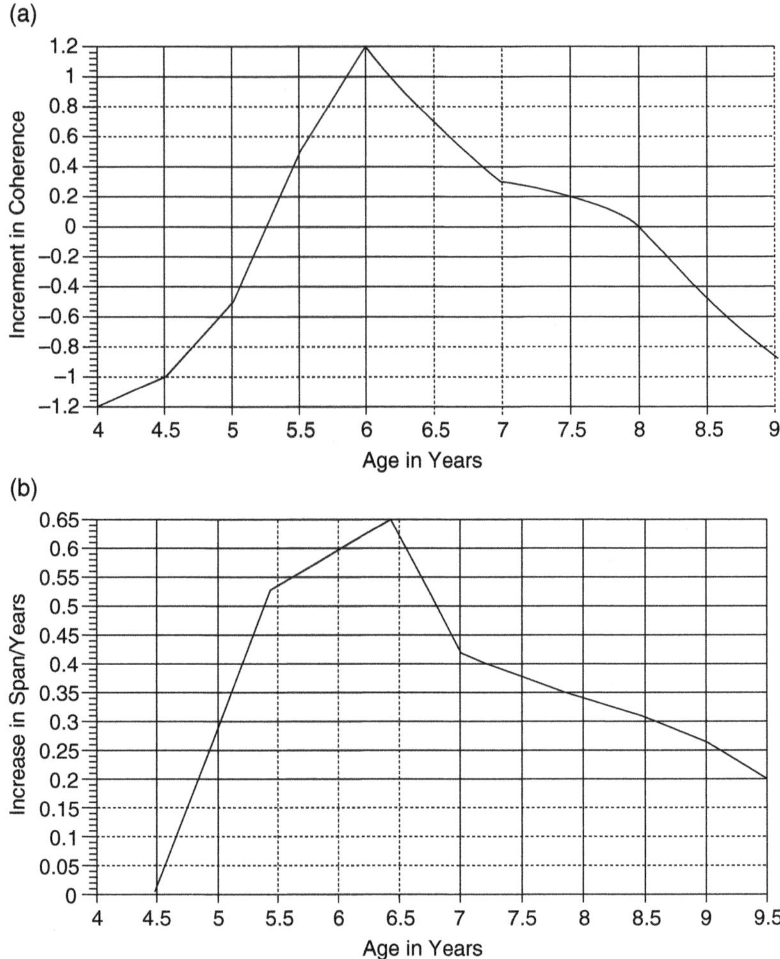

(b)

Figure 10.6 (a) Rate of growth of EEG coherence between frontal and posterior lobes during middle childhood (F 7–P 3 electrodes). Adapted from Thatcher, 1992. (b) Rate of growth of working memory (counting span and spatial span) during the same age range.

glance, there are a number of reasons why it cannot be regarded as more than suggestive at present. One of these is that there is no clear rationale for choosing the particular pair of electrodes presented—there are, after all, 56 possible pairs to select from, and there are always likely to be a few that show peaks in growth at similar ages to some developing cognitive function.

Even assuming that watertight correlations between the peaks in rates of increase in cognitive abilities and EEG coherence could be obtained, this association between brain and cognition would still be based only on temporal correlation, a form of evidence which, when taken in isolation, is not very compelling. Hopefully, in the future this approach will attempt to consider the fact that EEG coherence shows phases of significant decreases of coherence as well as increases, which may be associated with "dips" in performance at the behavioral level, and more dynamic, as opposed to stage-based, accounts of cognitive change.

Another potential criticism for the skill learning view is the findings from individuals with early prefrontal lesions. While there are many reports documenting long-term social and cognitive difficulties in such cases, intellectual abilities can develop in the normal range (Anderson et al., 2007). This would be unexpected if the PFC were critical for acquiring new knowledge and skills, and instead a more profound cognitive delay might be expected.

General Summary and Conclusions

Two different viewpoints on the development of the prefrontal cortex have been reviewed in this chapter. The view that the maturation of regions of the PFC enables particular cognitive functions is, of course, a *causal epigenesis* type of account (maturational view, Chapter 1), and such accounts will always be vulnerable to demonstrations of partial functioning of the region in question at earlier ages. On the other hand, the second approach discussed, in which the frontal lobes are a critical part of the cyclical self-organization of cortex, still remains relatively unspecified, with only preliminary evidence to support it at the moment. Nevertheless, bearing in mind the evidence for these two approaches, we can return to the unresolved issues raised at the start of the chapter.

The first of these issues concerned whether the specialized computations performed by the frontal cortex are due to a unique neuroanatomy and/or neurochemistry, or to other factors. The view put forward in Chapter 4 was that while the cerebral cortex provides architectural constraints on the representations that emerge during ontogeny, there are no innate representations. However, it seems likely that graded differences in neurotransmitter densities, and differences in patterns of interconnectivity to other regions of cortex, will provide initial biases that may subsequently lead to some regional differences in microcircuitry. In addition, however, the relatively delayed developmental trajectory of the frontal cortex means that it is likely to develop representations which capture invariances in the structure of the external world, and interactions with the external environment, at greater spatial and temporal distances than other regions of cortex. In other words, it will tend to integrate information over larger time and space intervals than will other regions of cortex. Thus, a combination of small intrinsic biases and relatively delayed development may give rise to the unique information-processing properties of this region of cortex. Whether this region plays a role in orchestrating the specialization of other regions of cortex, as Thatcher claims, is an exciting topic for research in the future.

The second unresolved issue concerned how to reconcile evidence for continuing neuroanatomical development in the frontal cortex until the teenage years, on the one hand, and evidence for some functioning in the region as early as the first few months of age, on the other. The evidence for early functioning of the PFC provides perhaps the biggest challenge to the maturational approach. One possible resolution to this issue is that representations that emerge within this region of cortex are initially weak, and sufficient only to control some types of output, such as saccades, but not others, such as reaching (Munakata et al., 1994). Other plausible resolutions of this issue come from Diamond's (1991) proposal that different regions of frontal cortex are differentially delayed in their development, and Thatcher's (1992) suggestion that prefrontal regions may have a continuing role in the cyclical reorganization of the rest of cortex. Whether these hypotheses work out or not, there is good reason why some degree of prefrontal cortical functioning is vital from the first weeks of postnatal life, or even earlier (Hykin et al., 1999; Moore et al., 2001; Fulford et al., 2003). The ability to form and retain goals, albeit for short periods, is vital for generating efforts to perform actions such as reaching for objects. Early, and often initially unsuccessful, attempts to perform motor actions provide the essential experience necessary for subsequent development.

Key Issues for Discussion

- How useful is the concept of "maturation" to account for the development of the prefrontal cortex, and what objective criteria could be used to establish functional or structural maturity?
- Why do several different developmental disorders involve symptoms of prefrontal cortex dysfunction?
- How do the different accounts of functional brain development (maturation, skill learning, and interactive specialization) account for the heightened sensitivity to social cues during adolescence?
- Is adolescence a social construct?

11

Educational Neuroscience

Learning to read and write (literacy) and to use numbers (numeracy) are key achievements for children, as these skills are important both for subsequent education and for everyday life in the majority of cultures. Findings from research on functional brain development have the potential to provide insight into how children acquire these skills and progress in formal educational settings, and into the source of individual differences in academic achievement, whether genetic or environmental. Educational neuroscience focuses on understanding the brain bases of acquiring academic skills, as well as the disorders involving specific difficulties in acquiring these skills. The field is also interested in understanding the neural mechanisms through which different ways of teaching (e.g., interleaving), interventions (e.g., working memory training), or specific practice (e.g., playing a musical instrument) may affect learning.

One factor contributing to the acquisition of skills in literacy and numeracy is the acquisition of domain-specific cognitive skills and knowledge within these areas. For example, literacy involves letter recognition and phonological awareness (being able to focus on the sound of words), while numeracy involves number recognition and counting. Brain areas have been identified that are involved in these skills, such as the visual word form area of the brain for reading and the intraparietal sulcus for numerosity processing. Another factor involved in acquiring academic skills is domain-general abilities, including executive skills, working memory (WM), and processing speed. Research documents predictive links between levels of domain-general abilities and later academic achievement, and shows that brain areas involved in domain-general processing are activated in tasks that require literacy and numeracy skills. Domain-general processes may be particularly important when new skills are being acquired, with a shift to more posterior cortical regions when skills have been consolidated. Difficulties in domain-specific and/or domain-general skills may contribute to specific disorders in literacy, known as dyslexia, and numeracy, known as dyscalculia. While scientists are making progress in translating knowledge of the developing human brain to applications in schools, educational neuroscience is a young field and in practice there still remains a big gap between basic research and effective applications of this knowledge in classroom settings.

Learning to read, write, and carry out mathematics can be viewed as a process whereby simpler skills (e.g., recognition of small shapes, the ability to track multiple objects) that are relevant to a domain of knowledge (domain-relevant skills) lead through education on to more complex ones that are specific for that domain (domain-specific skills), such as letter recognition or knowledge of symbolic numbers. These skills can be described as part of a domain-specific or crystallized knowledge base, built up based on past experiences at home, nursery, school, and so on. In addition to these domain-specific skills, there are also basic domain-general processing abilities that are important for learning but that may be less dependent on experience and prior knowledge (Gathercole et al., 2004). These domain-general skills are often described as some of the basic executive function skills that rely on prefrontal cortex, including WM (see Chapter 10), and also skills such as processing speed.

Acquiring academic skills can be considered an example of the skill learning view of human functional brain development (described in Chapters 3 and 12). From this perspective, we might expect that the brain regions involved in some domain-general, executive skills would overlap with brain activation seen during tasks like reading and mathematics, particularly during acquiring these skills when they are less routine and automatic. Brain regions involved in executive function may become less involved as the knowledge becomes crystallized and/or aspects of these skills become more routine and automatic. Since executive functions are particularly important in situations where routine, automatic processes are insufficient, the specific links between executive function and academic attainment are also likely to change over time, as some skills become automatic and other new skills are being acquired.

The idea of domain-specific skills does not necessarily mean that these specific skills are innately specified starting points. Instead, there are likely innately specified "domain relevant" starting points, which become domain-specific with the process of developmental and specific environmental interactions (Karmiloff-Smith, 1998). For example, the approximate number system (ANS), discussed in more detail later, can be thought of as a domain-relevant system for mathematics. It is clearly related to mathematics skills and may feed in to the development of mathematics-specific skills, yet it is also present in other species and thus in itself is not domain-specific to mathematics.

While this dual model of domain-specific and domain-general skills is a useful way of thinking about the neurocognitive components contributing to academic achievement, it also has some challenges:

- Acquiring literacy and numeracy likely requires both domain-general and domain-relevant skills, but how the related brain systems interact is not yet clear
- While studies have shown predictive links between domain-general and domain-relevant skills and later academic achievement, the causal direction of the brain–behavior relations is not always clear.
- Even when studies have shown a link between a skill, such as WM, and academic attainments in literacy or numeracy, there has not been compelling evidence that training WM improves academic progress, suggesting a more complicated relation among these skills.

Further Reading   Mareschal et al. (2013).

Numeracy

A basic skill underpinning numeracy is the ability to effortlessly perceive the number of objects in a set. Research by comparative psychologists and animal behavior experts reveals two domain-relevant systems for representing number in several animal species, including primates: (a) an ANS similar to that active in time or length judgments that is used to approximate numerosity of large sets of objects (see Dehaene, 1997; Gallistel, 1990, for reviews), and (b) a system engaged in tracking small numbers of objects (object-files). Both systems appear to be active in young human infants, and may underlie their ability to perform simple numerical computations with small numbers of objects.

The ANS is thought to represent quantity by reflecting the physical magnitude proportional to the items being enumerated, in the form of an analog magnitude representation. In this domain-specific system, "numerical" comparisons are made in a similar way to length or time comparisons. Three signatures of these approximate number representations are: (a) that discriminability is proportional to magnitude, in accord with Weber's Law (e.g., 1 and 2 are more discriminable than 7 and 8); (b) that successful representations are formed only when all members of a set are perceptually available at once (for visual/spatial arrays) or in immediate succession (for sequences of lights or sounds); and (c) that these representations can be transferred both across modalities (auditory and visual) and across formats (spatial and temporal). With regard to the latter point, it is striking that the same mental magnitude system may represent number, time, and the surface area of objects (Cordes & Brannon, 2008; Sokolowski et al., 2017; Walsh, 2003). Thus, while the system is number-relevant, it is probably not specific to the domain of number. Figure 11.1 shows some tasks used to evaluate numerical processing in children, with Figure 11.1a showing a non-symbolic task that can be computed by the ANS.

The second "object-file" system represents the exact numerosity of very small sets of objects (e.g., Hauser et al., 1996). Some have suggested that this is an "object-file" system that originally evolved to allow us to track up to about four moving objects at a time (Carey, 2001). Three signatures of these exact number representations are (a) that they are limited to set sizes of three or four, (b) that successful representations can be formed and maintained even when different members appear successively and then are occluded, and (c) that these representations are as abstract as large number representations, for they are impaired by simultaneous variation in continuous quantitative variables such as area or contour length (Carey, 2009).

Further Reading   Carey (2001); Cordes & Brannon (2008).

While both of the above systems are number-domain-relevant, they are not domain-specific in that they can be engaged by non-numerical tasks. Behavioral studies of human infants provide evidence that both these systems exist in humans and emerge early in development (see Cordes & Brannon, 2008, for review). By 6 months of age, infants discriminate between large numerosities when all other continuous variables are controlled, both in visual spatial arrays and in auditory temporal arrays. Like other animals, infants

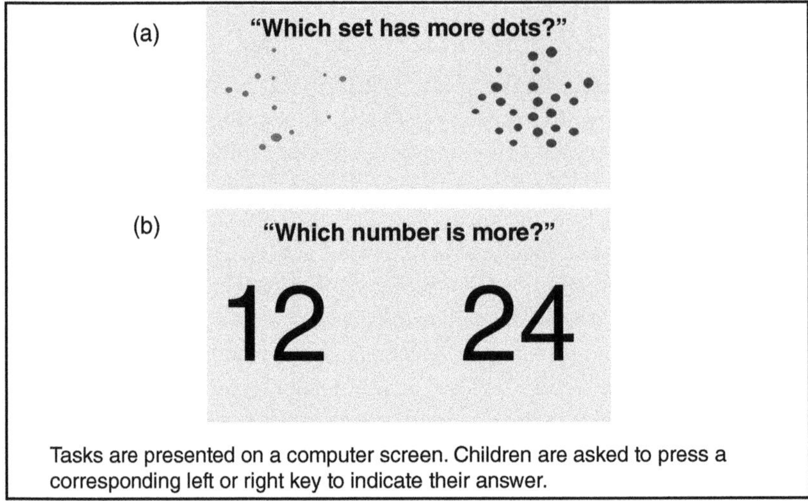

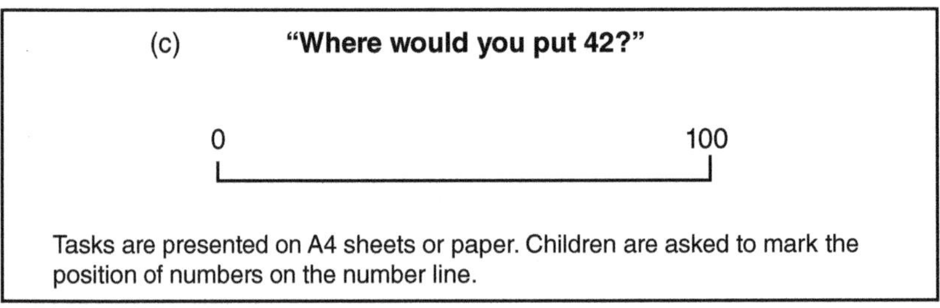

Figure 11.1 Examples of experimental tests used to assess (a) non-symbolic numerical representations, (b) symbolic representations, and (c) number line tasks.

discriminate between large numerosities only when the ratio difference is large: they discriminate arrays of 8 versus 16 dots or sequences of 8 versus 16 sounds but fail, in each case, to discriminate 8 from 12 elements (Lipton & Spelke, 2003; Xu & Spelke, 2000).

Experiments from several laboratories indicate that young infants discriminate between small numbers of objects exactly, both in visible arrays and in arrays in which each object is successively revealed and occluded (see Wynn, 1998, for review). For example, young infants and even newborns can discriminate between two and three dots, sounds, or objects, and 5-month-olds can track simple transformations of object arrays, such as addition and subtraction (Wynn, 1992). However, like monkeys, human infants fail to discriminate between large approximate numerosities when elements are successively revealed and occluded, fail to discriminate numerosities exactly for arrays of more than three to four objects (Chiang & Wynn, 2000), and form representations that are robust over variations in continuous quantities for large but not small numerosities (Feigenson et al., 2002). These findings suggest that both infants and other animals have indeed separate systems for representing large approximate numerosity and small exact numerosity. This idea is supported by findings from an event-related potential (ERP) experiment, where 6–7½-month-old infants showed different neural responses in tasks designed to engage the ANS and object file system (Hyde & Spelke, 2011). A positivity over parietal scalp sites was modulated by the ratio between successive large, but not small, numbers while an earlier positivity over occipital-temporal sites was modulated by the absolute cardinal value of small, but not large, numbers. These results provide evidence for two early developing systems of non-verbal numerical cognition: one that responds to small quantities as individual objects and a second that responds to large quantities as approximate numerical values.

It is important to note that some of the human infant experiments remain open to alternative (non-numerical) explanations (e.g., see Cohen & Marks, 2002; Cordes & Brannon, 2008), and that the capacity for successful behavior in small exact numerosity tasks may depend on a system that, while domain-relevant, is not necessarily domain-specific (see Carey, 2001).

There is evidence that both the approximate number and the object-file systems exist in human adults and are associated with bilateral activity in the inferior parietal lobes. This finding is consistent with the sensitivity of the dorsal pathway to the spatial-temporal aspects of object processing. Evidence for the ANS comes from experiments in which adults must rapidly (i.e., without counting) discriminate between numerosities in visual spatial arrays, visual sequences (light flashes), or auditory sequences. In these types of tasks, discriminability follows Weber's Law and is independent of stimulus modality, as found in animals (Barth et al., 2003; Whalen et al., 1999). Evidence for the object-file system comes from experiments in which adults must enumerate or attentively track several objects (Pylyshyn & Storm, 1988; Trick & Pylyshyn, 1994). In these studies of object-files there is usually a limit of four objects, and performance is unimpaired by occlusion, as in studies of infants and animals (see Carey, 2001).

In one of the first neuroimaging studies of number processing in children, Cantlon and colleagues (Cantlon et al., 2006) presented adults and 4-year-old children with sequences of visual arrays that varied either in the number of stimuli (with a minimum of eight items) or in their local shapes. During these presentations participants' brains were scanned using functional magnetic resonance imaging (fMRI). Adults showed greater activity around their intraparietal sulcus (IPS) when the arrays deviated in the number of elements than when they deviated in the shape of the elements, a result consistent with previous scanning studies that have linked non-symbolic and symbolic numerical processing. Interestingly, 4-year-old children with considerably less symbolic knowledge of number showed very similar patterns of activation, showing that the cortical tissue sensitive to non-symbolic numerical representations exists in early childhood and may provide a neural basis on which symbolic number representations could be mapped during their acquisition through education.

When children attend school and learn elementary arithmetic, however, they must work with a different system of number representation: a system that does not have an upper limit, that is not constrained by the Weber fraction, that is not constrained by perception, and that can be related to language including number words. This system has been termed an "integer-list," or natural number, representation (Carey, 2001). Studies of highly trained birds and primates so far provide no evidence that any non-human animal can acquire such a system (although some animals impressively expand their ability to use the two building-block systems described above; see, e.g., Brannon & Terrace, 2000; Matsuzawa, 1985, 1991; Pepperberg, 1987). How do human children construct this system?

One idea, advanced by Elizabeth Spelke and Susan Carey, is that children construct a new concept of number and gain arithmetic skills by bringing together their two initial systems of number representation, and that language—the number words and the verbal counting routine—plays a central role in orchestrating these systems. There are several lines of evidence in support of this view. First, when children learn the number words and the counting routine, they first map "one" to the object-file system and all other number words to the ANS indiscriminately: for example, they respond correctly when they are presented with arrays of one and of four objects and are asked to point to the array with "one" versus "four," but they respond at chance when presented with arrays of two versus four or eight objects and asked to point to the array with "two" versus "four" or "eight" (Wynn, 1992). Next, children coordinate the systems together to learn the meanings of "two" and "three," while continuing to use all other number words indiscriminately to mean, roughly, "some" (Condry et al., 2001; Condry & Spelke, 2008; Wynn, 1992). Finally, children surmise that each word in the count sequence refers to an array that includes one more entity, and a larger overall set size, than the array picked out by the previous word in the count sequence (Wynn, 1992).

Further evidence that the natural number concepts result from the coordination, through the language of counting, of the approximate number and object-file systems comes from studies of human adults. When adults make judgments about number words or symbols (e.g., whether a given number word is larger or smaller than five), they activate approximate number representation as well as exact numerosity, and therefore judge more quickly numbers that are more distant from the comparison number (Dehaene, 1997). This finding, and many other findings from behavioral studies of normal adults, behavioral studies of neuropsychological patients, and neuroimaging studies of normal adults, provides evidence that approximate number representations are activated in tasks requiring representations of the number concepts picked out by the counting words (see Dehaene, 1997, for review). Moreover, when bilingual adults learn new information about exact numerosity, their learning is language-specific: there is a cost in response time when the information is queried in the language not used in training. In contrast, when the same adults learn new information about approximate numerosity, their learning shows full transfer across languages. This finding, and parallel findings from neuroimaging studies, provides evidence that representations of large, exact numerosities depend on language for adults, whereas representations of large approximate numerosities do not (Dehaene et al., 1999).

A number of studies provide evidence in support of a link between number representation and mathematics learning. For example, children's performance on non-symbolic numerical tasks has been found to correlate with their scores on standardized or curriculum measures of mathematics. In this study, children were tested annually from age 5 to 10 years (Halberda et al., 2008). The results showed large individual differences in the non-verbal approximation abilities and that these current individual differences correlated with children's past scores on standardized mathematics achievement tests, extending all the way back to the younger ages.

In summary, research shows that there are two systems for representing numerosity present even in young infants, but while these are domain-relevant they are not specific to the domain of number. Work with children and adults using neuroimaging suggests that the intraparietal cortical regions are involved in numerical processing but also show developmental changes with a shift of relatively less frontal and more parietal activation triggered by basic and advanced numerical tasks (Ansari, 2008). For example, 8–12-year-olds differ from adults in the degree of recruitment of parietal and prefrontal networks while solving simple arithmetic tasks. Adults typically demonstrate activation primarily in the IPS, whereas children show significantly lower levels of activity in the IPS arithmetic network but more in frontal regions, areas responsible for attention and executive WM (Kucian et al., 2005). Thus, the functional specialization in the parietal cortex for mental arithmetic increases with age and is accompanied by a corresponding decrease of activity in prefrontal regions (Ansari et al., 2005; Rivera et al., 2005). The observed increases in the recruitment of parietal regions in numerical processing are taken as evidence for the specialization of parietal functioning during ontogeny (Ansari & Dhital, 2006; Holloway & Ansari, 2010). Another specialization observed over the course of development is that of the left IPS for ordinal number processing (Matejko et al., 2019). The decreasing involvement of prefrontal areas, on the other hand, is assumed to reflect a developmental disengagement of domain-general processes related to executive control and WM (Ansari et al., 2005; Rivera et al., 2005).

Literacy

Reading involves linking a visual written word with the sound structure of language and interpreting its meaning. The development of reading poses a somewhat different challenge than the development of the ability to perceive and produce spoken language (Chapter 9). Unlike the latter abilities, which seem to unfold without special effort as children develop, reading is a relatively recent development in human history that is acquired through explicit teaching and much practice. Thus, studying the acquisition of reading provides an interesting opportunity for investigating how specific experience impacts brain function during postnatal development. Both ERP and fMRI research have significantly contributed to our understanding of the neurocognitive processes involved in reading (Tong & McBride, 2020).

One brain area that has been intensively studied is the "visual word form area" (VWFA, Figure 11.2). The VWFA is a region in the left occipito-temporal cortex, centered on the mid-fusiform gyrus, that shows a preferential response to visual words compared to other complex visual stimuli. The VWFA appears to be involved in perceptual expertise for word recognition that allows words to be perceived and processed quickly and automatically in skilled readers. This echoes the role of another region of the fusiform gyrus that is also involved in perceptual expertise, the fusiform face area (see Chapter 7). Important changes have been observed in activation of the VWFA over the years when children begin to learn to read. Functional MRI studies have shown that VWFA is typically bilateral in beginning readers, but shifts to the mature pattern of left-lateralization with age and increasing reading skill (Schlaggar et al., 2002). One longitudinal fMRI study charted individual changes in cortical sensitivity to written words over a 4-year period as reading developed in a group of 7–12-year-olds. There were age-related changes in children's cortical sensitivity to word visibility in posterior left occipito-temporal sulcus nearby the anatomical location of the VWFA. The rate of increase in brain word sensitivity correlated with the rate of improvement in sight word efficiency (Ben-Shachar et al., 2011).

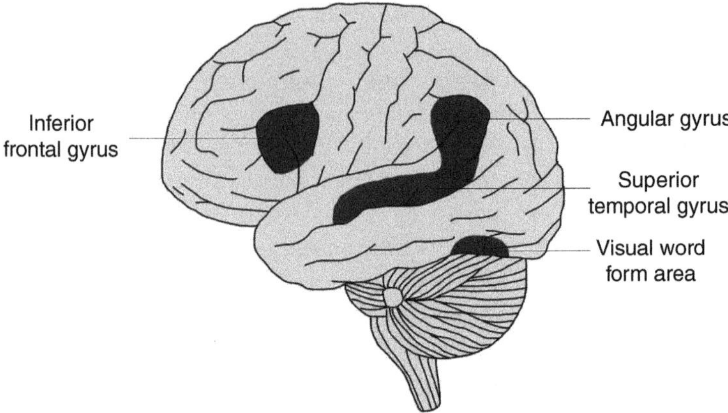

Figure 11.2 Key regions of the reading network. © Iroise Dumontheil.

ERP studies have also been employed to investigate the development of the VWFA, as this region is believed to contribute to the N170 ERP component elicited by visual words. In adults, the N170 for visual words is left-lateralized, and this spatial distribution of the response is believed to be a signature of perceptual expertise of visual words as opposed to other forms of perceptual expertise. Preschoolers' N170s for words are slow and do not show sensitivity to words or letters (Maurer et al., 2005); however, after a year and a half of reading instruction, reading fluency correlated with the degree to which the N170 showed an adult-like response (Maurer et al., 2006). Overall, these findings are consistent with the idea that perceptual expertise for recognizing visual words involves a process whereby occipito-temporal brain activity elicited by words becomes more specific to words and more focalized to the left hemisphere.

> **Further Reading** Schlaggar & McCandliss (2007); Tong & McBride (2020).

To summarize this section, language input from spoken and printed words plays an important role in shaping the neural bases of language. Generally speaking, the findings discussed above are consistent with the interactive specialization view that the brain activity underlying language functions becomes more focal with experience, as language skills become more efficient, automated, and specialized.

Domain-General Skills: Executive Functions and Processing Speed

There is general agreement that WM, inhibitory control, and switching are basic executive skills that are measurable in children and contribute to the emergence of more complex executive skills such as problem solving, planning, and reasoning (reviewed in Best & Miller, 2010; Diamond, 2011). It is easy to imagine how children who begin school with difficulties in these basic areas may struggle. For example, they may have problems remembering and carrying out instructions, inhibiting irrelevant information and staying focused on task, or planning and monitoring progress of individual steps of a task as it proceeds. The resulting missed opportunities for learning and practicing skills may then have a negative impact on the ability to achieve normal advances in complex skill domains such as reading and mathematics (Gathercole & Alloway, 2006).

Inhibition is the ability to stop an inappropriate response or thought (see Chapter 10). Several studies with preschoolers have found links between inhibitory control and mathematical skills, even when potential confounders like age, vocabulary, and mother's education are controlled for (Espy et al., 2004). Links between inhibition and academic skill can be indirect. For example, one study, examining whether executive function in 5–6-year-olds predicted reading abilities at 6 to 7 years, found that, once confounding variables were controlled, inhibition did not a have a direct effect on reading but did have an indirect effect on reading and writing via anger and aggression (Monette et al., 2011).

In middle-school-aged children, those who are poor at reading comprehension or at solving mathematical word problems have been found to have poorer inhibition, shown by poorer recall of critical task information and better recall of irrelevant information compared to their more able peers (DeBeni et al., 1998; Passolunghi et al., 1999). In 11-year-olds inhibition predicts curriculum attainment in mathematics and English, indicating that these skills support general academic learning rather than acquisition of skills and knowledge in specific domains (St. Clair-Thompson & Gathercole, 2006). Other research suggests that domain-specific aspects of inhibitory control may play a role in mathematics achievement, rather than putative domain-general inhibitory control processes. For example, Cragg and colleagues (Cragg et al., 2017) found that performance on an inhibition task involving numerical stimuli associated with individual differences in mathematics factual knowledge and procedural skills, while no association was observed with a non-numerical inhibition task.

Generally, the results of studies examining relations between inhibition and academic skills are consistent with the view that there is a more specific link between inhibitory control and mathematics in the preschool years, but that in older children there is a more general effect on learning. The early link with mathematics may reflect the fact that there are fewer automated processes for this skill during the initial learning stages; with skill acquisition, the intraparietal area implicated in numeracy in adults becomes increasingly active.

Further Reading   Holmes et al. (2010).

WM is our mental workspace, and as such is involved in many cognitive tasks (see Chapter 10). There have been several studies that have examined the relation of WM skills to development of reading and mathematics abilities. Visuospatial WM is correlated to mathematical abilities in children with (Henry & MacLean, 2003) and without (Cragg et al., 2017; Jarvis & Gathercole, 2003; Kyttälä et al., 2003) learning disabilities. This link has been interpreted as reflecting the numerous instances where visuospatial skills are needed in mathematics, from aligning columns of numbers to manipulating mental number lines. Holmes and Adams (2006) showed that visual spatial WM in a maze task uniquely predicted all aspects of 8-year-olds' mathematics achievements even after controlling for variance associated with age, phonological memory, and other measures of executive functioning. By contrast, for 10-year-olds visual spatial WM only predicted performance on more difficult mathematics questions. It may be that older children can use other strategies, such as direct retrieval, for simpler questions, but must revert to using visual-spatial strategies for more complex questions. This might suggest that as children become older the relations between WM and mathematics skills change, though other studies have reported a more consistent link over an even wider age span from 6 to 16 years (Dumontheil & Klingberg, 2011). Measures of WM can also predict future gains in mathematical skills. For example, one study of children aged 6–16 years found that visuospatial WM provided unique prediction of mathematics skills 2 years later even when age, verbal WM, and non-verbal intelligence quotient (IQ) were taken into account (Dumontheil & Klingberg, 2011). Another study showed that individual differences in a WM latent factor during adolescence predicted longitudinal progression in mathematics (as well as English and science) achievement between the ages of 11 and 16 years, while controlling for vocabulary and socioeconomic status (Donati et al., 2019). Additional work reported that both verbal and visuospatial WM predicted cross-sectionally individual differences in whole number arithmetic both directly and indirectly via influences on factual knowledge, procedural skills, and conceptual understanding (Cragg et al., 2017). Again, the association was found to be stable over late childhood (8–9 years), adolescence (11–14 years), and young adulthood.

Most studies of the relation between WM and reading have focused on verbal WM. Verbal WM skills at age 4 have been found to predict reading comprehension, writing, and spelling skills 2½ years later (Gathercole et al., 2003). One study found that the relation between verbal WM and reading in preschoolers was indirect and mediated through effects on anger and aggression (Monette et al., 2011). Generally, the most consistent links have been found between WM and reading comprehension (Cain et al., 2004; Swanson et al., 2006), and WM appears still relevant for English academic achievement in secondary school (Donati et al., 2019).

Are the links between development of executive functions and academic skills due to common neural correlates, whereby brain development mutually benefits both domains? The IPS is an area of convergence for both mathematical skills and visuospatial WM, suggesting that development of this area of cortex could underlie the links between these skills during ontogeny. A neuroimaging study found that greater activation of the left IPS relative to the rest of the brain during a visuospatial WM task is associated with poorer performance in mathematics 2 years later (Dumontheil & Klingberg, 2011). Left IPS activity was still a significant predictor of mathematics abilities when WM and reasoning abilities were taken into account.

In a meta-analysis of 52 studies involving over 800 children aged 4–17 years, Houdé et al. (2010) found both overlap and distinction when examining neural regions activated during mathematics, reading, and executive function (see Figure 11.3) in the color plate section. Mathematics tasks were related to the right inferior frontal gyrus, left superior frontal gyrus, and upper part of the left middle occipital gyrus near the junction with the parietal cortex; reading tasks were related to the left frontal, temporo-parietal, and occipito-temporal

(VWFA) regions, and executive functions were related to bilateral dorsolateral and inferior prefrontal cortex, extending to the insular cortex, as well as posterior cortical areas. Only for executive functions was the age contrast between children and adolescents significant. This was due to a greater involvement of the right anterior insula in adolescents compared to children. The main regions of overlap in the networks for reading and numeracy with executive skills were the middle and inferior frontal gyri. Generally, these findings were interpreted as suggesting that the brain network related to reading was relatively stable over the age range studied and similar to adults, whereas the brain network related to number skills showed more prefrontal and less parietal involvement compared to that typically observed in adults. A review of 38 studies of arithmetic in the developing brain also reported a decrease in activation in the prefrontal cortex and increase in posterior parietal cortex activation over development, which may reflect a change in strategy from calculation to fact retrieval (Peters & De Smedt, 2018). Another meta-analysis suggested that for children under 14 years, regions such as the insula and claustrum may play important roles with regards to motivation and integration of top-down and bottom-up processes (Arsalidou et al., 2018). Importantly, expertise does not mean that the involvement of the prefrontal cortex disappears completely. For example, a study reported that professional mathematicians, compared to non-mathematicians, recruited a network of bilateral intraparietal, inferior temporal, and dorsal prefrontal regions when they judged whether mathematics statements relating to a range of topics (e.g., algebra, topology, geometry) or non-mathematics statements were true or false (Amalric & Dehaene, 2016).

Individual Differences and Training Interventions

In addition to studying how academic skills are acquired, the field of educational neuroscience is also researching the source of individual differences in academic achievement, whether genetic or environmental, and how different ways of teaching, interventions, or specific types of practice may affect learning and possibly compensate for individual differences.

Decades of behavioral genetic research have shown that both environment and genetic influences play a role in academic attainment (years of education) and academic achievement (grades). As discussed before, genome-wide association studies (GWAS) of cognition and learning have demonstrated that these traits are highly polygenic, that is, individual differences in a trait are explained by variation in hundreds to thousands of genetic loci, and that there is significant pleiotropy for complex traits, that is, the same genetic variations affects two or more traits in a consistent direction (Visscher et al., 2017). A number of GWAS of length of time in education have been carried out, with increasing sample sizes. With data from approximately 1.1 million individuals, the consortium identified more than one thousand deoxyribonucleic acid (DNA) variants that each contributed to the number of years of schooling (Lee et al., 2018); however, the effect size of each association was very small, with a median effect size of 1.7 weeks of schooling per genetic variant. The advantage of such a broad phenotype is that it can be measured easily in a large number of individuals; however, how long one stays in school is influenced by a large number of facets of cognitive function, emotional behavior, personality, and social and economic circumstances. The small size of effects also precludes using these variations as individual predictors of risk and resilience in education and fails to give insight into relevant gene function or biological pathways.

Smaller-scale studies have tried to investigate genetic predictors of general intelligence or academic achievement in more specific subjects. For example, a GWAS of general intelligence identified 18 independent loci using a sample of more than 78,000 individuals (Sniekers et al., 2017), and a study found in approximately 3,000–5,000 individuals that single-nucleotide polymorphism (SNP)-based heritability of achievement in mathematics and science during adolescence remained present when removing shared variance between subjects (English, mathematics, and science) or IQ (Donati et al., 2021).

These results suggest that using more specific phenotypes of academic achievement may be necessary to help build more specific biological models of variance in learning processes and abilities, a conclusion that aligns with the finding that, for example, numeracy and literacy depend on both domain-relevant and domain-general skills.

Genetic studies have also shown that part of the observed associations between cognition and academic outcomes (e.g., the association between visuospatial WM and mathematics described above) are rooted in common genetic variation. For example, a study using the UK Biobank sample found that years in education and verbal-numerical reasoning were both phenotypically and genetically correlated, and that they both correlated genetically with cortical thickness measurements in bilateral primary motor cortex and parts of the temporal lobe (Ge et al., 2019). It has been argued, however, that the genetic association between cognition and academic outcome is amplified by environmental experience (Malanchini et al., 2020).

Further Reading   Donati & Meaburn (2020).

Socioeconomic status (SES) is a multidimensional construct which, for children and adolescents, is typically measured through family income, parental education, or parental occupation, but also encompasses neighborhood environment and school context. SES is a predictor of a range of psychosocial, environmental, and community exposures and opportunities, from whether the parent has the time and resources to invest in their child's development to toxins and pollution exposure. The interactionist model proposes that there is a dynamic relationship between SES, parenting behavior, and child development over time (Conger & Donnellan, 2007), and developmental cognitive neuroscience and educational research is studying how these relationships are implemented in changes in brain structure or brain function. Indeed, developmental cognitive neuroscience research has provided strong evidence of associations between SES and neurocognitive development (Farah, 2017), and SES has been found to moderate the association between neurocognitive development and academic outcome. For example, a study found that higher SES adolescents had higher WM and mathematics achievement than lower SES adolescents and that they showed greater WM activation at high load than lower SES adolescents, but that lower SES adolescents showed greater activation at low load. In addition, WM-related brain activity in the parietal cortex predicted mathematics achievement in both lower and higher SES groups, while activation in the prefrontal cortex predicted mathematics achievement in the higher SES group only (Finn et al., 2017). In a different dynamic relationship, there is also some evidence that brain structure and function may mediate the impact of SES on academic outcome. A study found that adolescents with higher SES had significantly greater cortical gray matter volume and cortical thickness, but not surface area or white matter volume, than adolescents with lower SES, and that greater cortical thickness, in particular in the temporal and occipital lobes, partially mediated the association between SES group and performance on a standardized test of academic achievement (Mackey et al., 2015). Limitations of this research are that there is less work done with younger children, due in parts to the difficulties in carrying out neuroimaging research in this population, and that there are also difficulties in recruiting participants from low SES. Nevertheless, the neuroscience approach may help better understand the origin of disparities and potentially inform policy (Farah, 2018). Importantly, the use of a neuroscience approach does not need to be based on a deficit-based view of the impact of SES but can help elucidate variation and adaptation; for example, consider the fact that SES differences may emerge as children and adolescents adapt to their environment. Neuroscience may also inform which interventions may be useful for specific groups, and provide insight into the mechanisms through which interventions may work.

Further Reading   Hackman & Kraemer (2020).

Indeed, educational neuroscience research has started investigating whether interventions could be designed, based on developmental cognitive neuroscience research, to improve academic outcome. One early example emerged from the finding that WM predicts mathematics and other academic achievement. Researchers predicted from this research that training WM skills, for example using repeated practice of increasingly difficult WM tasks in the form of video games (e.g., CogMed program), may help children achieve better at school. Unfortunately, the results showed that whilst WM could be improved, which was visible, for example, through improved performance on untrained WM tasks, there was no "transfer" to other skills, such as mathematics or reading comprehension (see Melby-Lervåg et al., 2016 for a review). In addition, it seemed that genetic variation may constraint the extent to which children may benefit from training. For example, a study in 7–19-year-olds found that SNPs located near the dopamine receptor 2 gene were associated with how much children's WM improved through training (Söderqvist et al., 2014). Another study with children with intellectual disabilities found that higher baseline performance predicted greater improvements after a training aiming to improve WM and non-verbal reasoning, suggesting again that not all children may benefit to the same extent (Söderqvist et al., 2012). Similar results have been found in studies of cognitive interventions aiming to improving sustained attention (Slattery et al., 2022).

Why are there no transfer effects from cognitive training? One possibility is that the phenotypic correlations between cognitive and academic measures are driven by shared underlying genetic pathways, and that training one component during childhood or adolescence will not affect the other. Similarly, the proposed benefits of musical practice, e.g., a link between music and language (Swaminathan & Schellenberg, 2020), may in fact be driven by genetic and SES differences in musical ability or between children who play and those who do not (Schellenberg, 2020). Another possibility is that rather than being domain-general skills, the neural networks supporting, for example, WM or reasoning may be intrinsically embedded into a particular domain. For example, increasing children's capacity to remember single-digit numbers or letters in a game may not improve their ability to remember the instructions given by their teachers, or the words of a long sentence they are reading. Interventions that target brain health (e.g., physical activity), cognition (e.g., mindfulness meditation), or parent–child interactions (e.g., Neville al., 2013), or that are implemented in a classroom or tutoring context closer to real life, may be more successful (Diamond & Ling, 2015; Slattery et al., 2022).

As suggested above, neuroscience may still play a role in identifying which children may be most susceptible to what type of training. For example, a study focused on determining the predictors of which children benefit most from academic tutoring in mathematics (Supekar et al., 2013). Children aged 8–9 years had assessments of IQ, WM, and mathematics skills as well as structural and resting state fMRI scans before having 8 weeks of one-to-one tutoring. With tutoring, the speed and accuracy of mathematics problem solving improved, but some children benefited more than others. None of the cognitive tests administered prior to tutoring predicted which children would benefit from tutoring, but pre-tutoring hippocampal volume and the functional connectivity of the hippocampus with dorsolateral and ventrolateral prefrontal cortices and the basal ganglia did predict improvements. By contrast, regions typically involved in arithmetic processing were not strong predictors of responsiveness to mathematics tutoring in children.

Dyscalculia and Dyslexia

Problems with understanding numbers come in several forms. Sometimes children with otherwise normal intelligence show particular problems with arithmetic. This is usually called "dyscalculia," and contrasts with cases where lower mathematical abilities co-occur with low IQ and other cognitive problems that result from genetic disorders. Some suggest that dyscalculia may not be as specific as originally supposed since it often co-occurs with dyslexia and attention deficit disorders (see Ansari & Karmiloff-Smith, 2002, for review). In genetic disorders such as Williams syndrome, numeracy problems are quite commonly observed (Ansari & Karmiloff-Smith, 2002). Further, performance in number tasks is often worse than reading scores, suggesting that numeracy may be a particularly vulnerable cognitive domain in the atypically developing child. Indeed, Peters and Ansari (2019) note the overlap between and heterogeneity within learning disorders and suggest that research should move away from studying specific learning disorders, such as dyslexia and dyscalculia, and instead try to understand interacting factors that are associated with overlapping domains of learning. Other authors suggest that dyscalculia may be a specific developmental syndrome (Butterworth, 2006) that can even be artificially induced in typical people by electrical stimulation to relevant parts of the parietal lobe (Cohen-Kadosh et al., 2007). Neuroimaging studies have shown structural and functional differences in parietal regions associated with number processing in children with developmental dyscalculia, which have been interpreted as reflecting a reduced ability to understand numerosities and mapping number symbols to number magnitude (see Butterworth et al., 2011, for review).

On this basis, a number of interventions have been designed to improve the arithmetic skills of children with dyscalculia by focusing on training their "number sense"; however, it is not evident so far that this type of intervention can turn children with developmental dyscalculia into "typical calculators" (Butterworth et al., 2011). Other evidence suggests that regions other than the parietal cortex are underactivated in children with dyscalculia, such as regions involved in verbal processing (Berteletti et al., 2014), and that while dyscalculia and dyslexia may have different patterns of performance, they show similar brain activity profiles during arithmetic (Peters et al., 2018). Iuculano (2016) proposes that in fact the evidence suggests that developmental dyscalculia is a multifaceted disorder, that multiple neurocognitive systems, supporting cognitive control, memory, and visuospatial skills in addition to numeracy, have been implicated in its etiology, and that therefore interventions would need to target the breadth of these cognitive processes.

Dyslexia was one of the first developmental disorders to be associated with a particular underlying neural atypicality. In 1907, Hinshelwood postulated that it was a developmental version of alexia (failure to recognize words) and associated it with atypicality of the left angular gyrus of the cortex (Hinshelwood, 1907). Key symptoms of dyslexia are difficulties in learning to read and spell, sometimes with letter and number reversals, and unusual errors. These correlated difficulties (such as in naming and verbal short-term memory) are thought to be due to primary cognitive difficulty in the phonological coding of spoken language. There is both structural and physiological evidence from the brains of dyslexics consistent with Hinshelwood's original proposal that developmental atypicalities in the left hemisphere are important.

The structural evidence comes from the work of Galaburda and colleagues (e.g., Galaburda et al., 1985), who conducted autopsies on the brains of several deceased dyslexic individuals. They examined the degree of symmetry (between the right and left side) of a particular part of the temporal lobe of the cortex, the planum temporale. This region is part of the so-called Wernicke's area, which has been implicated in phonological processing (Chapter 9). All brains showed a deviation from the standard pattern of symmetry of the planum temporale. Galaburda and colleagues also observed malformations in the clustering of neurons in this region and, to a lesser extent, elsewhere in the cortex of these individuals. These cellular atypicalities are not evident in structural neuroimaging studies, indicating that developmental atypicalities at the cellular or molecular level will not always be evident in terms of gross brain structure.

More recent efforts have also been made to identify structural asymmetries linked to dyslexia. In a study examining the brain images of a group of children and adults with reading and language disorders, Leonard and Eckert (2008) found two patterns of contrasting anatomical and reading profiles. One pattern was characterized by small, symmetrical brains and was related to difficulties in several domains of written and spoken language. The second pattern was characterized by larger, asymmetrical brain structures and was related to isolated phonological difficulties, as in dyslexia.

Turning to research on brain function, Wood et al. (1991) described studies from blood flow, positron emission tomography (PET), and scalp-recorded evoked responses in individuals with dyslexia. Consistent with the postmortem studies, they concluded that people with dyslexia showed atypical processing in the left temporal lobe in both phonemic discrimination and orthographic tasks. Similar conclusions were reached by Paulesu and colleagues (Paulesu et al., 1996), who compared dyslexics to a control group during a rhyming task and a short-term memory (for visually presented letters) task. The results showed that only a subset of the regions typically active during these tasks were activated in the individuals with dyslexia. Specifically, and unlike in the control group, Broca's area and Wernicke's area were never activated simultaneously, possibly because of dysfunction of the insula which normally provides the functional connection.

Research by Tallal and colleagues has suggested that the ability to process the rapid temporal information necessary for phonemic discrimination (see previous section) may be critical for normal language acquisition (e.g., Tallal & Stark, 1980; see Merzenich et al., 2002). Specifically, children with oral language delay often, but not always, have difficulties in the perception and discrimination of rapidly changing acoustic information (Tallal et al., 1980). The inability to discriminate some phonemes with rapid temporal transitions, such as /ba/ and /da/, may have consequences for oral speech recognition. Tallal, Merzenich, and colleagues have produced preliminary evidence that the inability to process rapid temporal transitions in children with language delay can be rectified by training with adapted speech in which the temporal transitions are extended (Merzenich et al., 2002; Tallal et al., 1996). Whether such remediation can be successfully extended to a wider range of language and reading disorders remains to be seen. However, some reports have indicated that children with language delay frequently develop reading problems symptomatic of some types of dyslexia.

Goswami (2020) has proposed a neurocognitive mechanistic theory accounting for the phonological processing difficulties of children with dyslexia. She proposes that brain oscillations that encode a phonological hierarchy of stressed syllables in a sentence, syllables within a word, and rhymes (Giraud & Poeppel, 2012) are less well synchronized to perceived speech in children with dyslexia than in typically developing children (Power et al., 2016). On the basis of this model of developmental dyslexia, interventions focusing on improving children's rhythm perception were developed and shown to have some success (Bhide et al., 2013), particularly for boys (Wilson et al., 2021) and for younger children and children with a family risk of dyslexia (Ahmed et al., 2020).

Researchers have also examined the role of the VWFA in dyslexia. Several studies have reported an under-activation of the VWFA in adults or adolescents with dyslexia (e.g., Maurer et al., 2007). A study investigated young children with dyslexia with just a few years' reading experience, and examined the VWFA as well as other regions of the visual word form pathway along the left occipito-temporal cortex (van der Mark et al., 2009). The results showed the children with dyslexia activated the same basic brain reading network as controls. However, there were differences in the degree of specialization of this activation. They found that only control children showed a differential pattern of activation for print compared to false fonts, with greater activation to print in anterior regions but the opposite in posterior regions. Moreover, they also found that only control children showed an "orthographic familiarity effect," whereby response in the visual word form pathway were greater for unfamiliar than familiar word forms. Consistent with these findings, Centanni et al. (2019) showed that a hypoactivation of left, but not right, occipito-temporal areas at kindergarten age was related to poorer reading outcomes 2 years later in those at familial risk for dyslexia. Together, these results thus show a dysfunction and lack of specialization of the VWFA and related regions in children with dyslexia even before they begin to read and continuing soon after, though they do not make clear to what extent this dysfunction is a cause, as opposed to a consequence, of the reading difficulties.

Further Reading   Eden & Flowers (2008); Merzenich et al. (2002).

The above studies are all broadly consistent with involvement of the left temporal lobe in both phonemic discrimination and dyslexia. However, it is important to stress that there is considerable controversy about the exact nature of the cognitive difficulties observed in dyslexia (e.g., Eden & Flowers, 2008; Pennington & Welsh, 1995) and that there is also evidence for atypicalities elsewhere in the brain (e.g., Livingstone et al., 1991). Further, the putative involvement of the left temporal lobe in this disorder need not be interpreted in terms of a causal epigenesis pathway (Chapter 1). It is tempting to infer that a neural atypicality in the left temporal lobe gives rise to a failure to process the rapid temporal transitions necessary for phonemic discrimination, and that in some cases this might be related to subtypes of dyslexia. However, the evidence on cortical development and plasticity discussed in Chapter 4, and the apparent initial success of the training studies by Merzenich et al. (2002), suggest a more probabilistic epigenetic view (Chapter 1).

Broadly speaking, this probabilistic epigenesis view is that regions of the left temporal lobe are engaged by inputs that require rapid temporal processing. This is due to having a neural architecture slightly different from the basic structure common to the rest of the cortex (architectural biases, Chapter 1). This, combined with constraints from inter-regional connectivity and sensory inputs, makes this region the best, but not the only possible, "home" for speech processing. If this region is damaged, other parts of cortex can also support these representations, though perhaps not quite as efficiently. Processing speech within this region influences its synaptic and dendritic microstructure through some of the mechanisms discussed in Chapter 4. Through interaction with a certain type of input it thus becomes increasingly specialized and different from other regions. Some degree of plasticity may remain during development, resulting in a reconfiguration of the microcircuitry of the region following specific training. Differences in neural structure between this region in adults with dyslexia and controls would reflect this deviated developmental pathway, some of which may have occurred postnatally. Thus, a slight initial bias can be enhanced through development into greater anatomical and functional specialization.

General Summary and Conclusions

Studies show that acquiring academic skills such as literacy and numeracy involves both domain-specific and domain-general skills. Young infants already have some basic skills that are relevant to this process, such as sensitivity to numerosity and to phonemes, and that may initially be more broad-based, but with experience develop into more specialized and domain-specific skills. There is evidence that, over childhood and as academic skills are acquired, there is increasing involvement of posterior brain regions and decreasing involvement of frontal brain regions. For example, functional specialization in the parietal cortex for mental arithmetic increases with age and is accompanied by a corresponding decrease of activity in prefrontal regions (Ansari et al., 2005; Rivera et al., 2005). The observed increases in the recruitment of parietal regions in numerical processing are taken as evidence for a specialization of parietal functioning during ontogeny (Ansari & Dhital, 2006; Holloway & Ansari, 2010). The decreasing involvement of prefrontal areas, on the other hand, is assumed to reflect a developmental disengagement of domain-general processes related to executive control and WM (Ansari et al., 2005; Rivera et al., 2005).

Studies of dyslexia and dyscalculia have investigated atypicalities in brain regions associated with reading and numerical processing respectively. When atypicalities have been observed, it is not always clear if these are the cause or consequence of the disorder.

Educational neuroscience is a new and emerging topic that promises much for the future. Success, however, will depend on more effective bridging between the laboratory and its associated brain scanning technologies, and the classroom.

Key Issues for Discussion

- In adolescence there are still immaturities of frontal executive functions such as cognitive control and error processing (Chapter 10). How might these limitations affect academic progress at this age?
- What cognitive mechanisms might underlie training effects on WM? How might this explain why training WM rarely translates into immediate improvements in academic skills?
- What key skills are essential for formal classroom education, and what brain systems support these skills?
- Overall, what are the prospects for a neuroscience approach to education, and what major barriers to progress need to be overcome?

12

Global and Cross-Cultural Perspectives

There is increasing awareness within developmental cognitive neuroscience research, and more generally in psychology and neuroscience, of the limitations of research focused mainly on Western, educated, industrialized, rich, democratic individuals. Conclusions based on these samples may not be generalizable to the majority of the global population. Greater inclusiveness can help determine which aspects of brain–behaviour relations are common across global/cultural settings, and which aspects are more specific adaptations to individual circumstances. For both of these aims, investigating a larger range of settings and experiences can inform on how experience shapes the brain, findings that are important for understanding developmental trajectories within settings, and for theories of the influence of experience on brain development, including interactive specialization.

A key paper published in 2008 pointed out that 96% of participants in psychology studies were from Western, educated, industrialized, rich, democratic (WEIRD) samples, even though these only represent 12% of the world's population (Arnett, 2008). This finding was highlighted by Henrich and colleagues, who pointed out the consequences and limitations of this approach in terms of generalizing from WEIRD to non-WEIRD samples, and the need for broader study (Henrich et al., 2010). The consequences for developmental psychology were elaborated on by Nielsen et al. (2017). A central point is that culturally specific findings are often interpreted as universally applicable "facts," when this may not necessarily be the case.

Few nowadays would argue with the potential value of studying a broader range of the global populations, and cross-cultural research is not new. However, it also presents some challenges both with respect to psychological/behavioral developmental assessments (e.g., Milosavljevic et al., 2019) and with respect to cognitive neuroscience approaches—which can involve specialized equipment and laboratories that may not be widely available (e.g., magnetic resonance imaging [MRI], magnetoencephalography [MEG], see Chapter 2). In spite of these challenges, there is value to understanding the aspects of brain–behavior

Developmental Cognitive Neuroscience: An Introduction, Fifth Edition. Michelle de Haan, Iroise Dumontheil, and Mark H. Johnson.
© 2023 John Wiley & Sons Ltd. Published 2023 by John Wiley & Sons Ltd.
Companion website: www.wiley.com/go/johnson/devneuro5e

relations that are common, i.e., universal (consistently observed across setting, culture, and conditions), and those that reflect adaptability for functioning optimally in a given environment, or in response to brain injury or disease. Some challenges to brain development are common across many different global settings (e.g., the COVID pandemic, nutrition, poverty), though the degree and qualitative aspects of these may still differ. Others are more region-specific (e.g., infection by cerebral malaria; exposure to war and experience as a refugee). Different measures may provide different information—for example, direct measurement of brain structure or resting state function may be more reliable and "culture free" across cultures; other measures, such as parental reports of child behavior, may be more culturally embedded. Below we will focus on studies that have applied cognitive neuroscience techniques in global, cross-cultural settings and discuss results and challenges for this research. First, we discuss studies that have used similar paradigms across settings, that can potentially identify common and more "universal" patterns. Then, we will turn to the question of factors that may contribute to differences in activation of these networks within and across settings.

Further Reading   Miller & Kinsbourne (2012); Jensen et al. (2019).

Developmental Cognitive Neuroscience: Factors to Consider in Global, Cross-Cultural Settings

As is clear from the chapters so far, developmental cognitive neuroscience often relies on technology such as eye tracking, event-related potentials (ERPs), functional near-infrared spectroscopy (fNIRS), or MRI/ functional MRI (fMRI). While these are increasingly available globally, they still are not available, or have very limited availability, in many settings. MRI/fMRI and MEG are costly and require a specific surrounding to operate. These kinds of factors present some barriers to application of techniques including MRI/fMRI or MEG more widely (although, some studies have shown that these barriers can be overcome, e.g., Turesky et al., 2021). Eye tracking, ERP, and fNIRS are in this regard more flexible and less costly techniques. Even so, obtaining good quality measures is also affected by the environments in which they are collected. Ambient temperature can both affect reliable operation of the equipment itself and introduce some challenges to measurement. For example, sweating can affect ERP measurements— temperature-controlled environments are therefore preferable. Cultural practices such as braiding of hair can also affect data collection as they can present challenges for placement of ERP or fNIRS sensors. Transport of equipment, including dismantling and re-assembly, can also be a challenge for data collection in remote areas or for collecting representative samples, regardless of global settings. However, these challenges can be overcome, as is illustrated in the studies described in the next section and in the Further Reading below.

Further Reading   Katus et al. (2019).

Behavioral genetics studies have also so far been mainly limited to Caucasian or localized populations. This raises the issue that polygenic risk scores that are currently available may not generalize outside the populations studied in the original samples (De La Vega & Bustamante, 2018; Popejoy & Fullerton, 2016). There is therefore a push to carry out this research in more varied samples, which raises analytical issues as populations can differ in their distribution of common genetic variations, and therefore genetic ancestry of different populations needs to be included in statistical models. Interestingly, research also suggests that there may be a genetic contribution to cultural differences. For example, variations within the genes of the serotonin and opioid neurotransmitter systems associate with individual differences in social sensitivity. The relative frequency of these variants in populations has been found to correlate across nations with their relative degree of individualism-collectivism. This pattern has been interpreted as suggesting that collectivity may have developed and persisted in populations with genetic variation that was more compatible with this cultural characteristic. A similar association was found with lifetime prevalence of major depression across nations, suggesting that reduced levels of depression in populations with a relatively higher frequency of social sensitivity alleles is mediated by greater collectivism (Way & Lieberman, 2010). Such correlations between phenotypes and genotypes between nations or cultures will be a challenge for behavioral genetic research.

Before moving to the next section, a factor to consider is the general aims of research in global, cross-cultural settings. Sometimes, researchers aim to find "culture-free" measures which can be applied similarly across settings. These are the universal/common aspects referred to above. Some brain methods, such as structural MRI or resting state MRI/ electroencephalography (EEG)/fNIRS might be considered culture-free compared to methods such as behavioral testing, where issues regarding language and cultural appropriateness of testing materials are a bigger factor (Milosavljevic et al., 2019). Another approach is to investigate the differences between cultures and explore how these affect the brain and development. This gives an idea of how the brain might adapt to different settings and challenges across the globe. Both approaches are valid but come from different perspectives in terms of the aims of what they hope to discover. We will discuss some examples of both approaches in the next sections, and provide reference to the theories discussed in Chapter 13.

Cognitive Neurosciences Approaches to Look at Commonalities Across Global/Cultural Settings

Infants' Response to Novelty

As discussed in Chapter 8, infants' responses to novelty have been considered an important marker of cognitive development in and of themselves, and have been used as a method to study many aspects of sensory, cognitive, and social development. However, this research has largely been based on WEIRD samples. Katus and her colleagues (2020) used ERPs to investigate the development of the novelty response in 1- and 5-month-old infants living in the United Kingdom and in rural The Gambia, West Africa. They presented infants with an intermixed soundtrack of a tone that repeated frequently (80% of the time), a burst of white noise that repeated infrequently (10% of the time—the most intense sound), and a category of "trial unique" sounds that were infrequent and never repeated (10% of the time). Some aspects of the response were common across the settings, e.g., overall morphology of ERP components; however, the modulation of these components differed across countries. A developmental change was observed in the UK infants, who showed a shift from a response based on perceptual intensity of the sounds at 1 month to a cognitive response to novelty at 5 months, whereas this latter developmental change was not observed in infants in The Gambia (see Figure 12.1).

Katus and her colleagues (2022) followed up these infants at 12 months of age and found that the ERP responses at 5 months of age predicted performance in deferred imitation memory tasks (see Chapter 8) at 12 months of age in *both* groups. This was true even when general neurodevelopment (measured by a standard assessment) was controlled for. These findings thus illustrate both commonalities (similar ERP components and prediction to later behavior) and differences (different early ERP development) in trajectories across settings. Future work can help identify the factors which contribute to these commonalities and differences—this will be discussed further later in this chapter.

Social Processing

Eye tracking is an approach that can give insight into early neurocognitive development in infants. Hernik and Broesch (2019) studied infants living in Vanuatu, in a culture where face-to-face interactions between parents and infants are less common than in Western populations, to test whether these cultural differences would impact gaze following. Gaze is an important social communicative cue, and the study showed that 5–7-month-olds in Vanuatu exhibited similar patterns of gaze following as previously studied 6-month-old Western infants, whereby they followed gaze shifts only if they were preceded by infant-directed speech, but not adult-directed speech. These results support the view that the early social communicative tendencies evidenced by gaze following are rooted in universal mechanisms and not dependent on specific sociocultural experience.

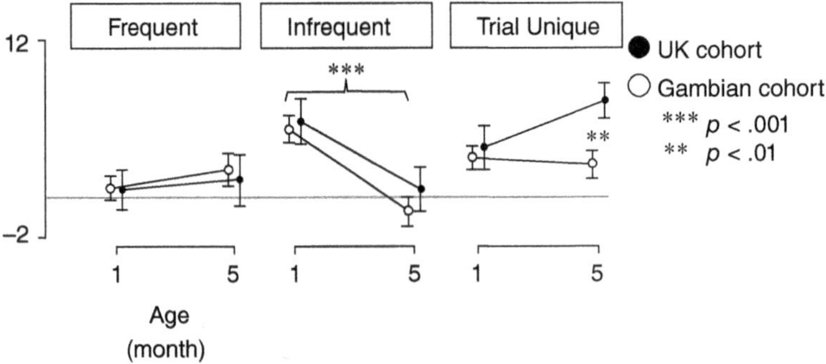

Figure 12.1 P3 ERP amplitude change between 1 and 5 months of age for frequent, infrequent, and trial unique sounds in a UK cohort and a cohort in the Gambian cohort. Error bars indicate 95% confidence intervals. From Katus et al. 2020.

In contrast, cultural norms can affect later development of face scanning. In a study with adults, British participants who observed avatar faces moving their mouth and then their eyes toward or away from them were less affected by the gaze shift and tended to fixate more on the mouth of the avatar, compared to Japanese participants, who fixated mainly on the eyes and followed the gaze of the avatar. This pattern of results was interpreted as consistent with cultural norms regarding eye contact in face-to-face interactions (Senju et al., 2013). In another study, East Asian participants, when asked to learn faces, recognize them, and categorize them according to race, attended to the center of the faces to a greater extent than Western Caucasians, who fixated more on facial features such as the eyes and the mouth (Blais et al., 2008). This cultural difference in face scanning is found by 7–8 years of age and does not appear to show major changes from 7 to 12 years of age, suggesting that aspects of social processing are culturally sensitive (Kelly et al., 2011).

fNIRS is another technique that has increasingly been applied across the globe. Lloyd-Fox, Bagus, et al. (2017) built on prior work in WEIRD settings using fNIRS to examine responses to social and non-social stimuli (Lloyd-Fox et al., 2014), showing that infants display differential activation to social compared to non-social stimuli in the inferior frontal, anterior temporal, and posterior superior temporal–temporo-parietal cortex (Lloyd-Fox et al., 2012; Minagawa Kawai et al., 2011). They applied this task to newborn to 2-year-old infants in rural The Gambia and urban UK. Overall, they found socially selective neural activity from 9 to 24 months for visual and auditory stimuli in Gambian infants similar to what was found previously, though there was no direct comparison in this study. Additionally, they reported that from birth to 8 months of age, Gambian infants showed non-social auditory selectivity. These results are similar to the ERP results discussed above, whereby infants initially are more responsive to sensory characteristics of inputs and then shift to more cognitive/social aspects. A similar finding, in terms of a change in fNIRS response around 8 months, was observed using this task in infants in Bangladesh (Perdue et al., 2019). Together, these findings help identify a core network involved in infant social information processing across settings. Factors that influence this network within and across settings will be discussed below.

Theory of mind is another aspect of social cognition that has been discussed in Chapter 7. As a brief recap, it involves understanding other people with respect to their point of view and what they are thinking. There is some evidence of cultural differences in the emergence of theory of mind—for example, children in Japan show a more protracted trajectory of development in this skill than Western children and focus more on socio-behavioral rules than mental states (Naito, 2003; Naito & Koyama, 2006). However, a study using the same paradigm in five different countries (Canada, India, Peru, Samoa, and Thailand) found children showed an understanding of false beliefs around 5 years of age in all five cultures. More broadly, developmental and social psychology studies have shown that the cultural context can affect the exhibition, meaning, and development of social behaviors such as sociability, shyness, cooperation or compliance, or aggression, as well as social competence (Chen & French, 2008). There are also different neural correlates—while some aspects of the social brain network are similarly activated (medial frontal and anterior cingulate cortex), the right temporo-parietal junction, a region known to be involved in mental state reasoning, was more activated in American than Japanese children (Kobayashi et al., 2007). A study modified the Reading the Mind in the

Eyes test to include an Asian version of the stimuli and tested 14 native white American participants and 14 native Japanese participants in the United States for a summer English language program (Adams et al., 2010). Each group completed both the Caucasian and Asian versions of the task, which requires participants to attribute one of four mental state terms to a picture cropped to the eye region of a face. Participants recruited a common set of regions including the superior temporal sulcus, temporal poles, and both lateral and medial aspects of the prefrontal cortex across the two types of stimuli. However, participants were more accurate and showed greater activation in the superior temporal sulcus and postcentral gyrus when attributing mental states to faces from their own culture than from the other culture. There was little difference between US and Japanese participants' brain activation. These results suggest while common brain systems are used across cultures, these systems have learned/adapted to individuals' experience of faces from their own culture.

Numeracy and Literacy

Little research has been done to study cultural similarities or differences in the neurocognitive correlates of numeracy. Tang and Liu (2009) asked 12 native Chinese speakers and 12 native English speakers to make judgments regarding the spatial orientation of non-numerical stimuli or Arabic numbers, and to carry out numerical additions or quantity comparisons, also with Arabic numbers, while being scanned with fMRI. The results suggest there may be a core system of number, associated with the bilateral intraparietal cortex, that is stable across cultures, and differential cortical representation in other circuits associated with language and education-specific strategies that differ between cultures. Notably, number words in Chinese are much shorter than in English and therefore are more easily stored in working memory. Reviewing previous work, Tang and Liu (2009) conclude that extracurricular culture-specific factors, rather than formal education, may drive difference in simple arithmetic skills between Asian and Canadian/US participants. Little is known of genetic differences that may drive differences in numeracy between cultures.

Somewhat more research has investigated reading and reading difficulties across cultures. For example, a meta-analysis of 25 studies compared word reading across different languages and cultures and found that the visual word form area (Figure 11.2) was central to word recognition across these studies. However, culture- and language-specific variation was also observed. For example, the Chinese writing system was associated with greater activation of visual processing brain areas than the Western alphabet (Bolger et al., 2005). Reading difficulties in dyslexia were compared in a study of English, Italian, and French adult readers. These languages differ in the shallowness of the orthography, and there was suggestion this led to different prevalence of dyslexia between countries. Paulesu and colleagues found that while Italian dyslexics performed better on reading tasks than English and French dyslexics, they performed similarly worse than their non-dyslexic peers. Positron emission tomography data showed a similar reduction in activity in a cluster in the temporal cortex in all three dyslexic groups, suggesting that there is a universal neurocognitive basis for dyslexia (Paulesu et al., 2014).

Developmental Cognitive Neuroscience: Examining Factors Affecting Similarities and Differences Among Global Settings and Cultures

There are a large number of factors, including social, medical, economic, education, and language, to name a few, that can potentially contribute to the differences in development of brain–behavior interactions and their relation across different settings (some of these issues are discussed more generally in Chapter 2). Here we will consider a few of these factors and give selected examples.

Maternal Stress, Caregiving and Education and Brain Responses to Social Stimuli

As discussed in Chapter 7, responses to social stimuli early in life are key in the development of the social brain network. Psychological studies suggest that some of the factors that affect social development in WEIRD samples also apply more broadly across cultural and global settings. For example, Pastorelli et al. (2021) found that in most of the 11 cultural groups they studied, greater parental warmth at 9–10 years was related to increases in prosocial behavior in later years. While they did not study the brain correlates, the findings suggest a universal role of warm relationships and shared prosocial values within the family context in fostering positive social development in the transition to adolescence. As in other fields, there is a push for research on adolescent neurocognitive development to focus to a greater extent on individual differences and to compare data across cultures, as cultural norms may particularly affect adolescent experiences in the transition to adulthood (Foulkes & Blakemore, 2018).

Considering brain networks, one study used fNIRS to examine how environmental factors influence the brain's response to social stimuli in children 6, 24, and 26 months of age in Bangladesh (Perdue et al., 2019). Infants were exposed to social and non-social video stimuli (social short movies of adults moving their eyes, mouths, and hands as if performing nursery rhymes or non-social movies of modes of transportation) and auditory stimuli (social crying, laughing sounds, or non-social sounds such as a fan whirring). Cortical specialization to sounds within the social brain network was found at all ages, and was similar to previous studies in WEIRD and other non-WEIRD settings. Moreover, the strength of the socially selective response was related to maternal education, maternal stress, and the caregiving environment. These findings suggest some common factors in the influence of environmental factors on the social brain network across samples, though it is important to note that the study did not directly investigate differences across global settings.

Interactions Between Factors Predicting Individual Differences in Neurocognition

One approach to investigate the potential impact of race and culture is to compare them by recruiting different participant groups within a country. For example, the Adolescent Brain Cognitive Development (ABCD) study is a US longitudinal study which is collecting a range of measures related to culture, race, and socioeconomic status (SES) (Gonzalez et al., 2021), in addition to neurocognitive and health measures over the course of adolescent development. One study using this dataset has demonstrated that neurocognitive differences are due not merely to additive effects of race and socioeconomic status but also to multiplicative effects. Researchers compared Black and white 9–10-year-olds—with a total sample size of 4,290 children—and found that in white, but not Black, children parental education, a measure of socioeconomic status, associated with greater left orbito-frontal cortex activity during a working memory task (N-back task) (Assari et al., 2021a). One suggestion proposed through the marginalization-related diminished returns (MDRs) framework is that structural racism and social stratification mean the benefits of parental education are smaller for Black and other racial/ethnic minority children than for white children. In a separate study, using 10,871 participants from the ABCD data, Assari and colleagues (Assari et al., 2021b) followed up previous findings that parental education is associated with larger superior temporal cortical surface area and higher reading skills in children. While they replicated these findings overall, there were interactions with race/ethnicity, indicating that high parental education led to a lesser increase in superior temporal cortical surface area bilaterally in Black than white children. The reverse was observed for behavior, whereby parental education had a greater effect on reading ability in Black than white children. Future studies looking at the longitudinal, rather than cross-sectional, data from the ABCD sample may further our understanding of this type of interaction between neurocognition and cultural and environmental factors.

Brain Responses in the Context of Global-Specific Risk

Environmental factors can be common across many settings—e.g., parenting, socioeconomic factors. To some extent these factors may support common development, but may also, depending on the range of differences within these factors across cultures, contribute to differences. There are also factors that are relatively unique to certain cultures that can affect brain development. For example, malaria is a parasitic infection that can affect the central nervous system. Malaria that affects the brain can cause neurocognitive deficits in up to one quarter of cases; it is common in some geographic locations (e.g., sub-Saharan Africa) but rarer in other geographic locations (e.g., North America, Europe). Kihara, de Haan, et al. (2010) compared children exposed to malaria with aged-matched unexposed children using a novelty-response paradigm similar to that described above. Children exposed to severe malaria had significantly smaller responses to novelty compared to unexposed children for both auditory and visual modalities. The overall pattern suggested that severe malaria affected prefrontal and temporal cortices normally activated by stimulus novelty.

General Summary and Conclusions

There is increasing interest in including broader global and culture samples in developmental cognitive neuroscience research. To some extent, there is already a history of looking at this in terms of certain skills such as speech processing and face processing among cultures. However, there remain many areas to be considered, and in global health settings.

Identifying common/universal patterns and the environmental factors that support these, and considering the factors driving differences, are two key approaches. It is encouraging that neuroimaging methods have been applied across global settings, though there is still much work to be done. Findings, particularly with respect to differences in environmental factors, are relevant to understanding theories of the development of brain–behavior relations. Global settings help to understand these relations beyond WEIRD samples.

Key Issues for Discussion

- What factors might contribute to common patterns in neurocognitive development across cultures and global settings, and which to differences? Is there overlap in these factors?
- What are the strengths and weaknesses of using neuroimaging in global settings?
- How do theories of brain development inform study in global/cross-cultural settings?
- What are the ethical challenges of global cognitive neuroscience research?
- What might the influence of the global COVID pandemic on brain–behavior relations be across global settings?

13

Toward an Integrated Developmental Cognitive Neuroscience

This final chapter ties together some themes of the book and points to future directions. First, we return to the three viewpoints on human functional brain development raised in Chapter 1. We contend that interactive specialization provides the best framework for understanding the majority of data currently available. Interactive specialization provides an account of developmental changes in the postnatal functional development of cortex: localization (the extent of cortex activated in a given task situation) and specialization (how finely tuned the functionality of a given cortical area is). Mechanisms underlying interactive specialization are reviewed and some of their functional consequences explored in more detail. To date, most research has investigated the emergence of functions in particular regions or areas. The major challenge for the next decade will be to reveal how interacting networks of regions emerge during postnatal development (network specialization). We review current attempts to understand the emergence of functional brain networks during development. The next issue discussed is the value of the molecular genetic analysis of cognitive development. It is suggested that the role of genes in functional development can only be adequately interpreted from within a developmental cognitive neuroscience approach. Next, we consider the value of neural network modeling that can make contact with both neural and cognitive data. This approach can also be extended to developmental disorders and to assessing the effects of developmental changes in levels of neuromodulators. General guidelines for developing theory within the field are proposed. Implications and applications of developmental cognitive neuroscience for issues of societal and educational importance are presented. Finally, we stress the need for applying multiple methods to cognitive transitions in development and for training a future generation of developmental cognitive neuroscientists.

Introduction

The earlier chapters of this book have introduced the reader to the field of developmental cognitive neuroscience. While there is a way to go to resolve debates and unanswered questions in the field, the extent to which previously disparate fragments of information are coming together encourages optimism for the future. In this final chapter we draw together some overall conclusions and make recommendations for future studies which may help progress the field.

Three Viewpoints on Human Functional Brain Development

In Chapter 1 we discussed three different perspectives on human functional brain development: the maturational view, the skill learning view, and interactive specialization (IS). In subsequent chapters, a number of themes emerged from our review of several domains of cognitive development. From our review of the available evidence on cortical development in Chapter 4 we concluded that while large-scale regions of cerebral cortex show graded differential patterns of gene expression, many of the small-scale functional areas of interest to the cognitive neuroscientist (such as number, language-related, and face areas) require activity-dependent processes for their specialization to emerge. However, the nature and location of these emerging functional specializations are heavily constrained by intrinsic brain connectivity, architectural and timing factors, and the nature of the environment that the child experiences.

In several previous chapters evidence from human infants and children was consistent with the IS framework. In some domains examined there was evidence that neural representations underpinning cognitive functions emerged postnatally, and partially under the influence of the nature of the information in the domain under investigation. In some cases where there was evidence of specific "prior" information about the world in the newborn, such as the preferential responses to face patterns (Chapter 7), subcortical circuits appeared to play an important contributory role in controlling the behavior. Thus, while the current state of the evidence does not allow definitive conclusions, of the three perspectives on functional brain development discussed in this book, the IS approach appears most consistent with the majority of data currently available.

Table 13.1 illustrates some of the assumptions that underlie the three approaches outlined earlier. It is a defining feature of the maturational approach that it assumes deterministic epigenesis; region-specific gene expression is assumed to effect changes in intra-regional connectivity that, in turn, allow new functions to emerge. A related assumption commonly made within the maturational approach is that there is a one-to-one mapping between brain and cortical regions and particular cognitive functions, such that specific computations come "on-line" following that maturation of circuitry intrinsic to the corresponding cortical region. In some respects, this view parallels "mosaic" development at the cellular level in which simple organisms (such as *Caenorhabditis elegans*) are constructed through cell lineages that are largely independent of each other (Elman et al., 1996). Similarly, different cortical regions are assumed to have different maturational timetables, thus enabling new cognitive functions to emerge in a relatively isolated fashion at different ages.

In contrast to the maturational approach, IS (Johnson, 2001, 2002, 2011) has a number of different underlying assumptions. Specifically, a probabilistic epigenesis assumption is coupled with the view that cognitive functions are the emergent product of interactions between different brain regions. With regard to the latter of these assumptions, IS follows trends in adult functional neuroimaging. It is no longer considered that cortical regions function as isolated "modules"; leading authors point out that it may be an error to assume that

Table 13.1 Three Viewpoints on Human Functional Brain Development

	Brain-cognitive mapping	Primary locale of changes	Plasticity	Cause
Maturational	One-to-one mapping Static over development	Intra-region connectivity matures	A specialized mechanism invoked by stroke or injury	Brain changes cause cognitive development
Skill learning	Changes during acquisition of skill		Life-long; no clear sensitive period	
Interactive specialization	Networks/neural systems	Inter-regional connectivity changes and shapes intra-regional connectivity	An inherent property; the state of having not yet specialized	Bidirectional relations between structure and function
	Dynamic changes during development		Sensitive; period determined by state of specialization	

particular functions can be localized within a certain cortical region (Friston & Price, 2001). Rather, the response properties of individual regions are determined by their patterns of connectivity to other regions, as well as by their current activity states. By this view, "the cortical infrastructure supporting a single function may involve many specialized areas whose union is mediated by the functional integration among them" (Friston & Price, 2001, p. 276). Similarly, in discussing the design and interpretation of adult functional magnetic resonance imaging (fMRI) studies, Carpenter and collaborators have argued that: "In contrast to a localist assumption of a one-to-one mapping between cortical regions and cognitive operations, an alternative view is that cognitive task performance is subserved by large-scale cortical networks that consist of spatially separate computational components, each with its own set of relative specializations, that collaborate extensively to accomplish cognitive functions" (Carpenter et al., 2001, p. 360). Extending these ideas to development, the IS approach emphasizes changes in inter(between)-regional connectivity, as opposed to the maturation of intra(within)-regional circuitry. Specifically, inter-regional connectivity influences intra-regional connectivity to a large extent, including the shaping of smaller-scale areas. While the maturational approach may be analogous to mosaic cellular development, the IS view corresponds to the "regulatory" development seen in higher organisms in which cell–cell interactions are critical in determining developmental fate. While mosaic development can be faster than regulatory in building simple animals, the latter has several advantages. Namely, regulatory development is more flexible and better able to respond to damage, and it is much more efficient in terms of genetic coding. In regulatory development genes need only orchestrate cellular-level interactions to yield more complex structures (see Elman et al., 1996).

As well as the mapping between structure and function at one age, we can also consider how this mapping might change during development. When discussing functional imaging of developmental disorders, many researchers have assumed that the relation between brain structure and cognitive function is unchanging during development. Specifically, in accordance with a maturational view, when new structures come on-line, the existing (already mature) regions continue to support the same functions that they did at earlier developmental stages. The "static assumption" is partly why it is acceptable to study developmental disorders in adulthood and then extrapolate back in time to early development. Contrary to this view, the IS viewpoint suggests that when a new computation or skill is acquired, there is a reorganization of interactions between different brain structures and regions (Johnson, 2001, 2011). This reorganization process could even change how previously acquired cognitive functions are represented in the brain. Thus, the same behavior could potentially be supported by different neural substrates at different ages during development.

Stating that structure–function relations can change with development is all very well, but it lacks the specificity required to make all but the most general predictions. Fortunately, the view that there is competitive specialization of regions during development gives rise to expectations about the types of changes in structure–function relations that should be observed. Specifically, as regions become increasingly selective in their response properties during infancy, patterns of cortical activation during behavioral tasks may therefore change dynamically.

The basic assumption underlying the skill learning approach is that there is a continuity of the circuitry underlying skill acquisition from birth through to adulthood. This circuit is likely to involve a network of structures that retains the same basic function across developmental time (a static brain-cognition mapping). However, other brain regions may respond to training with dynamic changes in functionality similar or identical to those hypothesized within the IS framework. Another way in which the skill learning view differs from the other perspectives is with regard to the notion of "plasticity."

Plasticity in brain development is a phenomenon that has generated much controversy, with several different conceptions and definitions having been presented (Thomas & Johnson, 2008). The three perspectives we have discussed provide different viewpoints on plasticity. According to the maturational framework, plasticity is a specialized mechanism that is activated following brain injury. According to the IS approach, plasticity is simply the state during which a region's function is not yet fully specialized. That is, there is still remaining scope for developing more finely tuned responses. This definition corresponds well to the view of developmental biologists that development involves the increasing "restriction of fate." Finally, according to the skill learning hypothesis, plasticity is at least the result of specific circuitry that remains in place throughout the life span. This hypothesis, unlike the IS approach, does not claim that plasticity necessarily reduces during development.

Interactive Specialization (IS)

IS specifically addresses two of the most fundamental issues in cognitive neuroscience: *localization* and *specialization*. In this context, *localization* refers to the extent to which a given function is associated with a region or area of cortex. Specifically, the extent (quantity) of cortex activated following the presentation of a given task or perceptual stimulus may change during development. *Specialization* refers to the degree of specificity of function of a given region or area of cortex. Functions may be finely tuned, such as an area that is activated only by a restricted category of visual objects or under a very narrow range of task circumstances, or broadly tuned, in that they are activated under a wide range of circumstances. According to the IS view, the issues of localization and specialization are two sides of the same coin, and both are consequences of the same common underlying mechanisms.

To briefly review some of the evidence consistent with this approach, in the development of face processing (Chapter 7) we discussed event-related potential (ERP) and behavioral evidence consistent with the idea of increasingly finely tuned cortical processing of faces (also referred to by Nelson [2003] as "perceptual narrowing"). For example, this narrowing process resulted in better recognition of non-human faces by younger infants (Pascalis et al., 2002). Along with these changes in specialization, fMRI work showed increasingly restricted localization of face processing when children were compared to adults in a face-matching task. Some similar findings were reported for reading acquisition in Chapter 11. For example, Schlaggar and McCandliss (2007) marshal evidence from neuroimaging and other sources to argue that the emergence of the cortical area specifically activated by reading words (the left hemisphere "visual word form area") occurs through the IS processes of increased specialization and localization, in close association with the child learning to read. In another example, from Chapter 5, we discussed evidence that changes in visual orienting abilities as assessed by fMRI revealed multiple sites of change throughout a network of different pathways involved in oculomotor control, and not just the activation of one or two "new" functional areas. This evidence is consistent with adjustments throughout a network of regions as it accommodates to a new function. In atypical development resulting from genetic disorders, structural and functional imaging has usually revealed widespread atypical patterns of activation, and changes in the extent of white matter (connectivity) (Johnson et al., 2002). These latter findings are consistent with processes of IS that have gone awry, or specialized differently in an attempt to compensate for an earlier deviation or deficit.

Further Reading   Johnson (2011); Johnson, Jones, & Gliga (2015).

In summary, according to the IS view, small-scale areas of cortex become tuned for certain functions as a result of a combination of factors, including (a) the suitability or otherwise of the biases within the large-scale region (e.g., transmitter types and levels, synaptic density, patterns of inter-regional connectivity, etc.), (b) the information within the sensory inputs (sometimes partly determined by other brain systems), and (c) competitive interactions with neighboring regions (so that functions are not redundant). Two significant challenges for IS going forward are to better specify the cellular and neural mechanisms that underlie the processes of specialization and localization, and to go from an understanding of regional specialization to that of whole networks of regions acting collectively to process information.

One of the cellular processes that may underlie IS is the marked postnatal loss of synaptic contacts within the cerebral cortex we reviewed in Chapter 4. This widespread pruning of synapses has led some authors to speculate on the functional consequence of this selective loss process (Changeux, 1985; Ebbesson, 1980; Edelman, 1987; Gazzaniga, 1983). One of the first "selectionist" accounts was provided by Jean-Pierre Changeux, who argued that connections between classes of neurons are prespecified, but are initially labile (plastic). Synapses in their labile state may either become stabilized or regress, depending on the total activity of the post-synaptic cell. The activity of the post-synaptic cell is, in turn, dependent upon its input. Initially, this input may be the result of spontaneous activity in the network, but rapidly it is evoked by input to the circuit. The critical concept here is that of *selective stabilization.* In short, Changeux and colleagues proposed that "to learn is to eliminate," as opposed to learning taking place by instruction or new growth.

Selectionist-type mechanisms of functional brain development vary along a number of dimensions. One of these dimensions is the scale of the unit selected. For example, Edelman (1987) proposed that particular "neuronal groups" are the unit of selection rather than single synapses (Crick, 1989, suggests that such neuronal groups will be composed of around 200–1,000 neurons, or a "mini-column"). The larger-scale cognitive level of selection posited by Changeux and Dehaene (1989) could be implemented as the dynamic selection of large-scale neural circuits or pathways according to particular task demands. Most likely, selection occurs at multiple scales and time courses.

A number of computer-based neural network modelers have investigated the functional consequences of different variations of link or connection loss (Barto et al., 1983; Jacobs, 2002; Kerszberg et al., 1992; Thomas & Johnson 2006; Thomas et al., 2011), including the interaction between activity-dependent link loss and overall network decreases in "trophic" factors (Shrager & Johnson, 1995). However, it is important to note that selective loss is only one of several aspects of postnatal neural development (see Sanes et al., 2006), and so a selectionist account alone can never provide a complete account of neurocognitive development.

Further Reading   Bourgeois (2001); Greenough et al. (2002); Sanes et al. (2006); Thomas & Johnson (2008).

As discussed above, the IS view predicts that cortical areas will gain increasing functional specialization during development. Another facet of this specialization may be that processing in any given area becomes increasingly isolated from that of neighboring regions, a process sometimes referred to as "parcellation" (Ebbesson 1984). Ebbesson (1984) argued that the increasing differentiation of the brain into separate processing streams and structures during both phylogeny and ontogeny occurs as a result of the selective loss of synapses and dendrites. The result of this parcellation process is the creation of "informationally encapsulated" processing streams and structures. At the cognitive level, such systems may be akin to "modules" in the adult mind/brain (Fodor, 1983).

One example of parcellation is the emergence of ocular dominance columns, discussed in Chapter 5. Recall that ocular dominance columns that received inputs exclusively from one eye emerged from a situation in which most layer 4 cells (in the primary visual cortex) received afferents from both eyes. The parcellation process would lead us to expect that the selective loss of afferents will result in components of the network becoming increasingly encapsulated and inaccessible to other parts. Held and his colleagues (see Held, 1993) tested this idea with the experiment described earlier in which they demonstrated that infants under 4 months had a form of integration between the two eyes not observed in older infants. The older infants did not show this integration between the eyes because by this age neurons in cortical layer 4 of primary visual cortex only receive inputs from one eye.

The example of ocular dominance column formation also demonstrates that parcellation (or segregation) at one level of processing can allow for more adaptive or accurate recombination at another level of processing. Thus, while parcellation may result in a loss of a certain level of integration between systems (e.g., binocular summation in the case of ocular dominance columns), it may be followed by the acquisition of a new level of integration (e.g., binocular vision in the current example).

Another example of increased segregation with development comes from the separation of color- and motion-processing channels in adult primates (Dobkins & Anderson, 2002). Specifically, in adults, brain pathways that encode direction of object motion do not also process color (e.g., Merigan & Maunsell, 1993). This dissociation of processing means that adults are surprisingly poor at detecting the direction of motion (as measured by direction-ally appropriate eye movements) of red- and green-striped color gratings. In contrast, infants of 2, 3, and 4 months were much better at detection than adults, suggesting that color input to motion processing is relatively stronger in the immature visual system. Furthermore, younger infants showed more evidence of this integration than did older ones, demonstrating a dynamic process of increasing segregation of these streams of information (Dobkins & Anderson, 2002).

The mechanisms underlying IS can also be investigated through computational neural network modeling. For example, several groups have used simple "cortical matrix" models to investigate the factors and mechanisms responsible for cortical specialization (e.g., Kerszberg et al., 1992; Oliver et al., 2000; Shrager & Johnson, 1995). In these artificial neural networks, connections between nodes are pruned according to variations of Hebbian learning: links between nodes that are often active together are strengthened, whereas links between nodes that are not often co-active get weaker and are pruned. In some of these models, the degree of pruning of connections during learning approximately matches that seen during the course of human brain development. During exposure to patterned input (roughly equivalent to sensory stimulation), nodes become more selective in their response properties, and under certain conditions clusters of nodes with similar response properties emerge. Thus, in these computational models selective pruning plays a role in the emergence of clusters of nodes (localization) that share common specific response properties (specialization) (see Figure 13.1). More recently, a new class of models, termed "generative network models," examine the growth of simulated neural networks based on the trade-off between minimizing the energy of brain wiring and maximization of their typology (Akarca et al., 2021). Such models can be anchored by the physical neuroanatomy observed in structural MRI at different age points, and as such offer a directly testable implementation of frameworks such as IS.

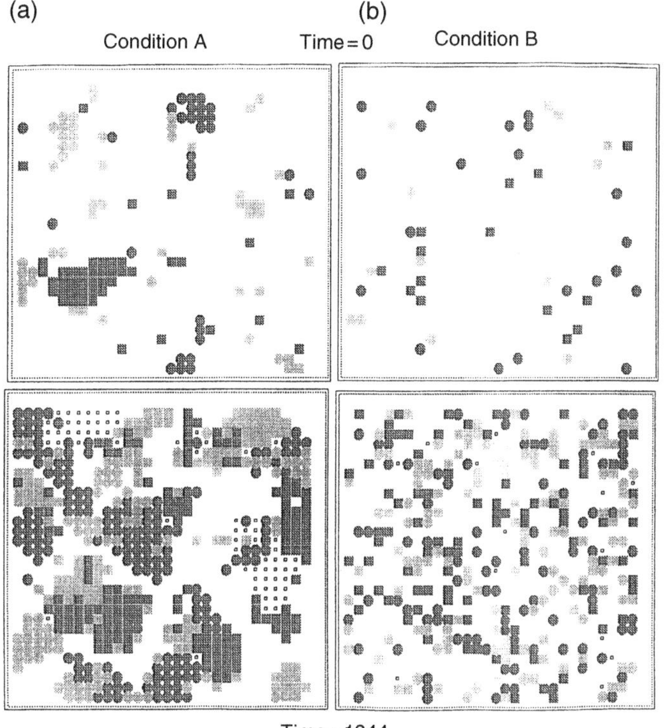

Figure 13.1 The formation of representations in the cortical matrix model under two different architectural conditions. The left upper panel shows the starting state and the left lower the final state. In the final state, "structured" representations emerge in which stimuli that have features in common tend to be clustered together (spatially aligned). With just a minor change in the architecture of the network (changing the relevant average lengths of intrinsic excitatory and inhibitory links: the right-hand side), nodes in the network fail to form structured clustered representations.

Emerging Networks

To date, the majority of the research on the emergence of specialized functions in human cortex has focused on specific regions. However, it is clear from the IS viewpoint that the next step is to understand how *networks* involving different regions, each with their own different specializations, emerge. In other words, while we are beginning to understand functional brain development at the level of individual cortical regions, we are still somewhat in the dark about how the larger scale of cortical function in terms of networks of regions develops (Johnson & Munakata, 2005). In this section we advance some initial evidence and speculation that addresses this intriguing issue.

Before considering the empirical evidence, we need to consider what makes a network of functional nodes more or less successful. A branch of mathematics called "graph theory" concerns itself with the relative efficiency of different kinds of networks (see Figure 13.2). While it may seem at first that a lattice or grid pattern is the best design for a network,

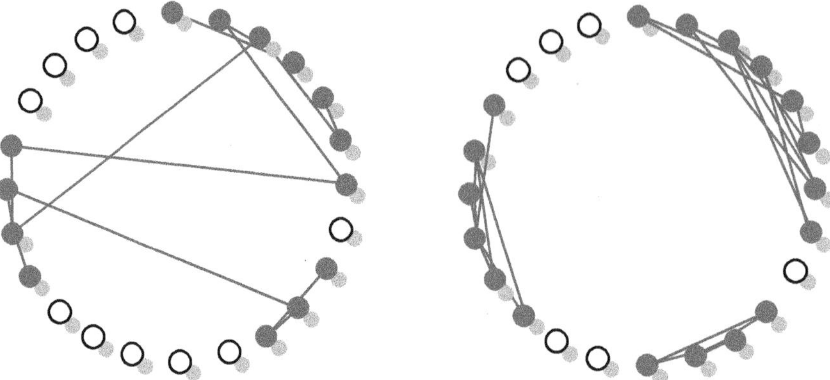

Figure 13.2 The illustration of different kinds of brain connectivity. The right-hand panel shows dense local connectivity without long-range connections. The left panel show the more optional arrangement that balances local connectivity with some long-range connectivity (a "small world") network.

formal analysis of measures of local network connectivity, and the average path length from one node to another, show that so-called "small world" networks are the most efficient. In contrast to the grid pattern of streets found in many American cities, small world networks are more like the clusters of small streets in a village that is then linked to other such villages by fast highways. Although the overall balance of the small local streets and highways can vary, most biological systems (and even the World Wide Web) are small world networks. Several studies have shown the regional interconnectivity of the adult brain is a highly efficient small world network, but how does this efficient network emerge?

The first piece of the jigsaw comes from work by Fair et al. (2007, 2009), who used functional connectivity analyses in fMRI to study resting state "control" networks in school-age children and adults. Their analysis allows them to infer the nature and strength of functional connections between 39 different cortical regions. They found that development entailed both *segregation* (i.e., decreased short-range connectivity) and *integration* (i.e., increased long-range connectivity) of brain regions that contribute to a network. In a similar study, the general developmental transition from more local connectivity to greater and stronger long-range network connectivity was confirmed using slightly different methods and 90 different cortical and subcortical regions (Supekar et al., 2009).

The decrease in short-range inter-regional functional connectivity is readily explicable in terms of the IS view. As neighboring regions of cortical tissue become increasingly specialized for different functions (e.g., objects versus faces), they will less commonly be co-activated. This process may also involve synaptic pruning and, as we heard in the last section, has been simulated in neural network models of cortex in which nodes with similar response properties cluster together spatially away from nodes with other response properties (Oliver et al., 1996). Thus, decreasing functional connectivity between neighboring areas of cortex is readily predicted by models implementing the IS view. More challenging from the current perspective is to account for the increase in long-range functional connections.

A maturational explanation of the increase in long-range functional connectivity would suggest that this increase is due to the establishment or strengthening of the relevant fiber bundles. However, the increase in functional connectivity during development may occur after the relevant long-range fiber bundles are in place (see Fair et al., 2009; Supekar et al., 2009 for discussion). While increased myelination is likely to be a contributory factor, (a) myelination itself can be a product of the activity/usage of a connection (Markham & Greenough, 2004) and (b) a general increase in myelin does not in itself account for the specificity of inter-regional activity into functional networks that support particular computations (but see Nagy et al., 2004). Thus, the strengthening and maintenance of long-range brain connections is likely to also be an activity-dependent aspect of brain development. This raises the question of why and how particular anatomically distant brain regions begin to cooperate in a functional network.

A key to answering this question may lie in scaling up the basic mechanisms of Hebbian learning. Instead of "cells that fire together wire together," we are seeing regions that tend to be co-activated in a given task context strengthening or maintaining the neural pathways between them. While each region is becoming individually specialized for a particular function, this intra-region change in tuning is modulated and influenced by its presence within an emerging network of co-activated structures (Akarca et al., 2021). For example, in a task that requires visually guided action, a variety of visual and motor areas will be co-activated along with multimodal integration areas. If the task is repeated sufficiently often then these patterns of co-activation will be strengthened, and specialization of individual regions will proceed within this context of overall patterns of activation.

A second source of co-activation in the developing human brain is commonly overlooked: spontaneous activity during the resting state (with no task demands). Although there has been great interest in the resting state or "default network" in adults, only more recently has this been studied using fMRI in children (although, as mentioned in Chapter 2, there is a long history of studying resting electroencephalography [EEG] in children). We learned earlier (Chapter 4) about the importance of spontaneous activity during prenatal development in shaping aspects of cortical structure and function. It seems likely that the oscillatory resting activity of the brain, which possibly occupies more waking hours than any specific tasks, may play a key role in strengthening and pruning the basic architecture of long-range connections.

A third reason why anatomically distant regions may strengthen and maintain their connectivity relates to the fact that most of the long-range functional connections studied by Fair et al. (2007) involved links to parts of the prefrontal cortex (PFC). As mentioned earlier (Chapter 10), this part of the cortex is generally considered to have a special role during development in childhood and skill acquisition in adults (Gilbert & Sigman, 2007; Thatcher, 1992). In earlier chapters, we reviewed a number of studies consistent with the idea that PFC may play a role in orchestrating the collective functional organization of other cortical regions during development. While there are several neural network models of PFC functioning in adults (e.g., O'Reilly, 2006), few if any of these have addressed development. However, another class of model intended to simulate aspects of development may be relevant both to PFC and to the issue of how networks of specialized regions come to coordinate their activity to support cognition. Knowledge-based cascade correlation (KBCC; Shultz et al., 2007) involves an algorithm and architecture that recruits previously learned functional networks when required during learning. Computationally, this dynamic neural network architecture has a number of advantages over other learning systems. Put simply, it can learn many tasks faster, or learn tasks that other networks cannot, because it can recruit the "knowledge" (computational abilities) of other self-contained networks as and when required. In a sense, it selects from a library of available computational systems to orchestrate the best combination for the learning problem at hand. While this class of model is not intended to be a detailed model of brain circuits (Shultz & Rivest, 2001; Shultz et al., 2007), it has been used to characterize frontal systems (Thivierge et al., 2005) and may capture important elements of the emerging interactions between PFC and other cortical regions at an abstract level. In addition, it offers initially attractive accounts of (a) why PFC is required for the acquisition of new skills, (b) why PFC is active from early in development but also shows prolonged developmental change, and (c) why early damage to PFC can have widespread effects over many domains.

Although much work remains to be done to understand in more detail the factors that lead to the emergence of long-range networks, graph theory analyses of changes during the school-age years are generating important insights. While, as described earlier, there are differences in the balance of short and long connections between children and adults, it is important to note that the network organization of children's brains is as efficient as that of adults (Yates et al., 2021). In other words, while children's brains are wired differently from those of adults (see Figure 13.3 in the color plate section), they are still optimally geared for the rapid and high-fidelity transmission of information. Whether the same is true in infancy and early childhood remains unknown.

Aside from the shift from local to long-range connectivity, another change in network structure observed using graph theory analysis during development is in the hierarchical structure. Adult networks have a more hierarchical structure that is optimally connected to support top-down relations between one part of the network and another (Supekar et al., 2009). While hierarchical networks have a number of computational advantages, to be discussed below, they are known to be less plastic and more vulnerable to damage or noise in the particular nodes at the top of the hierarchy. Thus, the network arrangement of children may be more flexible and plastic in response to unusual or atypical sensory input or environmental context. Further, the response to focal brain damage, particularly in the PFC (Chapter 10), may be more clearly understood in the light of these different network structures.

One of the features of a hierarchical network is the capacity for one region to feed back highly processed sensory or motor input to the earlier stages of processing. In much the same way as we hypothesized that lateral inter-regional interactions help shape the intrinsic connectivity of areas to result in functional specialization, interactions between regions connected by feedback and feed-forward connections may also help shape the specialization of the areas involved. Top-down effects play an important role in sensory information processing in the adult brain (e.g., Siegel et al., 2000). For example, during perception, information propagates through the visual processing hierarchy from primary sensory areas to higher cortical regions, while feedback connections convey information in the reverse direction. In a neurocomputational model of feedback in visual processing in the adult brain, Spratling and Johnson (2004) demonstrated that a number of different phenomena associated with visual attention, figure/ground segmentation, and contextual cueing could all be accounted for by a common mechanism underlying cortical feedback. Extending these ideas to development, there are potentially two important implications of feedback that will benefit from future exploration. The first of these will be to examine how the specialization of early sensory areas is shaped by top-down feedback, and vice versa, during development. The second topic for investigation will be to examine the consequences of relatively poor or diffuse cortical feedback in the immature cortex. For example, possibly some of the failures in object processing in infants (Chapter 6) can be explained by a lack of adequate top-down feedback.

Top-down feedback from PFC may also have a direct role in shaping the functional response properties of posterior cortical areas. In cellular recording studies from both humans and animals, evidence has accrued that the selectivity of response of neurons in areas such as the fusiform cortex may increase in real time following the presentation of a stimulus. For example, McCarthy and colleagues (McCarthy et al., 1997) measured local field potentials in face-selective regions of lateral fusiform cortex in human adults and found that responses of these neurons go from being face-selective at around 200 ms after stimulus presentation, to being face-identity- or emotion-selective at later temporal windows. This suggests that top-down cortical feedback pathways, in addition to their importance in attention and object processing (Spratling & Johnson, 2004, 2006), may increase the degree of specialization and localization in real time, as well as in developmental time. Thus, some of the changes in functional specialization and localization seen in face-sensitive regions may reflect the increasing influence of inter-regional coordination with other regions, including the PFC.

A final aspect of the transition from child brain network to the adult one is the greater connectivity between cortical and subcortical structures seen at younger ages (Supekar et al., 2009). This observation may be fundamental for our understanding of the emergence of the social brain (Chapter 7) and memory systems (Chapter 8) as it implies that the specialization of some cortical areas may be initially more dominated by structures such as the amygdala and hippocampus. As we approach adulthood, more networks become intrinsic to the cortex and develop a complex hierarchical structure more dominated by PFC.

Genes and Cognitive Development

In Chapter 3 we reviewed how molecular biological techniques will inevitably have a major impact on our understanding of genetic contributions to cognitive change. However, it is important to remember that genes do not "code for" functional components of cognition in any direct sense. We are never likely to discover a "gene for language" for the same reason as there is not a single gene for the big toe. Both language and big toes are the result of complex interactions between many genes, their products, and multiple levels of environments. Similarly, with regard to brain structure, it is likely that most relevant genes have widespread effects throughout several or all brain regions and, commonly, other organs (such as the heart). As attractive as the notion may seem at first sight, patterns of gene expression rarely neatly localize to small-scale functional areas of cerebral cortex (though there are examples related to larger scale structure-defined regions). Since brain structure is partly the product of complex self-organizing interactive processes, providing causal accounts of cognitive change in development purely in terms of genetics is certain to be inadequate. Rather, expression of particular genes will have to be located within a developmental cognitive neuroscience framework that includes some account of the interactions at molecular, cellular, and organism–environment levels. Ultimately, development is the path from genotype to phenotype, and no cognitive "function" for a gene can be attributed without providing some account of this mapping. In other words, attempting to explain a cognitive developmental change purely in terms of gene expression will miss out much that should be of core interest to the developmental cognitive neuroscientist.

Relations Between Brain Structure and Function in Development

Another issue that surfaced in several of the preceding chapters was the apparent discrepancy between developmental data on structural as compared to functional development. For example, with regard to the PFC, there is evidence of neuroanatomical changes occurring until the teenage years, yet infants as young as 6 months appear to pass some behavioral marker tasks of PFC functioning and show activation in functional imaging experiments. These findings are, however, only a problem for a causal epigenesis (maturational) view (see earlier) in which cause operates in a single direction from brain development to cognitive change. The data can be resolved by taking a probabilistic epigenesis view, in which there are two-way interactions between brain and cognitive development. In fact, there are several ways that they can be reconciled. One way is to invoke the graded development of representations in which input helps to tune the fine structure of a network over time (Chapter 10). Another is the view advanced by Thatcher (1992) and others that there are dynamic changes in the connectivity between cortical regions which lead to a reorganization of representations at several points in development. Finally, the IS view predicts that detailed structural changes in postnatal development will be partly a consequence of increasing functional specialization.

In this book we have made use of comparisons to computer-based neural network models. The use of these models does not necessarily commit one to an empiricist (behaviorist, Chapter 1) viewpoint on development, as some have suggested. Rather, as we have seen in this book, these models are excellent research tools for exploring interactions between intrinsic and extrinsic information and the emergence of representations, and for interpreting the functional consequence of partial or weak representations. When information about brain structure is added to the network architecture, neural network models can potentially provide a theoretical bridge between neurobiology and cognitive psychology (Akarca et al., 2021).

Further Reading   Mareschal et al. (2007).

Combining these two applications of connectionist modeling, we can devise models in which aspects of postnatal developmental neuroanatomy can be simulated, such as the selective loss of synapses, and the interacting effects of different variables studied. However, for such models to be useful they need to be pitched at an appropriately abstract level to make contact with both neural and cognitive data. Such models are only just beginning to be developed. Similarly, some network models of neural development are pitched at a detailed cellular level and thus do not allow any inferences about cognitive representations.

In order to study the possible functional consequence of developments in neural structure, we need models that make contact with both sets of data. Other types of non-linear models, such as those derived from dynamic systems theory (Thelen & Smith, 1994), may be useful in certain contexts, such as motor development. However, while these models tend to give good descriptions of the shape of development change in a given task or behavior, they neglect changes in representations (see Karmiloff-Smith & Johnson, 1994). Consequently, they may as yet be of limited utility for studying some types of cognitive transitions during development (but see Spencer et al., 2009).

Further Reading   Spencer et al. (2009).

One of the general assumptions that pervade this book is that an understanding of cognitive development will require consideration of computations of neurons and neural networks. Looking for mechanisms of cognitive change at this level does not imply a maturationist, empiricist, or reductionist viewpoint. It does, however, require a belief that a level of explanation closer to neural mechanisms will bring benefits in understanding. While this belief remains largely untested, it is no less plausible than the widely held assumption that cognitive models, unconstrained by neural evidence, can provide the best explanation of behavioral change.

Neuroconstructivism

We have argued that the IS approach to human functional brain development is likely to be more helpful in explaining developmental change than the maturational or skill learning views. IS is a specific example of a broader approach to brain and cognitive development sometimes called neuroconstructivism (Mareschal et al., 2007). As outlined in Chapter 1, constructivism, and its brain-based cousin neuroconstructivism, differs in several ways from more traditional approaches to behavioral and cognitive development. One difference is that the organism's interaction with its current external environment is taken into account. Taking this view of cognitive development changes the types of representations we consider in our models of cognitive development in two ways. First, it inclines the researcher to consider the entire neurocognitive path from (sensory) input to (motor) output. This is because the structure of the sensory information and the nature of the representations necessary for planning motor actions severely restrict the possible intermediate representations. In other words, there is a need for computer models that, while less detailed, attempt to capture the functioning of several brain systems within a given domain. The second, and related, difference is that taking into account the structure of the external environment in which the organism develops means that the information contained in mental representations can be relatively impoverished, but still be sufficient to produce adaptive behavior. In other words, the representation need only be as detailed as that which is sufficient for adaptive behavior in a given context. For example, in Chapter 7 we reviewed the hypothesis that the face representation possessed by newborn infants is merely a primitive "sketch" of the face (something like three blobs corresponding to the eyes and mouth) (Johnson & Morton, 1991a). While this is an impoverished representation of a face, it is sufficient for adaptive behavior given that faces engage this representation, and that the early environment of the infant is guaranteed to have faces present. Few naturally occurring stimuli (other than those generated by developmental psychologists for experiments!) are likely to engage the same representation.

Neuroconstructivism also provides a particular perspective on developmental disorders. The developmental disorders discussed in this book are those in which the most research has been done in attempting to link cognitive and neural deficits (see Chapter 2). However, even in these cases the mapping from brain to cognitive system remains far from clear. In some cases, multiple different brain regions have been implicated despite apparently focal cognitive deficits. A number of factors contribute to the complexity of understanding the brain basis of developmental disorders:

- If a region that develops relatively early in the course of brain development is atypical, then this is likely to have knock-on consequences for later developing regions. For example, if the projections from the thalamus to the cortex are atypical, then the subsequent parcellation of the cortex into areas will be disturbed. Further, it is likely that cognitive deficits result from a combination of primary and secondary neural atypicalities. Indeed, it is possible that the primary cognitive deficits arise from secondary neural effects.
- It is likely that the most devastating forms of brain injury in development are those that affect a major brain system, rather than, for example, a particular cortical area. In most cases of genetic atypicality, or insult early in gestation, the neural consequences are likely to be widespread. Thus, they are more likely to disturb whole brain systems than are focal lesions acquired in later life. Focal lesions commonly result in compensation at neural and cognitive levels in primate infants.
- It is plausible that some common developmental disorders result from widespread mild atypicalities in synaptic functioning occurring during prenatal and early postnatal development. Such atypical synaptic function may change the balance of excitatory and inhibitory influences on neural processing. Such disturbances to early development will be difficult to accommodate with neuropsychological analysis as applied to adults with acquired brain damage (Karmiloff-Smith, 1998). Indeed, many of the developmental consequences of such an event may be the result of brain adaptations. Importantly, a variety of different low-level disturbances to neural function may result in a common adaptive response, much in the same way as the body responds to a variety of different factors (viruses, bacterial infection, etc.) by raising its core temperature (fever). This common adaptive response may be why there are multiple different causal factors implicated in developmental disorders such as autism (Johnson, Jones, & Gliga, 2015).

Within the same general constructivist viewpoint we can also consider developmental disorders in terms of Waddington's intuitive conceptualization of developmental trajectories in the form of an epigenetic landscape (see Chapter 1; Figure 1.2). In the epigenetic landscape, a perturbation to the trajectory early in development (possibly corresponding to the prenatal period) will lead to an entirely different pathway (valley) being taken. However, a variety of different sources of perturbation could lead to the same alternative path being taken, and therefore the same behavioral phenotype could result. This corresponds to the fact that there are probably several sources of perturbation that can result in autistic symptoms, for example. Waddington's analysis would, however, predict that these perturbations occur around the same developmental stage. Perturbations to development that occur later in development when the organism is in a chreod (possibly corresponding to perinatal or early postnatal) are compensated for by self-regulatory (adaptive) processes which ensure that the same behavioral phenotype results. This may correspond to the effects of perinatal and early postnatal acquired cortical damage. As we have seen, if perinatal damage is limited to regions of the cortex, there is often a degree of functional compensation not seen in adults and older children. For example, lesions to the left temporal lobe early in life do not have a devastating effect on language acquisition (Chapter 9). The effects are subtle and affect other domains of cognition also. Thus, in these cases the injured brain adapts to preserve the patterns of specialization that arise from typical development. Obviously, major perturbations, such as the loss of most of the cortex, or prolonged rearing in darkness or social isolation, are likely to push the child into a different phenotype.

Criticisms of Developmental Cognitive Neuroscience

Now that development cognitive neuroscience has become established as an interdisciplinary field in its own right, it is time to consider and question the directions we are going in. One criticism leveled at our emerging field is that it is primarily being driven forward by the powerful new methods for imaging brain structure and function in an infant- and child-friendly way, in addition to new techniques for genetic analyses, and that it therefore lacks the theory-driven approach that characterizes much of the best work in the neighboring field of cognitive development. Similar concerns are expressed, albeit less directly, by students who can be daunted by the exciting but fragmentary islands of data we have acquired to date about human functional brain development. Where is the overarching theory or framework within which they can make sense of disparate observations? A related concern sometimes expressed by those in cognitive science is that the hypotheses which are presented in developmental cognitive neuroscience are reductionist, or are otherwise impoverished as a proper psychological explanation of infant or child behavior. In other words, this criticism is that what hypotheses and theories there are in the field are of the wrong type and do not offer a satisfactory explanation of behavioral change in development.

Starting with the criticism of a relative lack of theories in developmental cognitive neuroscience, we have to acknowledge that, at least compared to the parent discipline of cognitive development, work in developmental cognitive neuroscience generally tends to be less theory-driven (with some notable exceptions). Why is this? A large part of the explanation we believe to be due to the sudden increase in the volume and diversity of data from the new methods that have become available. Many cognitive theories that successfully accounted for sets of behavior observations in child development founder on the rocks when we also try to account for neuroscience data relating to the same behavioral tasks. For one thing, when you more than double the *quantity of data* to be accounted for, then many previously successful theories will no longer offer a satisfactory explanation, simply because the chance of observing refuting evidence is much higher. Bringing powerful new methods into a field is analogous to a catastrophic environmental change during evolution—the majority of species (theories) simply cannot adapt and therefore die off. It then takes generations for the better-adapted species to emerge.

This leads us to the second common criticism of theory in developmental cognitive neuroscience. This criticism is that the theories inspired by neural development are of the *wrong type* to be of relevance for explaining the development of human behavior. The view is sometimes expressed that theories in developmental cognitive neuroscience are reductionist and therefore do not offer good explanations of cognitive change. Cognition, it is argued following Marr (1982), is a level of explanation independent from the underlying neuroscience. A simple analogy sometimes given is that computer software could, in theory, be run on a variety of different hardware from microchips to wooden cogwheels. Recent directions in neuroscience suggest that, to the contrary, there is a large degree of interdependence between levels in real complex biological systems such as the brain. This has led to the proposal that we should be seeking theories that are *consistent* between different levels of explanation (Mareschal et al., 2007). Ultimately, theories that are consistent with both behavioral and brain development evidence will have greater explanatory power than those confined to one level of observable only.

In considering the issues above, the current dearth of plausible theories in developmental cognitive neuroscience seems less surprising. After all, new fields in the biological sciences (in contrast to some physical sciences) often go through a phase in which collection of basic data is the priority—a kind of natural history phase. However, given the importance of theory in directing hypothesis-driven empirical research programs, we offer three hallmarks of a good theory in developmental cognitive neuroscience.

1) The theory advanced should relate neural observations to behavioral ones, and be equally well tested (and refuted) by either brain- or behavioral-level observations. Although a variety of different types of theories could make this bridge, they are unlikely to resemble existing cognitive development theories based purely on behavioral observations.

2) Theories should center on mechanisms of change. This suggestion is not new (e.g., Mareschal & Thomas, 2007), but it is still surprisingly common to see theories that explain the state of affairs before and after a developmental transition but that do not specify the mechanisms of the transition itself (other than invoking the terms "maturation" or "learning"). Theories of development need to be focused on change.

3) Given that theories in developmental cognitive neuroscience are accounting for several levels of observable, and that they also need be compatible with undoubtedly complex and dynamic aspects of neural processing, we need to find ways to elucidate and present those theories so that they are both comprehensible and clarifying. This is the attraction and importance of computational modeling, be it symbolic, connectionist, or hybrid-style (see Mareschal et al., 2007). While theories may initially develop as informal ideas, ultimately we should aim to implement them as working computer models.

Finally, in the long term it is probably good for the field to have a heterogeneous mix of different types of theories and let the data, and time, select those with the best fit to reality. After all, despite their prolonged domination, the dinosaurs did not inherit the globe.

Applications of Developmental Cognitive Neuroscience

Developmental cognitive neuroscience has been identified by several national and international grant-funding agencies as one of the most significant growth areas in all of the neurosciences, partly because an understanding of human functional brain development has fundamental implications for social, educational, and clinical policies and strategy. In several chapters of this book we have heard how some developmental disorders can be characterized as a deviation from the typical trajectory of postnatal human brain development (Karmiloff-Smith, 1998). For example, according to the IS view, biases in attention and processing in early infancy are reinforced by differential patterns of experience subsequently resulting in the patterns of specialization observed in adults. Thus, some adult patterns of cortical functional specialization are an inevitable consequence of interacting factors in typical development (Johnson, 2001). A corollary to this is that disruption of one or more of these factors can lead to failure to develop the typical degree or patterns of cortical specialization. Further, an initial slight deviance can lead to increasingly compounded patterns of deviance as others around the affected individual alter their own natural behavior. However, the positive side of the IS view of atypical development is that remediation strategies may be effective in alleviating some symptoms in at least some cases as long as it is started early in life, before the compounding of symptoms into a more complex syndrome. Thus, on the agenda for the next decade will be attempts to intervene in the course of development before major symptoms in some disorders (such as autism) have become compounded.

Further Reading   Green (2017).

While in many cases developmental disorders arise from atypical genetics, atypical early environments can also lead to adverse outcomes. Evidence from several sources indicates that childhood, and particularly infancy, within a low-income (low socioeconomic status, or SES) family context has a surprisingly strong relationship to later cognitive ability as measured by intelligence quotient (IQ) and school achievement (see Chapters 2, 11, and 12). While there are multiple correlated factors such as diet, parental drug abuse, and life stress, being raised in a poor family itself correlates with a variety of cognitive differences in language, working memory, spatial cognition, and attention. In particular, lower language abilities and specific differences in prefrontal and executive functions have been consistently identified (for review see Hackman & Farah, 2009). Governments have acknowledged how crucial the level of cognitive development during the early years is for the later educational and life outcomes (Allen & Duncan Smith, 2008). As a consequence of this, a number of pilot intervention programs have been funded, some of which involved weekly home visits by well-trained teachers and extra school hours, over periods of 1 to 3 years. Most of these programs proved that there are economic gains of early interventions, in terms of a decreased number of individuals requiring benefits or ending up in prison (Heckman, 2007). However, to date, few of these programs have been specifically designed based on current knowledge about cognitive and brain development.

Because most early intervention programs have been generic (e.g., targeting both cognitive and non-cognitive factors, like access to better health care) and have measured different outcomes at different ages, it is difficult to make a comparative analysis of their results and to asses which were the intervening factors that led to their success or otherwise. Nevertheless, a challenge for the application of developmental cognitive neuroscience over the next decade will be developing theoretically motivated and evidence-based early interventions for children raised in poverty.

Many of the issues related to environments apply both within and across cultures (e.g., poverty); other factors are more culture-specific (e.g., diseases that are prevalent that might affect the brain). Further study across different settings and cultures across the globe can help identify both more universal and more setting/culture-specific aspects of development in brain–behavior interactions. While there are challenges to applying developmental cognitive neuroscience methods across the globe, these are increasingly being overcome with advances in technology and research techniques (see Chapter 12).

Concluding Remarks

The past decades have seen developmental cognitive neuroscience grow from newborn to adolescent. The next steps will be dependent on successful collaboration between developmental neuroscientists, cognitive developmentalists, and computational modelers. New methods and new theoretical approaches will be equally important. In the longer term we will depend on the next generation, who will need some familiarity with all of these fields and will become a new generation of true developmental cognitive neuroscientists.

Key Issues for Discussion

- How would the three different viewpoints on human functional brain development account for the emergence of integrated functional networks of brain regions?
- How might graph theory analyses be applied to developmental disorders?
- What are the prospects for applying a neuroscience approach to societal issues such as poverty and education, and what major barriers to progress need to be overcome?
- What domains of cognition have yet to be approached by a developmental cognitive neuroscience approach, and why have these been neglected to this point?
- What new (or old) methods will be most useful for advancing developmental cognitive neuroscience in the future, and why?

References

Aamodt, C. M., Farias-Virgens, F., & White, S. A. (2020). Birdsong as a window into language origins and evolutionary neuroscience. *Philosophical Transactions of the Royal Society of London. Series B, Biological Sciences, 375*(1789), 20190060. https://doi.org/10.1098/rstb.2019.0060

Abrahams, B. S., & Geschwind, D. H. (2008). Advances in autism genetics: On the threshold of a new neurobiology. *Nature Reviews Genetics, 9*, 3410355.

Adams, R. B., Rule, N. O., Franklin, R. G., Wang, E., Stevenson, M. T., Yoshikawa, S., Nomura, M., Sato, W., Kveraga, K., & Ambady, N. (2010). Cross-cultural reading the mind in the eyes: An fMRI investigation. *Journal of Cognitive Neuroscience, 22*(1), 97–108.

Adlam, A. L., Vargha-Khadem, F., Mishkin, M., & de Haan, M. (2005). Deferred imitation of action sequences in developmental amnesia. *Journal of Cognitive Neuroscience, 17*, 240–248.

Adolphs, R. (2003). Cognitive neuroscience of human social behaviour. *Nature Reviews Neuroscience, 4*(3), 165–178.

Ahmed, H., Wilson, A., Mead, N., Noble, H., Richardson, U., Wolpert, M. A., & Goswami, U. (2020). An evaluation of the efficacy of GraphoGame Rime for promoting English phonics knowledge in poor readers. *Frontiers in Education, 5*, 132.

Akarca, D., Vértes, P. E., Bullmore, E. T., CALM team, & Astle, D. E. (2021). A generative network model of neurodevelopmental diversity in structural brain organization. *Nature Communications, 12*, 4216.

Akhtar, N., & Enns, J. T. (1989). Relations between covert orienting and filtering in the development of visual attention. *Journal of Experimental Child Psychology, 48*(2), 315–334.

Alexander, G. E., DeLong, M. R., & Strick, P. L. (1986). Parallel organization of functionally segregated circuits linking basal ganglia and cortex. *Annual Review of Neuroscience, 9*, 357–382.

Allen, G., & Duncan Smith, I. (2008). *Early intervention: Good parents, great kids, better citizens.* The Centre for Social Justice and the Smith Institute.

Amalric, M., & Dehaene, S. (2016). Origins of the brain networks for advanced mathematics in expert mathematicians. *Proceedings of the National Academy of Sciences of the United States of America, 113*(18), 4909–4017. https://doi.org/10.1073/pnas.1603205113

Developmental Cognitive Neuroscience: An Introduction, Fifth Edition. Michelle de Haan, Iroise Dumontheil, and Mark H. Johnson.
© 2023 John Wiley & Sons Ltd. Published 2023 by John Wiley & Sons Ltd.
Companion website: www.wiley.com/go/johnson/devneuro5e

Andersen, R. A., Batista, A. P., Snyder, L. H., Buneo, C. A., & Cohen, Y. E. (2000). Programming to look and reach in the posterior parietal cortex. In M. S. Gazzaniga (Ed.), *The new cognitive neurosciences* (2nd ed., pp. 515–524). MIT Press.

Andersen, R. A., & Zipser, D. (1988). The role of the posterior parietal cortex in coordinate transformations for visual-motor integration. *Canadian Journal of Physiology and Pharmacology, 66*, 488–501.

Anderson, S. W., Aksan, N., Kochanska, G., Damasio, H., Wisnowski, J., & Afifi, A. (2007). The earliest behavioural expression of focal damage to human prefrontal cortex. *Cortex, 43*, 806–816.

Andre, J., Picchioni, M., Zhang, R., & Toulopoulou, T. (2015). Working memory circuit as a function of increasing age in healthy adolescence: A systematic review and meta-analyses. *NeuroImage: Clinical, 12*, 940–948. https://doi.org/10.1016/j.nicl.2015.12.002

Ansari, D. (2008). Effects of development and enculturation on number representation in the brain. *Nature Reviews Neuroscience, 9*, 278–291.

Ansari, D., & Dhital, B. (2006). Age-related changes in the activation of the intraparietal sulcus during nonsymbolic magnitude processing: An event-related functional magnetic resonance imaging study. *Journal of Cognitive Neuroscience, 18*, 1820–1828.

Ansari, D., Garcia, N., Lucas, E., Hamon, K., & Dhital, B. (2005). Neural correlates of symbolic number processing in children and adults. *Neuroreport, 16*, 1769–1773.

Ansari, D., & Karmiloff-Smith, A. (2002). Atypical trajectories of number development. *Trends in Cognitive Sciences, 6*(12), 511–516.

Apperly, I. A. (2021). Cognitive basis of mindreading in middle childhood and adolescence. In R. T. Devine & S. Lecce (Eds.), *Theory of mind in middle childhood and adolescence: Integrating multiple perspectives* (pp. 37–54). Routledge.

Arnett, J. J. (2008). The neglected 95%: Why American psychology needs to become less American. *American Psychologist, 63*(7), 602–614. https://doi.org/10.1037/0003-066X.63.7.602

Arsalidou, M., Pawliw-Levac, M., Sadeghi, M., & Pascual-Leone, J. (2018). Brain areas associated with numbers and calculations in children: Meta-analyses of fMRI studies. *Developmental Cognitive Neuroscience, 30*, 239–250.

Aslin, R. N. (1981). Development of smooth pursuit in human infants. In D. F. Fisher, R. A. Monty, & J. W. Senders (Eds.), *Eye movements: Cognition and visual perception* (pp. 31–51). Erlbaum.

Aslin, R. N. (2007). What's in a look? *Developmental Science, 10*, 48–53.

Aslin, R. N., Clayards, M. A., & Bardhan, N. P. (2008). Mechanisms of auditory reorganization during development from sounds to words. In C. A. Nelson & M. Luciana (Eds.), *The handbook of developmental cognitive neuroscience* (2nd ed., pp. 97–116). MIT Press.

Aslin, R. N., Shukla, M., & Emberson, L. (2015). Hemodynamic correlates of cognition in human infants. *Annual Review of Psychology, 3*(66), 349–379.

Assari, S., Boyce, S., Bazargan, M., Thomas, A., Cobb, R. J., Hudson, D., Curry, T. J., Nicholson, H. L., Jr., Cuevas, A. G., Mistry, R., Chavous, T. M., Caldwell, C. H., & Zimmerman, M. A. (2021a). Parental educational attainment, the superior temporal cortical surface area, and reading ability among American children: A test of marginalization-related diminished returns. *Children, 8*(5), 412. https://doi.org/10.3390/children8050412

Assari, S., Boyce, S., Saqib, M., Bazargan, M., & Caldwell, C. H. (2021b). Parental education and left lateral orbitofrontal cortical activity during N-back task: An fMRI study of American adolescents. *Brain Sciences, 11*(3), 401. https://doi.org/10.3390/brainsci11030401

Atkinson, J. (1984). Human visual development over the first six months of life: A review and a hypothesis. *Human Neurobiology, 3*, 61–74.

Atkinson, J. (1998). The 'where and what' or 'who and how' of visual development. In F. Simion & G. Butterworth (Eds.), *The development of sensory, motor and cognitive capacities in early infancy: From perception to cognition* (pp. 3–20). Psychology Press.

Atkinson, J., & Braddick, O. (2003). Neurobiological models of normal and abnormal visual development. In M. de Haan & M. H. Johnson (Eds.), *The cognitive neuroscience of development* (pp. 43–71). Psychology Press.

Atkinson, J., & Braddick, O. (2020). Visual development. *Handbook of Clinical Neurology, 174*, 121–142.

Avidan, G., Hasson, U., Malach, R., & Behrmann, M. (2005). Detailed exploration of face-related processing in congenital prosopagnosia: Functional neuroimaging findings. *Journal of Cognitive Neuroscience, 17*, 1150–1167.

Bachevalier, J. (2008). Nonhuman primate models of memory development. In C. A. Nelson & M. Luciana (Eds.), *The handbook of developmental cognitive neuroscience* (2nd ed., pp. 499–508). MIT Press.

Bachevalier, J., Brickson, M., & Hagger, C. (1993). Limbic-dependent recognition memory in monkeys develops early in infancy. *Neuroreport, 4*, 77–80.

Bachevalier, J., & Mishkin, M. (1984). An early and a late developing system for learning and retention in infant monkeys. *Behavioral Neuroscience, 98*, 770–778.

Bachevalier, J., & Vargha-Khadem, F. (2005). The primate hippocampus: Ontogeny, early insult and memory. *Current Opinion in Neurobiology, 15*, 168–174.

Baillargeon, R. (1993). The object concept revisited: New directions in the investigation of infants' physical knowledge. In C. E. Granrud (Ed.), *Visual perception and cognition in infancy* (pp. 265–315). Lawrence Erlbaum.

Baird, A. A., Kagan, J., Gaudette, T., Walz, K. A., Hershlag, N., & Boas, D. A. (2002). Frontal lobe activation during object permanence: Data from near-infrared spectroscopy. *NeuroImage, 16*, 1120–1126.

Balaban, C. D., & Weinstein, J. M. (1985). The human pre-saccadic spike potential: Influences of a visual target, saccade direction, electrode laterality and instruction to perform saccades. *Brain Research, 347*, 49–57.

Ball, G., Aljabar, P., Zebari, S., Tusor, N., Arichi, T., Merchant, N., Robinson, E. C., Ogundipe, E., Rueckert, D., Edwards, A. D., & Counsell, S. J. (2014). Rich-club organization of the newborn human brain. *Proceedings of the National Academy of Sciences, 111*, 7456–7461.

Banerjee, S., Riordan, M., & Bhat, M. A. (2014). Genetic aspects of autism spectrum disorders: Insight from animal models. *Frontiers in Cellular Neuroscience, 8*, 58.

Banks, M. S., & Shannon, E. (1993). Spatial and chromatic visual efficiency in human neonates. In C. E. Granrud (Ed.), *Visual perception and cognition in infancy* (pp. 1–46). Lawrence Erlbaum.

Baron-Cohen, S. (1995). *Mindblindness: An essay on autism and theory of mind*. MIT Press.

Baron-Cohen, S., Leslie, A. M., & Frith, U. (1985). Does the autistic child have a "theory of mind"? *Cognition, 21*, 37–46.

Baron-Cohen, S., Leslie, A. M., & Frith, U. (1986). Mechanical, behavioural and intentional understanding of picture stories in autistic children. *British Journal of Developmental Psychology, 4*, 113–125.

Barth, H., Kanwisher, N., & Spelke, E. (2003). The construction of large number representations in adults. *Cognition, 86*(3), 201–221.

Barto, A. G., Sutton, R. S., & Anderson, C. W. (1983). Neuronlike adaptive elements that can solve difficult learning control problems. *Institute of Electrical Engineers Transactions on System, Man and Cybernetics, 15*, 835–846.

Bate, S., & Bennetts, R. (2015). The independence of expression and identity in face-processing: Evidence from neuropsychological case studies. *Frontiers in Psychology, 9*(6), 770.

Bates, E., Thal, D., & Janowsky, J. S. (1992). Early language development and its neural correlates. In I. Rapin & S. Segalowitz (Eds.), *Handbook of neuropsychology* (Vol. 7, pp. 2–62). Elsevier.

Bates, E. A., & Roe, K. (2001). Language development in children with unilateral brain injury. In C. A. Nelson & M. Luciana (Eds.), *The handbook of developmental cognitive neuroscience* (pp. 281–307). MIT Press.

Bathelt, J., Dale, N., & de Haan, M. (2017). Event-related potential response to auditory social sitmuli, parent-reported social communicative deficits and autism risk in school-aged children with congenital visual impairment. *Developmental Cognitive Neuroscience, 27*, 10–18.

Bathelt, J., Dale, N., de Haan, M., & Clark, C. (2020). Brain structure in children with congenital visual disorders and visual impairment. *Developmental Medicine and Child Neurology, 62*, 125–131.

Bathelt, J., O'Reilly, H., Clayden, J. D., Cross, J. H., & de Haan, M. (2013). Functional brain network organisation of children between 2 and 5 years derived from reconstructed activity of cortical sources of high-density EEG recordings. *NeuroImage, 15*(82), 595–604.

Batki, A., Baron-Cohen, S., Wheelwright, S., Connellan, J., & Ahluwalia, J. (2001). How important are the eyes in neonatal face perception? *Infant Behavior and Development, 23*, 223–229.

Batterink, L. J., Paller, K. A., & Reber, P. J. (2019). Understanding the neural bases of implicit and statistical learning. *Cognitive Science, 11*, 482–503.

Bauer, P. J. (2008). Toward a neuro-developmental account of the development of declarative memory. *Developmental Psychobiology, 50*, 19–31.

Bauer, P. J., Wiebe, S. A., Carver, L. J., Lukowski, A. F., Haight, J. C., Waters, J. M., & Nelson, C. A. (2006). Electrophysiological indexes of encoding and behavioural indexes of recall: Examining relations and developmental change late in the first year of life. *Developmental Neuropsychology, 29*, 293–320.

Bear, M. F., & Singer, W. (1986). Modulation of visual cortical plasticity by acetylcholine and noradrenaline. *Nature, 320*, 172–176.

Beauchamp, M. H., Thompson, D. K., Howard, K., Doyle, L. W., Egan, G. F., Inder, T. E., & Anderson, P. J. (2008). Preterm infant hippocampal volumes correlate with later working memory deficits. *Brain, 131*, 2986–2994.

Bechara, A., Damasio, A. R., Damasio, H., & Anderson, S. W. (1994). Insensitivity to future consequences following damage to human prefrontal cortex. *Cognition, 50*, 7–15.

Bechara, A., Damasio, H., Tranel, D., & Damasio, A. R. (1997). Deciding advantageously before knowing the advantageous strategy. *Science, 275*, 1293–1295.

Becker, L. E., Armstrong, D. L., Chan, F., & Wood, M. M. (1984). Dendritic development on human occipital cortex neurons. *Brain Research, 315*, 117–124.

Bednar, J. A., & Miikkulainen, R. (2003). Learning innate face preferences. *Neural Computation, 15*, 1525–1557.

Behrmann, M., & Avidan, G. (2005). Congenital prosopagnosia: Face blind from birth. *Trends in Cognitive Sciences, 9,* 180–187.

Bell, M.A. (1992a). *A not B task performance is related to frontal EEG asymmetry regardless of locomotor experience.* Paper presented at the Proceedings of the VIIIth International Conference on Infant Studies, Miami Beach, FL.

Bell, M.A. (1992b). *Electrophysiological correlates of object search performance during infancy.* Paper presented at the Proceedings of the VIIIth International Conference on Infant Studies, Miami Beach, FL.

Bell, M. A., & Fox, N. A. (1992). The relations between frontal brain electrical activity and cognitive development during infancy. *Child Development, 63*(5), 1142–1163.

Bell, M. A., & Wolfe, C. D. (2007). Brain reorganization from infancy to early childhood: Evidence from EEG power and coherence during working memory tasks. *Developmental Neuropsychology, 31,* 21–38.

Bellugi, U., Bihrle, A., Neville, H., Jernigan, T., & Doherty, S. (1992). Language, cognition and brain organization in a neurodevelopmental disorder. In M. Gunnar & C. Nelson (Eds.), *Developmental behavioral neuroscience* (pp. 201–232). Lawrence Erlbaum.

Bellugi, U., Poizner, H., & Klima, E. S. (1989). Language, modality and the brain. *Trends in Neurosciences, 12,* 380–388.

Ben-Ari, Y. (2002). Excitatory actions of GABA during development: The nature of the nurture. *Nature Reviews Neuroscience, 3*(9), 728–739.

Ben-Ari, Y., Khalilov, I., Kahle, K. T., & Cherubini, E. (2012). The GABA excitatory/inhibitory shift in brain maturation and neurological disorders. *The Neuroscientist, 18*(5), 467–486.

Benes, F. M. (1994). Development of the corticolimbic system. In G. Dawson & K. W. Fischer (Eds.), *Human behavior and the developing brain* (pp. 176–206). Guilford Press.

Benes, F. M. (2001). The development of prefrontal cortex: The maturation of neurotransmitter systems and their interactions. In C. A. Nelson & M. Luciana (Eds.), *The handbook of developmental cognitive neuroscience* (pp. 79–92). MIT Press.

Ben-Shachar, M., Dougherty, R. F., Deutsch, G. K., & Wandell, B. A. (2011). The development of cortical sensitivity to visual word forms. *Journal of Cognitive Neuroscience, 23*(9), 2387–2399.

Berenbaum, S. A., Moffat, S., Wisniewski, A., & Resnick, S. (2003). Neuroendocrinology: Cognitive effects of sex hormones. In M. de Haan & M. H. Johnson (Eds.), *The cognitive neuroscience of development* (pp. 207–236). Psychology Press.

Berman, S., & Friedman, D. (1995). The development of selective attention as reflected by event-related brain potentials. *Journal of Experimental Child Psychology, 59,* 1–31.

Berteletti, I., Prado, J., & Booth, J. R. (2014). Children with mathematical learning disability fail in recruiting verbal and numerical brain regions when solving simple multiplication problems. *Cortex, 57,* 143–155.

Best, J. R., & Miller, P. H. (2010). A developmental perspective on executive function. *Child Development, 81,* 1641–1660.

Bethlehem, R. A. I., Seidlitz, J., White, S. R., Vogel, J. W., Anderson, K. M., Adamson, C., Adler, S., Alexopoulos, G. S., Anagnostou, E., Areces-Gonzalez, A., Astle, D. E., Auyeung, B., Ayub, M., Bae, J., Ball, G., Baron-Cohen, S., Beare, R., Bedford, S. A., Benegal, V., ... Alexander-Bloch, A. F. (2022). Brain charts for the human lifespan. *Nature, 2022,* 1–11. https://doi.org/10.1038/s41586-022-04554-y

Bhide, A., Power, A. J., & Goswami, U. (2013). A rhythmic musical intervention for poor readers: A comparison of efficacy with a letter-based intervention. *Mind, Brain, and Education, 7*(2), 113–123.

Bialystok, E. (2017). The bilingual adaptation: How minds accommodate experience. *Psychological Bulletin, 143*, 233–262.

Birnbaum, R., & Weinberger, D. R. (2017). Genetic insights into the neurodevelopmental origins of schizophrenia. *Nature Reviews Neuroscience, 18*(12), 727–740.

Bishop, D. V. M. (1983). Linguistic impairment after hemidecortication for infantile hemiplegia? A reappraisal. *Quarterly Journal of Experimental Psychology, 35A*, 199–207.

Bishop, D. V. M. (1997). *Uncommon understanding: Development and disorders of language comprehension in children.* Psychology Press.

Blais, C., Jack, R. E., Scheepers, C., Fiset, D., & Caldara, R. (2008). Culture shapes how we look at face. *PLoS One, 3*(8), e3022.

Blakemore, S. J. (2008). The social brain in adolescence. *Nature Reviews Neuroscience, 9*(4), 267–277.

Blakemore, S.-J., & Choudhury, S. (2006). Development of the adolescent brain: Implications for executive function and social cognition. *The Journal of Child Psychology and Psychiatry, 47*, 296–312.

Blakemore, S.-J., den Ouden, H., Choudhury, S., & Frith, C. (2007). Adolescent development of the neural circuitry for thinking about intentions. *Social Cognitive and Affective Neuroscience, 2*, 130–139.

Blakemore, S. J., & Mills, K. L. (2014). Is adolescence a sensitive period for sociocultural processing? *Annual Review of Psychology, 65*, 187–207.

Blasi, A., Lloyd-Fox, S., Katus, L., & Elwell, C. E. (2019). fNIRS for tracking brain development in the context of global health projects. *Photonics, 6*(3), 89.

Blass, E. (1992). Linking developmental and psychobiological research. *Society for Research in Child Development Newsletter*, 3–10.

Blokland, G. A. M., de Zubicaray, G. I., McMahon, K. L., & Wright, M. J. (2012). Genetic and environmental influences on neuroimaging phenotypes: A meta-analytical perspective on twin imaging studies. *Twin Research and Human Genetics, 15*(3), 351–371.

Blokland, G. A. M., McMahon, K. L., Thompson, P. M., Martin, N. G., de Zubicaray, G. I., & Wright, M. J. (2011). Heritability of working memory brain activation. *Journal of Neuroscience, 31*(30), 10882–10890.

Bolger, D. J., Perfetti, C. A., & Schneider, W. (2005). Cross-cultural effect on the brain revisited: Universal structures plus writing system variation. *Human Brain Mapping, 25*(1), 92–104.

Bolhuis, J. J. (1991). Mechanisms of avian imprinting: A review. *Biological Reviews, 66*, 303–345.

Born, A. P., Rostrup, E., Miranda, M. J., Larsson, H. B. W., & Lou, H. C. (2002). Visual cortex reactivity in sedated children examined with perfusion MRI (FAIR). *Magnetic Resonance Imaging, 20*(2), 199–205.

Born, P., Rostrup, E., Leth, H., Peitersen, B., & Lou, H. C. (1996). Change of visually induced cortical activation patterns during development. *Lancet, 347*(9000), 543.

Bourgeois, J. P. (2001). Synaptogenesis in the necortex of the newborn: The ultimate frontier for individuation? In C. A. Nelson & M. Luciana (Eds.), *The handbook of developmental cognitive neuroscience* (pp. 23–34). MIT Press.

Bourne, S. V., Korom, M., & Dozier, M. (2022). Consequences of inadequate caregiving for children's attachment, neurobiological development and adaptive functioning. *Clinical Child and Family Psychology Review, 25*, 166–181.

Braddick, O., Atkinson, J., & Wattam-Bell, J. (2003). Normal and anomalous development of visual motion processing: Motion coherence and 'dorsal-stream vulnerability'. *Neuropsychologia, 41*, 1769–1784.

Braddick, O. J., Atkinson, J., Hood, B., Harkness, W., Jackson, G., & Vargha-Khadem, F. (1992). Possible blindsight in infants lacking one cerebral hemisphere. *Nature, 360*, 461–463.

Brain Development Cooperative Group. (2012). Total and regional brain volumes in a population-based normative sample from 4 to 18 years: The NIH MRI study of normal brain development. *Cerebral Cortex, 22*(1), 1–12.

Brannon, E. M., & Terrace, H. S. (2000). Representation of numerosities 1–9 by rhesus macaques (*Macaca mulatta*). *Journal of Experimental Psychology – Animal Behaviour Processes, 26*, 31–49.

Brodeur, D. A., & Boden, C. (2000). The effects of spatial uncertainty and cue predictability on visual orienting in children. *Cognitive Development, 15*, 367–382.

Brodmann, K. (1909). *Vergleichende Lokalisationslehre der Grosshirnrinde in ihren Prinzipien dargestellt auf Grund des Zellenbaues*. Barth.

Brodmann, K. (1912). Neue Ergebnisse über die vergleichende histologische Lokalisation der Grosshirnrinde mit besonderer Berücksichtigung des Stirnhirns. *Anatomischer Anzeiger (Suppl.), 41*, 157–216.

Bronson, G. W. (1974). The postnatal growth of visual capacity. *Child Development, 45*, 873–890.

Bronson, G. W. (1982). *The scanning patterns of human infants: Implications for visual learning*. Ablex.

Brooksbank, B. W. L., Atkinson, D. J., & Balasz, R. (1981). Biochemical development of the human brain: II. Some parameters of the GABAergic system. *Developmental Neuroscience, 1*, 267–284.

Bruinink, A., Lichtensteinger, W., & Schlumpf, M. (1983). Pre- and postnatal ontogeny and characterization of dopaminergic D2, serotonergic S2, and spirodecanone binding sites in rat forebrain. *Journal of Neurochemistry, 40*, 1227–1237.

Butterworth, B. (2006). *The mathematical brain*. Macmillan.

Butterworth, B., Varma, S., & Laurillard, D. (2011). Dyscalculia: From brain to education. *Science, 332*(6033), 1049–1053.

Butterworth, G., & Jarrett, N. (1991). What minds have in common is space: Spatial mechanisms serving joint visual attention in infancy. *British Journal of Developmental Psychology, 9*, 55–72.

Byers-Heinlein, K., Fennell, C. T., & Werker, J. F. (2013). The development of associative word learning in monolingual and bilingual infants. *Bilingualism: Language and Cognition, 16*(1), 198–205. https://doi.org/10.1017/S1366728912000417

Bystron, I., Blakemore, C., & Rakic, P. (2008). Development of the human cerebral cortex: Boulder Committee revisited. *Nature Reviews Neuroscience, 9*, 110–122.

Cahalane, D. J., Clancy, B., Kingsbury, M. A., Graf, E., Sporns, O., & Finlay, B. L. (2011). Network structure implied by initial axon outgrowth in rodent cortex: Empirical measurement and models. *PLoS One, 6*(1), e16113.

Cain, K., Oakhill, J., & Bryant, P. (2004). Children's reading comprehension ability: Concurrent prediction by working memory, verbal ability, and component skills. *Journal of Educational Psychology, 96*(1), 31–42.

Callaghan, T., Rochat, P., Lillard, A., Claux, M. L., Odden, H., Itakura, S., Tapanya, S., & Singh, S. (2005). Synchrony in the onset of mental-state reasoning: Evidence from five cultures. *Psychological Science, 16*(5), 378–384.

Calvo-Merino, B., Glaser, D. E., Grèzes, J., Passingham, R. E., & Haggard, P. (2005). Action observation and acquired motor skills: And fMRI study with expert dancers. *Cerebral Cortex, 15*, 1243–1249.

Cameron, J. L. (2001). Effects of sex hormones on brain development. In C. A. Nelson & M. Luciana (Eds.), *The handbook of developmental cognitive neuroscience* (pp. 59–78). MIT Press.

Canfield, R. L., & Haith, M. M. (1991). Young infants' visual expectations for symmetric and asymmetric stimulus sequences. *Developmental Psychology, 27*, 198–208.

Canfield, R. L., Smith, E. G., Brezsnyak, M. P., & Snow, K. L. (1997). Information processing through the first year of life: A longitudinal study using the visual expectation paradigm. *Monographs of the Society for Research in Child Development, 62*(2), 1–145.

Cantlon, J. F., Brannon, E. M., Carter, E. J., & Pelphrey, K. A. (2006). Functional imaging of numerical processing in adults and 4-y-old children. *PLoS Biology, 4*, e125.

Carey, S. (2001). Bridging the gap between cognition and developmental neuroscience: The example of number presentation. In C. A. Nelson & M. Luciana (Eds.), *The handbook of developmental cognitive neuroscience* (pp. 415–432). MIT Press.

Carey, S. (2009). The making of abstract concepts: A case study of natural number. In D. Mareschal (Ed.), *The making of abstract concepts* (pp. 265–294). Oxford University Press.

Carlin, J. D., & Calder, A. J. (2013). The neural basis of eye gaze processing. *Current Opinion in Neurobiology, 23*, 450–455.

Carpenter, P. A., Just, M. A., Keller, T., Cherkassky, V., Roth, J. K., & Minshew, N. (2001). Dynamic cortical systems subserving cognition: fMRI studies with typical and atypical individuals. In J. L. McClelland & R. S. Siegler (Eds.), *Mechanisms of cognitive development* (pp. 353–386). Lawrence Erlbaum.

Carroll, S. B. (2005). Evolution at two levels: On genes and form. *PLoS Biology, 3*, e245.

Carter, E. J., & Pelphrey, K. (2006). School-aged children exhibit domain-specific responses to biological motion. *Social Neuroscience, 1*, 396–411.

Carver, L. J., & Bauer, P. J. (1999). When the event is more than the sum of its parts: 9-month-olds' long-term ordered recall. *Memory, 7*, 147–174.

Carver, L. J., & Bauer, P. J. (2001). The dawning of a past: The emergence of long-term explicit memory in infancy. *Journal of Experimental Psychology General, 130*, 726–745.

Casanova, M. F., & Trippe, J., II (2006). Regulatory mechanisms of cortical maniar development. *Brain Research Reviews, 51*, 72–84.

Case, R. (1992). The role of the frontal lobes in the regulation of human development. *Brain and Cognition, 20*, 51–73.

Casey, B. J., Getz, S., & Galvan, A. (2008). The adolescent brain. *Developmental Review, 28*, 62–77.

Cassia, V. M., Turati, C., & Simion, F. (2004). Can a non-specific bias towards top-heavy patterns explain newborns' face preference? *Psychological Science, 5*(6), 379–383.

Cassotti, M., Ania, A., Osmont, A., Houdé, O., & Borst, G. (2014). What have we learned about the processes involved in the Iowa Gambling Task from developmental studies? *Frontiers in Psychology, 5*, 915. https://doi.org/10.3389/fpsyg.2014.00915

Castelhano, M. S., & Henderson, M. J. (2008). Stable individual differences across images in human saccadic eye movements. *Canadian Journal of Experimental Psychology, 62*, 1–14.

Cech, T. R., & Steitz, J. A. (2014). The noncoding RNA revolution – trashing old rules to forge new ones. *Cell, 157*(1), 77–94.

Centanni, T. M., Norton, E. S., Ozernov-Palchik, O., Park, A., Beach, S. D., Halverson, K., Gaab, N., & Gabrieli, J. D. E. (2019). Disrupted left fusiform response to print in beginning kindergartners is associated with subsequent reading. *NeuroImage: Clinical, 22*, 101715. https://doi.org/10.1016/j.nicl.2019.101715

Changeux, J.-P. (1985). *Neuronal man: The biology of mind*. Pantheon Books.

Changeux, J.-P., & Dehaene, S. (1989). Neuronal models of cognitive functions. *Cognition, 33*, 63–109.

Chapman, M. (1981). Dimensional separability or flexibility of attention? Age trends in perceiving configural stimuli. *Journal of Experimental Child Psychology, 31*, 332–349.

Chein, J., Albert, D., O'Brien, L., Uckert, K., & Steinberg, L. (2011). Peers increase adolescent risk taking by enhancing activity in the brain's reward circuitry. *Developmental Science, 14*(2), F1–F10. https://doi.org/10.1111/j.1467-7687.2010.01035.x

Chen, X., & French, D. C. (2008). Children's social competence in cultural context. *Annual Review of Psychology, 59*, 591–616.

Cheour-Luhtanen, M., Alho, K., Kujala, T., Sainio, K., Reinikainen, K., Renlund, M., Aitonen, O., Eorela, O., & Naatanen, R. (1995). Mismatch negativity indicates vowel discrimination in newborns. *Hearing Research, 82*, 53–58.

Chiang, W.-C., & Wynn, K. (2000). Infants' representation and tracking of multiple objects. *Cognition, 77*, 169–195.

Chien, S. H. (2011). No more top-heavy bias: Infants and adults prefer upright faces but not top-heavy geometric or face-like patterns. *Journal of Vision, 11*(6), 13. https://doi.org/10.1167/11.6.13

Chinello, A., Cattani, V., Bonfiglioli, C., Dehaene, S., & Piazza, M. (2013). Objects, numbers, fingers, space: Clustering of ventral and dorsal functions in young children and adults. *Developmental Science, 16*(3), 377–393.

Chiu, C. Y., Schmithorst, V. J., Brown, R. D., Holland, S. K., & Dunn, S. (2006). Making memories: A cross-sectional investigation of episodic memory encoding in childhood using fMRI. *Developmental Neuropsychology, 29*, 321–240.

Chiu, Y.-C., Huang, J. T., Duann, J. R., & Lin, C. H. (2018). Editorial: Twenty years after the Iowa Gambling Task: Rationality, emotion, and decision-making. *Frontiers in Psychology, 8*, 2353. https://doi.org/10.3389/fpsyg.2017.02353

Chorghay, Z., Káradóttir, R. T., & Ruthazer, E. S. (2018). White matter plasticity keeps the brain in tune: Axons conduct while glia wrap. *Frontiers in Cellular Neuroscience, 12*, 428. https://doi.org/10.3389/fncel.2018.00428

Chou, S.-J., Babot, Z., Leingärtner, A., Studer, M., Nakagawa, Y., & O'Leary, D. D. M. (2013). Geniculocortical input drives genetic distinctions between primary and higher-order visual areas. *Science, 340*, 1239–1242.

Chugani, H. T. (1994). Development of regional brain glucose metabolism in relation to behavior and plasticity. In G. Dawson & K. W. Fischer (Eds.), *Human behavior and the developing brain* (pp. 153–175). Guilford Press.

Chugani, H. T., Hovda, D. A., Villablanca, J. R., Phelps, M. E., & Xu, W. F. (1991). Metabolic maturation of the brain: A study of local cerebral glucose utilization in the developing cat. *Journal of Cerebral Blood Flow and Metabolism, 11*(1), 35–47.

Chugani, H. T., & Phelps, M. E. (1986). Maturational changes in cerebral function in infants determined by [18] FDG positron emission tomography. *Science, 231*, 840–843.

Chugani, H. T., Phelps, M. E., & Mazziotta, J. C. (1987). Positron emission tomography study of human brain functional development. *Annals of Neurology, 22*(4), 487–497.

Chugani, H. T., Phelps, M. E., & Mazziotta, J. C. (2002). Positron emission tomography study of human brain functional development. In M. H. Johnson, Y. Munakata, & R. Gilmore (Eds.), *Brain development and cognition: A reader* (2nd ed., pp. 101–116). Blackwell.

Clancy, B., Darlington, R. B., & Finlay, B. L. (2000). The course of human events: Predicting the timing of primate neural development. *Developmental Science, 3*(1), 57–66.

Cicchetti, D., & Cohen, D. J. (1995). Perspectives on developmental psychopathology. In D. Cicchetti & D. J. Cohen (Eds.), *Developmental psycholopathology, Vol. 1. Theory and methods* (pp. 3–20). John Wiley & Sons.

Clohessy, A. B., Posner, M. I., Rothbart, M. K., & Vecera, S. P. (1991). The development of inhibition of return in early infancy. *Journal of Cognitive Neuroscience, 3*, 345–350.

Cobos, I., Calcagnotto, M. E., Vilaythong, A. J., Thwin, M. T., Noebels, J. L., Baraban, S. C., & Rubenstein, J. L. R. (2005). Mice lacking Dlx1 show subtype-specific loss of interneurons, reduced inhibition and epilepsy. *Nature Neuroscience, 8*(8), 1059–1068.

Cohen Kadosh, K., Cohen Kadosh, R., Dick, F., & Johnson, M. H. (2011). Developmental changes in effective connectivity in the emerging core face network. *Cerebral Cortex, 21*(6), 1389–1394. https://doi.org/10.1093/cercor/bhq215

Cohen, L. B., & Marks, K. S. (2002). How infants process addition and subtraction events. *Developmental Science, 5*(2), 186–201.

Cohen, N. J., & Squire, L. R. (1980). Preserved learning and retention of pattern analyzing skill in amnesia: Dissociation of knowing how and knowing what. *Science, 210*, 207–209.

Cohen-Kadosh, K., & Johnson, M. H. (2007). Developing a cortex specialized for face perception. *Trends in Cognitive Science, 11*, 367–369.

Cohen-Kadosh, R., Cohen-Kadosh, K., Schuhmann, T., Kaas, A., Goebel, R., Henik, A., & Sack, A. T. (2007). Virtual dyscalculia induced by parietal lobe TMS impairs automatic magnitude processing. *Current Biology, 17*, 689–693.

Colantuoni, C., Lipska, B. K., Ye, T., Hyde, T. M., Tao, R., Leek, J. T., Colantuoni, E. A., Elkahloun, A. G., Herman, M. M., Weinberger, D. R., & Kleinman, J. E. (2011). Temporal dynamics and genetic control of transcription in the human prefrontal cortex. *Nature, 478*, 519–523.

Conboy, B. T., & Kuhl, P. K. (2011). Impact of second-language experience in infancy: Brain measures of first- and second-language speech perception. *Developmental Science, 14*(2), 242–248. https://doi.org/10.1111/j.1467-7687.2010.00973.x

Condry, K. F., Smith, W. C., & Spelke, E. S. (2001). Development of perceptual organization. In F. Lacerda, C. V. Hofsten, & M. Heimann (Eds.), *Emerging cognitive abilities in early infancy* (pp. 1–28). Erlbaum.

Condry, K. F., & Spelke, E. S. (2008). The development of language and abstract concepts: The case of natural number. *Journal of Experimental Psychology: General, 137*(1), 22–38. https://doi.org/10.1037/0096-3445.137.1.22

Conel, J. L. (1939–1967). *The postnatal development of the human cerebral cortex* (Vol. I–VIII). Harvard University Press.

Conger, R. D., & Donnellan, M. B. (2007). An interactionist perspective on the socioeconomic context of human development. *Annual Review of Psychology, 58*, 175–199.

Conte, S., Richards, J. E., Guy, M. W., Xie, W., & Roberts, J. E. (2020). Face-sensitive brain responses in the first year of life. *NeuroImage, 211*, 116602. https://doi.org/10.1016/j.neuroimage.2020.116602

Conway, M. A., Wang, Q. I., Hanyu, K., & Haque, S. (2005). A cross-cultural investigation of autobiographical memory on the universality and cultural variation of the reminiscence bump. *Journal of Cross-Cultural Psychology, 36*(6), 739–749.

Cooper, N. G. F., & Steindler, D. A. (1986). Lectins demarcate the barrel subfield in the somatosensory cortex of the early postnatal mouse. *Journal of Comparative Neurology, 249*, 157–169.

Cordes, S., & Brannon, E. M. (2008). Quantative competencies in infancy. *Developmental Science, 11*, 803–808.

Cornish, K., & Wilding, J. (2010). *Genes, cognition and early brain development.* Oxford University Press.

Courchesne, E., Yeung-Courchesne, R., Press, G. A., Hesselink, J. R., & Jernigan, T. L. (1988). Hypoplasia of cerebellar vermal lobules VI and VII in autism. *The New England Journal of Medicine, 318*, 1349–1354.

Coyle, J. T., & Molliver, M. (1977). Major innervation of newborn rat cortex by monoaminergic neurons. *Science, 196*, 444–447.

Craft, S., White, D. A., Park, T. S., & Figiel, G. (1994). Visual attention in children with perinatal brain injury: Asymmetric effects of bilateral lesions. *Journal of Cognitive Neuroscience, 6*, 165–173.

Cragg, L., Keeble, S., Richardson, S., Roome, H. E., & Gilmore, C. (2017). Direct and indirect influences of executive functions on mathematics achievement. *Cognition, 162*, 12–26.

Crick, F. (1989). Neural Edelmanism. *Trends in Neurosciences, 12*, 240–248.

Cromer, R. E. (1992). A case study of dissociation between language and cognition. In H. Tager-Flusberg (Ed.), *Constraints on language acquisition: Studies of atypical children* (pp. 141–153). Lawrence Erlbaum.

Crone, E. A., & Dahl, R. E. (2012). Understanding adolescence as a period of social-affective engagement and goal flexibility. *Nature Reviews Neuroscience, 13*(9), 636–650.

Crone, E. A., & Steinbeis, N. (2017). Neural perspectives on cognitive control development during childhood and adolescence. *Trends in Cognitive Sciences, 21*(3), 205–215. https://doi.org/10.1016/j.tics.2017.01.003

Crone, E. A., & van Duijvenvoorde, A. C. K. (2021). Multiple pathways of risk taking in adolescence. *Developmental Review, 62*, 100996.

Crone, E. A., & Westenberg, P. M. (2009). A brain-based account of developmental changes in social decision making. In M. de Haan & M. R. Gunnar (Eds.), *Handbook of developmental social neuroscience* (pp. 378–398). The Guilford Press.

Crystal, J. D., & George Wilson, A. (2015). Prospective memory: A comparative perspective. *Behavioural Processes, 112*, 88–99.

Csibra, G. (2003). Teleological and referential understanding of action in infancy. *Philosophical Transactions of the Royal Society of London. Series B, Biological Sciences, 358*, 447–458.

Csibra, G. (2007). Action mirroring and action interpretation: An alternative account. In P. Haggard, Y. Rosetti, & M. Kawato (Eds.), *Sensorimotor foundations of higher cognition. Attention and performance XXII* (pp. 435–459). Oxford University Press.

Csibra, G., Davis, G., Spratling, M. W., & Johnson, M. H. (2000). Gamma oscillations and object processing in the infant brain. *Science, 290*, 1582–1585.

Csibra, G., & Johnson, M. H. (2007). Investigating event-related oscillations in infancy. In M. de Haan (Ed.), *Infant EEG and event-related potentials* (pp. 289–304). Psychology Press.

Csibra, G., Johnson, M. H., & Tucker, L. A. (1997). Attention and oculomotor control: A high-density ERP study of the gap effect. *Neuropsychologia, 35*(6), 855–865.

Csibra, G., Tucker, L. A., & Johnson, M. H. (1998). Neural correlates of saccade planning in infants: A high-density ERP study. *International Journal of Psychophysiology, 29*, 201–215.

Csibra, G., Tucker, L. A., & Johnson, M. H. (2001). Differential frontal cortex activation before anticipatory and reactive saccades in infants. *Infancy, 2*(2), 159–174.

Dale, N., & Sonksen, P. (2002). Developmental outcome, including setback, in young children with severe visual impairment. *Developmental Medicine and Child Neurology, 44*(9), 613–622.

Darki, F., & Klinberg, T. (2014). The role of fronto-parietal and fronto-striatal networks in the development of working memory: A longitudinal study. *Cerebral Cortex, 25*(6), 1587–1595. https://doi.org/10.1093/cercor/bht352

Dawson, G., Carver, L. J., Meltzoff, A. N., Panagiotides, H., McPartland, J., & Webb, S. J. (2002). Neural correlates of face and object recognition in young children with autism spectrum disorder, developmental delay, and typical development. *Child Development, 73*, 700–717.

De Bie, H. M. A., Boersma, M., Wattjes, M. P., Adriaanse, S., Vermeulen, R. J., Oostrom, K. J., Huisman, J., Veltman, D. J., & Delemarre-Van De Waal, H. A. (2010). Preparing children with a mock scanner training protocol results in high quality structural and functional MRI scans. *European Journal of Pediatrics, 169*(9), 1079–1085.

de Boysson-Bardies, B., de Schonen, S., Jusczyk, P., McNeilage, P., Morton, J., & (Eds.). (1993). *Developmental neurocognition: Speech and face processing in the first year of life.* Kluwer.

de Haan, M. (2008). Neurocognitive mechanisms for the development of face processing. In C. A. Nelson & M. Luciana (Eds.), *The handbook of developmental cognitive neuroscience* (2nd ed., pp. 509–520). MIT Press.

de Haan, M. (2014). Neuroscientific methods with children. In W. Overton, P. Molenaar, & R. Lerner (Eds.), *Handbook of child psychology: Relational, developmental systems theories and methods* (7th ed., Vol. 1, pp. 683–712). Wiley.

de Haan, M., & Johnson, M. H. (2003). *The cognitive neuroscience of development.* Psychology Press.

de Haan, M., Johnson, M. H., & Halit, H. (2003). Development of face-sensitive event-related potential components during infancy. *International Journal of Psychophysiology, 51*, 45–58.

de Haan, M., Mishkin, M., Baldeweg, T., & Vargha-Khadem, F. (2006). Human memory development and its dysfunction after early hippocampal injury. *Trends in Neurosciences, 29*, 374–381.

de Haan, M., Pascalis, O., & Johnson, M. H. (2002). Specialization of neural mechanisms underlying face recognition in human infants. *Journal of Cognitive Neuroscience, 14*, 199–209.

De La Vega, F. M., & Bustamante, C. D. (2018). Polygenic risk scores: A biased prediction? *Genome Medicine, 10*(1), 1–3.

DeBeni, R., Palladino, P., Pazzaglia, F., & Cornoldi, C. (1998). Increases in intrusion errors and working memory deficits of poor comprehenders. *Quarterly Journal of Experimental Psychology, 51*, 305–320.

DeGutis, J. M., Bentin, S., Robertson, L. C., & D'Esposito, M. (2007). Functional plasticity in ventral temporal cortex following cognitive rehabilitation of a congenital prosopagnosic. *Journal of Cognitive Neuroscience, 19*, 1790–1802.

Dehaene, S. (1997). *The number sense: How the mind creates mathematics.* Oxford University Press.

Dehaene, S., Spelke, E., Pinel, P., Stanescu, R., & Tsivkin, S. (1999). Sources of mathematical thinking: Behavioural and brain-imaging evidence. *Science, 284*, 970–974.

Dehaene-Lambertz, G. (2017). The human infant brain: A neural architecture able to learn language. *Psychonomic Bulletin & Review, 24*, 48–55.

Dehaene-Lambertz, G., & Dehaene, S. (1994). Speed and cerebral correlates of syllable discrimination in infants. *Nature, 370*, 292–295.

Dehaene-Lambertz, G., Dehaene, S., & Hertz-Pannier, L. (2002). Functional neuroimaging of speech perception in infants. *Science, 298*(5600), 2013–2015.

Dehay, C., & Kennedy, H. (2007). Cell-cycle control and cortical development. *Nature Reviews Neuroscience, 8*, 438–450.

Dehay, C., & Kennedy, H. (2009). Transcriptional regulation and alternative splicing make for better brains. *Neuron, 62*, 455–457.

Dekker, T., Mareschal, D., Sereno, M. O., & Johnson, M. H. (2011). Dorsal and ventral stream activation and object recognition performance in school-aged children. *NeuroImage, 57*, 659–670.

Delgado-Reyes, L., Wijeakumar, S., Magnotta, V. A., Forbes, S. H., & Spencer, J. H. (2020). The functional brain networks that underlie visual working memory in the first two years of life. *Neuroimage, 219*, 116971. https://doi.org/10.1016/j.neuroimage.2020.116971

Dennis, M., & Whitaker, H. (1976). Language acquisition following hemidecortication: Linguistic superiority of the left over the right hemisphere. *Brain and Language, 3*, 404–433.

Dennison, M., Whittle, S., Yücel, M., Vijayakumar, N., Kline, A., Simmons, J., & Allen, N. B. (2013). Mapping subcortical brain maturation during adolescence: Evidence of hemisphere- and sex-specific longitudinal changes. *Developmental Science, 16*(5), 772–791.

Diamond, A. (1985). Development of the ability to use recall to guide action, as indicated by infants' performance on AB. *Child Development, 56*, 868–883.

Diamond, A. (1991). Neuropsychological insights into the meaning of object concept development. In S. Carey & R. Gelman (Eds.), *The epigenesis of mind: Essays on biology and cognition* (pp. 67–110). Lawrence Erlbaum.

Diamond, A. (2001). Looking closely at infants' performance and experimental procedures in the A-not-B task. *The Behavioral and Brain Sciences, 24*, 38–41.

Diamond, A. (2011). Biological and social influences on cognitive control processes dependent on prefrontal cortex. In O. Braddick, J. Atkinson, & G. Innocenti (Eds.), *Progress in brain research* (Vol. 189, pp. 319–339). Elsevier.

Diamond, A. (2013). Executive functions. *Annual Review of Psychology, 64*(1), 135–168.

Diamond, A., & Doar, B. (1989). The performance of human infants on a measure of frontal cortex function, the delayed response task. *Developmental Psychobiology, 22*(3), 271–294.

Diamond, A., & Goldman-Rakic, P. S. (1986). Comparative development of human infants and infant rhesus monkeys of cognitive functions that depend on prefrontal cortex. *Neuroscience Abstracts, 12,* 274.

Diamond, A., & Goldman-Rakic, P. S. (1989). Comparison of human infants and infant rhesus monkeys on Piaget's AB task: Evidence for dependence on dorsolateral prefrontal cortex. *Experimental Brain Research, 74,* 24–40.

Diamond, A., & Ling, D. S. (2015). Conclusions about interventions, programs, and approaches for improving executive functions that appear justified and those that, despite much hype, do not. *Developmental Cognitive Neuroscience, 18,* 34–48.

Diamond, A., Werker, J. F., & Lalonde, C. (1994). Toward understanding commonalities in the development of object search, detour navigation, categorization and speech perception. In G. Dawson & K. W. Fisher (Eds.), *Human behavior and the developing brain* (pp. 380–426). Guilford Press.

Diamond, A., Zola-Morgan, S., & Squire, L. R. (1989). Successful performance by monkeys with lesions of the hippocampal formation on AB and object retrieval, two tasks that mark developmental changes in human infants. *Behavioral Neuroscience, 103*(3), 526–537.

Diebler, M. F., Farkas-Bargeton, E., & Wehrle, R. (1979). Developmental changes of enzymes associated with energy metabolism and synthesis of some neurotransmitters in discrete areas of human neorcortex. *Journal of Neurochemistry, 32,* 429–435.

Dobkins, K. R., & Anderson, C. M. (2002). Color-based motion processing is stronger in infants than in adults. *Psychological Science, 13*(1), 75–79.

Donati, G., Dumontheil, I., Pain, O., Asbury, K., & Meaburn, E. L. (2021). Evidence for specificity of polygenic contributions to attainment in English, maths and science during adolescence. *Scientific Reports, 11*(1), 3851. https://doi.org/10.1038/s41598-021-82877-y

Donati, G., & Meaburn, E. (2020). What has behavioural genetic research told us about the origins of individual differences in educational abilities and achievements? In M. S. C. Thomas, D. Mareschal, & I. Dumontheil (Eds.), *Educational neuroscience: Development across the life span* (pp. 53–87). Routledge.

Donati, G., Meaburn, E. L., & Dumontheil, I. (2019). The specificity of associations between cognition and attainment in English, Maths and Science during adolescence. *Learning and Individual Differences, 69,* 84–93.

Doria, V., Beckmann, C. F., Arichi, T., Merchant, N., Groppo, M., Turkheimer, F. E., Counsell, S. J., Murgasova, M., Aljabar, P., Nunes, R. G., Larkman, D. J., Rees, G., & Edwards, A. D. (2010). Emergence of resting state networks in the preterm human brain. *Proceedings of the National Academy of Sciences, 107,* 20015–20020.

Driver, J., Davis, G., Ricciardelli, P., Kidd, P., Maxwell, E., & Baron-Cohen, S. (1999). Gaze perception triggers reflexive visuo-spatial orienting. *Visual Cognition, 6,* 509–540.

Drummey, A., & Newcombe, N. (1995). Remembering versus knowing the past: Children's explicit and implicit memories for pictures. *Journal of Experimental Child Psychology, 59,* 549–565.

Drummey, A. B., & Newcombe, N. S. (2002). Developmental changes in source memory. *Developmental Science, 5,* 502–513.

Duchaine, B., & Nakayama, K. (2006). Developmental prosopagnosia: A window to content-specific face processing. *Current Opinion in Neurobiology, 16*, 166–173.

Duchaine, B., & Yovel, G. (2015). A revised neural framework for face processing. *Annual Review of Vision Science, 1*, 393–416.

Duell, N., Steinberg, L., Icenogle, G., Chein, J., Chaudhary, N., Di Giunta, L., Dodge, K. A., Fanti, K. A., Lansford, J. E., Oburu, P., Pastorelli, C., Skinner, A. T., Sorbring, E., Tapanya, S., Uribe Tirado, L. M., Alampay, L. P., Al-Hassan, S. M., Takash, H. M. S., Bacchini, D., & Chang, L. (2018). Age patterns in risk taking across the world. *Journal of Youth and Adolescence, 47*(5), 1052–1072.

Dumontheil, I. (2014). Development of abstract thinking during childhood and adolescence: The role of rostrolateral prefrontal cortex. *Developmental Cognitive Neuroscience, 10*, 57–76. https://doi.org/10.1016/j.dcn.2014.07.009

Dumontheil, I., Apperly, I. A., & Blakemore, S. J. (2010). Online usage of theory of mind continues to develop in late adolescence. *Developmental Science, 13*, 331–338.

Dumontheil, I., Hillebrandt, H., Apperly, I. A., & Blakemore, S.-J. (2012). Developmental differences in the control of action selection by social information. *Journal of Cognitive Neuroscience, 24*(10), 2080–2095. https://doi.org/10.1162/jocn_a_00268

Dumontheil, I., & Klingberg, T. (2011). Brain activity during a visuospatial working memory task predicts arithmetical performance 2 years later. *Cerebral Cortex, 22*, 1078–1085.

Dumontheil, I., Roggeman, C., Ziermans, T., Peyrard-Janvid, M., Matsson, H., Kere, J., & Klingberg, T. (2011). Influence of the COMT genotype on working memory and brain activity changes during development. *Biological Psychiatry, 70*(3), 222–229.

Duncan, J. (2001). An adaptive coding model of neural function in prefrontal cortex. *Nature Reviews Neuroscience, 2*, 820–829.

Dziurawiec, S. (1996). Blink reflex modification in neonates and its relationship to maturational indices. *Infant Behavior & Development, 19*, 67.

Ebbesson, S. O. (1980). The parcellation theory and its relation to interspecific variability in brain organization, evolutionary and ontogenetic development, and neuronal plasticity. *Cell and Tissue Research, 213*, 179–212.

Ebbesson, S. O. (1984). Evolution and ontogeny of neural circuits. *Behavioral and Brain Sciences, 7*, 321–366.

Edelman, G. M. (1987). *Neural Darwinism: The theory of neuronal group selection.* Basic Books.

Eden, G. F., & Flowers, D. L. (2008). Learning, skill acquisition, reading and dyslexia. *Annals of the New York Academy of Sciences, 1145*, ix–xii.

Edin, F., Macoveanu, J., Olesen, P., Tegner, J., & Klingberg, T. (2007). Stronger synaptic connectivity as a mechanism behind development of working memory-related brain activity during childhood. *Journal of Cognitive Neuroscience, 19*, 750–760.

Eichenbaum, H. (2012). What H.M. taught us. *Journal of Cognitive Neuroscience, 25*, 14–21.

Elison, J. T., Paterson, S. J., Wolff, J. J., Reznick, J. S., Sasson, N. J., Gu, H., Botteron, K. N., Dager, S. R., Estes, A. M., Evans, A. C., Gerig, G., Hazlett, H. C., Schultz, R. T., Styner, M., Zwaigenbaum, L., Piven, J., & IBIS Network. (2013). White matter microstructure and atypical visual orienting in 7-month-olds at risk for autism. *American Journal of Psychiatry, 170*(8), 899–908. https://doi.org/10.1176/appi.ajp.2012.12091150

Elman, J., Bates, E., Johnson, M. H., Karmiloff-Smith, A., Parisi, D., & Plunkett, K. (1996). *Rethinking innateness: A connectionist perspective on development.* MIT Press.

Elsabbagh, M., Fernandes, J., Webb, S. J., Dawson, G., Charman, T., Johnson, M. H., & The BASIS Team. (2013). Disengagement of visual attention in infancy is associated with emerging autism in toddlerhood. *Biological Psychiatry, 74*, 189–194.

Elsabbagh, M., & Johnson, M. H. (2007). Infancy and autism: Progress, prospects, and challenge. *Progress in Brain Research, 164*, 355–382.

Elsabbagh, M., & Johnson, M. H. (2010). Getting answers from babies about autism. *Trends in Cognitive Science, 14*, 81–87.

Engelhardt, L. E., Harden, K. P., Tucker-Drob, E. M., & Church, J. A. (2019). The neural architecture of executive functions is established by middle childhood. *NeuroImage, 185*, 479–489.

Enns, J. T., & Brodeur, D. A. (1989). A developmental study of covert orienting to peripheral visual cues. *Journal of Experimental Child Psychology, 48*, 171–189.

Enns, J. T., & Girgus, J. S. (1985). Developmental changes in selective and integrative visual attention. *Journal of Experimental Child Psychology, 48*, 315–334.

Espy, K. A., McDiarmid, M. M., Cwik, M. F., Stalets, M. M., Hamby, A., & Senn, T. E. (2004). The contribution of executive functions to emergent mathematic skills in preschool children. *Developmental Neuropsychology, 26*(1), 465–486.

Fabiani, M., & Wee, E. (2001). Age-related changes in working memory function: A review. In C. Nelson & M. Luciana (Eds.), *The handbook of developmental cognitive neuroscience* (pp. 473–488). MIT Press.

Fair, D. A., Cohen, A. L., Power, J. D., Dosenbach, N. U. F., Church, J. A., Miezin, F. M., Schlaggar, B. L., & Petersen, S. E. (2009). Functional brain networks develop from a "local to distributed" organization. *PLoS Computational Biology, 5*, e1000381.

Fair, D. A., Dosenbach, N. U. F., Church, J. A., Cohen, A. L., Brahmbhatt, S., Miezin, F. M., Barch, D. M., Raichle, M. E., Petersen, S. E., & Schlaggar, B. L. (2007). Development of distinct control networks through segregation and integration. *Proceedings of the National Academy of Sciences of the United States of America, 104*, 13507–13512.

Fan, J., Fossella, J., Sommer, T., Wu, Y., & Posner, M. I. (2003). Mapping the genetic variation of executive attention onto brain activity. *Proceedings of the National Academy of Sciences of the United States of America, 100*, 7406–7411.

Fantz, R. L. (1964). Visual experience in infants: Decreased attention to familiar patterns relative to novel ones. *Science, 46*, 668–670.

Farah, M. J. (2017). The neuroscience of socioeconomic status: Correlates, causes, and consequences. *Neuron, 96*(1), 56–71.

Farah, M. J. (2018). Socioeconomic status and the brain: Prospects for neuroscience-informed policy. *Nature Reviews. Neuroscience, 19*, 428–438.

Farroni, T., Csibra, G., Simion, F., & Johnson, M. H. (2002). Eye contact detection in humans from birth. *Proceedings of the National Academy of Sciences of the United States of America, 99*, 9602–9605.

Farroni, T., Johnson, M. H., Brockbank, M., & Simion, F. (2000). Infants' use of gaze direction to cue attention: The importance of perceived motion. *Visual Cognition, 7*(6), 705–718.

Feigenson, L., Carey, S., & Spelke, E. S. (2002). Infants' discrimination of number vs. continuous extent. *Cognitive Psychology, 44*, 33–66.

Fetita, R., Hillary, R. F., Price, D. J., & Lawrie, S. M. (2021). The neuropathology of autism: A systematic review of post-mortem studies of autism and related disorders. *Neuroscience & Biobehavioral Reviews, 129*, 35–62. https://doi.org/10.1016/j.neubiorev.2021.07.014

Filipek, P. A. (1999). Neuroimaging in the developmental disorders: The state of the science. *Journal of Child Psychology and Psychiatry, 40*(1), 113–128.

Filipek, P. A., Kennedy, D. N., & Caviness, V. S. J. (1992). Neuroimaging in child neuropsychology. In I. Rapin & S. J. Segalowitz (Eds.), *Handbook of neuropsychology* (Vol. 6, pp. 301–309). Elsevier Science.

Finlay, B. L., & Darlington, R. B. (1995). Linked regularities in the development and evolution of mammalian brains. *Science, 268*(5217), 1578–1584.

Finn, A. S., Minas, J. E., Leonard, J. A., Mackey, A. P., Salvatore, J., Goetz, C., West, M. R., Gabrieli, C. F. O., & Gabrieli, J. D. E. (2017). Functional brain organization of working memory in adolescents varies in relation to family income and academic achievement. *Developmental Science, 20*(5), e12450.

Finn, A. S., Sheridan, M. A., Kam, C. L., Hinshaw, S., & D'Esposito, M. (2010). Longitudinal evidence for functional specialisation of the neural circuitry supporting working memory in the human brain. *Journal of Neuroscience, 30*, 11062–11067.

Fischer, B., & Breitmeyer, B. (1987). Mechanisms of visual attention revealed by saccadic eye movements. *Neuropsychologia, 25*, 73–83.

Fisher, S. E., & Scharff, C. (2009). FOXP2 as a molecular window into speech and language. *Trends in Genetics, 25*(4), 166–177. https://doi.org/10.1016/j.tig.2009.03.002

Fodor, J. A. (1983). *The modularity of mind.* MIT Press.

Foldiak, P. (1996). Learning constancies for object perception. In V. Walsh & J. Kulikovski (Eds.), *Perceptual constancy: Why things look as they do* (pp. 144–172). Cambridge University Press.

Fosse, V. M., Heggelund, P., & Fonnum, F. (1989). Postnatal development of glutamatergic, GABAergic and cholinergic neurotransmitter phenotypes in the visual cortex, lateral geniculate nucleus pulvinar and superior colliculus in cats. *Journal of Neuroscience, 9*, 426–435.

Foulkes, L., & Blakemore, S.-J. (2018). Studying individual differences in human adolescent brain development. *Nature Neuroscience, 21*(3), 315–323.

Fox Keller, E. (2002). *The century of the gene.* First Harvard University Press.

Fox, N. A., & Bell, M. A. (1990). Electrophysiological indices of frontal lobe development. In A. Diamond (Ed.), *The development and neural bases of higher cognitive functions* (Vol. 608, pp. 677–698). New York Academy of Sciences.

Fraguas, D., Díaz-Caneja, C. M., Pina-Camacho, L., Janssen, J., & Arango, C. (2016). Progressive brain changes in children and adolescents with early-onset psychosis: A meta-analysis of longitudinal MRI studies. *Schizophrenia Research, 173*(3), 132–139.

Franchuk, J. M., & Yu, C. (2022). Beyond screen time: Using head-mounted eye tracking to study natural behavior. *Advances in Child Development and Behavior, 62*, 61–91.

Fransson, P., Skiöld, B., Horsch, S., Nordell, A., Blennow, M., Lagercrantz, H., & Aden, U. (2007). Resting-state networks in the infant brain. *Proceedings of the National Academy of Sciences of the United States of America, 104*, 15531–15536.

Fray, P. J., Robbins, T. W., & Sahakian, B. J. (1996). Neuropsychiatric applications of CANTAB. *International Journal of Geriatric Psychiatry, 11*, 329–336.

Friederici, A. D. (2008). Brain correlates of language processing during the first years of life. In C. A. Nelson & M. Luciana (Eds.), *The handbook of developmental cognitive neuroscience* (2nd ed., pp. 117–126). MIT Press.

Friederici, A. D. (2009). Pathways to language: Fiber tracts in the human brain. *Trends in Cognitive Sciences, 13*(4), 175–181. https://doi.org/10.1016/j.tics.2009.01.001

Friesen, C. K., & Kingstone, A. (1998). The eyes have it! Reflexive orienting is triggered by nonpredictive gaze. *Psychonomic Bulletin and Review, 5*, 490–495.

Friston, K. J., & Price, C. J. (2001). Dynamic representation and generative models of brain function. *Brain Research Bulletin, 54*(3), 275–285.

Frith, U. (2003). *Autism: Explaining the enigma* (2nd ed.). Blackwell.

Frodl, T., & Skokauskas, N. (2012). Meta-analysis of structural MRI studies in children and adults with attention deficit hyperactivity disorder indicates treatment effects. *Acta Psychiatrica Scandinavica, 125*(2), 114–126.

Fulford, J., Vadeyar, S. H., Dodampahala, S. H., Moore, R. J., Young, P., Baker, P. N., James, D. K., & Gowland, P. A. (2003). Fetal brain activity in response to visual stimulus. *Human Brain Mapping, 20*, 239–245.

Funahashi, S., Bruce, C. J., & Goldman-Rakic, P. S. (1989). Mnemonic coding of visual space in the monkey's dorsolateral prefrontal cortex. *Journal of Neurophysiology, 61*(2), 331–349.

Funahashi, S., Bruce, C. J., & Goldman-Rakic, P. S. (1990). Visuospatial coding in primate prefrontal neurons revealed by oculomotor paradigms. *Journal of Neurophysiology, 63*(4), 814–831.

Gagliardi, C., Arrigoni, F., Nordio, A., De Luca, A., Peruzzon, D., Decio, A., Leemans, A., & Borgati, R. (2018). A different brain: Abnormalities of funtional and structural connections in Williams Syndromd. *Frontiers in Neurology, 9*, 721. https://doi.org/10.3389/fneur.2018.00721

Galaburda, A. M., & Bellugi, U. (2000). Multi-level analysis of cortical neuroanatomy in Williams syndrome. *Journal of Cognitive Neuroscience, 12*(Supplement), 74–88.

Galaburda, A. M., Sherman, G. F., Rosen, G. D., Aboitiz, F., & Geschwind, N. (1985). Development dyslexia: Four consecutive patients with cortical anomalies. *Annals of Neurology, 18*, 222–232.

Galaburda, A. M., Wang, P. P., Bellugi, U., & Rosen, M. (1994). Cytoarchitectonic anomalies in a genetically based disorder: Williams syndrome. *Cognitive Neuroscience and Neuropsychology, Neuroreport, 5*, 753–757.

Gallistel, C. R. (1990). *The organization of learning*. MIT Press.

Galvan, A., Hare, T., Voss, H., Glover, G., & Casey, B. J. (2006). Risk-taking and the adolescent brain: Who is at risk? *Developmental Science, 10*, F8–F14.

Gardner, M., & Steinberg, L. (2005). Peer influence on risk taking, risk preference, and risky decision making in adolescence and adulthood: An experimental study. *Developmental Psychology, 41*(4), 625–635.

Gathercole, S. E., & Alloway, T. P. (2006). Short-term and working memory impairments in neurodevelopmental disorders: A diagnosis and remedial support. *Journal of Child Psychology and Psychiatry, 47*, 4–15.

Gathercole, S. E., Brown, L., & Pickering, S. J. (2003). Working memory assessments at school entry as longitudinal predictors of National Curriculum attainment levels. *Educational and Child Psychology, 20*, 109–122.

Gathercole, S. E., Pickering, S. J., Knight, C., & Stegman, Z. (2004). Working memory skills and educational attainment: Evidence from National Curriculum assessments at 7 and 14 years of age. *Applied Cognitive Psychology, 18*, 1–16.

Gathers, A. D., Bhatt, R., Corbly, C. R., Farley, A. B., & Joseph, J. E. (2004). Developmental shifts in cortical loci for face and object recognition. *Neuroreport, 15,* 1549–1553.

Gauthier, I., & Nelson, C. (2001). The development of face expertise. *Current Opinion in Neurobiology, 11,* 219–224.

Gauthier, I., Tarr, M. J., Anderson, A. W., Skudlarski, P., & Gore, J. C. (1999). Activation of the middle fusiform "face area" increases with expertise in recognizing novel objects. *Nature Neuroscience, 2,* 568–573.

Gazzaniga, M. (1983). Right-hemisphere damage following brain bisection. *American Psychologist, 25,* 549.

Ge, T., Chen, C. Y., Doyle, A. E., Vettermann, R., Tuominen, L. J., Holt, D. J., Sabuncu, M. R., & Smoller, J. W. (2019). The shared genetic basis of educational attainment and cerebral cortical morphology. *Cerebral Cortex, 29*(8), 3471–3481.

Ge, X., Zhang, K., Gribizis, A., Hamodi, A. S., Sabino, A. M., & Crair, M. C. (2021). Retinal waves prime visual motion detection by simulating future optic flow. *Science, 373*(6553), eabd0830. https://doi.org/10.1126/science.abd0830

Ghetti, S., & Bunge, S. A. (2012). Neural changes underlying the development of episodic memory during middle childhood. *Developmental Cognitive Neuroscience, 13,* 5017–5026.

Ghetti, S., DeMaster, D. M., Yonelinas, A. P., & Bunge, S. A. (2010). Developmental differences in medial temporal lobe function during memory encoding. *Journal of Neuroscience, 30,* 9548–9546.

Ghetti, S., & Fandakova, Y. (2020). Neural development of memory and metamemory in childhood and adolescence: Toward an integrative model of the development of episodic recollection. *Annual Review of Developmental Psychology, 2*(1), 365–388.

Gibson, J. J. (1979). *The ecological approach to visual perception.* Houghton Mifflin.

Giedd, J. N., Blumenthal, J., Jeffries, N. O., Castellanos, F. X., Lui, H., Zijdenbos, A., Paus, T., Evans, A. C., & Rapoport, J. L. (1999). Brain development during childhood and adolescence: A longitudinal MRI study. *Nature Neuroscience, 2,* 861–863.

Gilbert, C., & Sigman, M. (2007). Brain states: Top-down influences in sensory processing. *Neuron, 54,* 677–696.

Gilles, F. H., Shankle, W., & Dooling, E. C. (1983). Myelinated tracts: Growth patterns. In F. H. Gilles, A. Leviton, & E. C. Dooling (Eds.), *The developing human brain: Growth and epidemiological neuropathology* (pp. 117–183). John Wright-PSG.

Gilmore, R. O., & Johnson, M. H. (1995). Working memory in infancy: Six-month-olds' performance on two versions of the oculomotor delayed response task. *Journal of Experimental Child Psychology, 59,* 397–418.

Gilmore, R. O., & Johnson, M. H. (1997). Egocentric action in early infancy: Spatial frames of reference for saccades. *Psychological Science, 8,* 224–230.

Giraud, A. L., & Poeppel, D. (2012). Cortical oscillations and speech processing: Emerging computational principles and operations. *Nature Neuroscience, 15,* 511–517.

Goddings, A.-L., Beltz, A., Peper, J. S., Crone, E. A., & Braams, B. R. (2019). Understanding the role of puberty in structural and functional development of the adolescent brain. *Journal of Research on Adolescence, 29*(1), 32–53. https://doi.org/10.1111/jora.12408

Goddings, A.-L., Burnett Heyes, S., Bird, G., Viner, R. M., & Blakemore, S.-J. (2012). The relationship between puberty and social emotion processing. *Developmental Science, 15*(6), 801–811. https://doi.org/10.1111/j.1467-7687.2012.01174.x

Goddings, A.-L., Mills, K. L., Clasen, L. S., Giedd, J. N., Viner, R. M., & Blakemore, S.-J. (2014). The influence of puberty on subcortical brain development. *NeuroImage, 88*, 242–251.

Gogtay, N., Giedd, J. N., Lusk, L., Hayashi, K. M., Greenstein, D., Vaituzis, C., Nugent, T. F., III, Herman, D. H., Clasen, L. S., Toga, A. W., Rapoport, J. L., & Thompson, P. M. (2004). Dynamic mapping of human cortical development during childhood through early adulthood. *Proceedings of the National Academy of Science of the United States of America, 101*, 8174–8179.

Gogtay, N., Nugent, T. F., Herman, D. H., Ordonez, A., Greenstein, D., Hayashi, K. M., Clasen, L., Toga, A. W., Rappoport, J. L., & Thompson, P. M. (2006). Dynamic mapping of normal human hippocampal development. *Hippocampus, 16*, 664–672.

Golarai, G., Ghahrmani, D. G., Whitfield-Gabrieli, S., Reiss, A., Eberhardt, J. L., Gabrieli, D. E., & Grill-Spector, K. (2007). Differential development of high-level visual cortex correlates with category-specific recognition memory. *Nature Neuroscience, 10*, 512–522.

Goldman-Rakic, P. S. (1987). Development of cortical circuitry and cognitive function. *Child Development, 58*, 601–622.

Goldman-Rakic, P. S. (1994). Introduction. In G. Dawson & K. W. Fischer (Eds.), *Human behavior and the developing brain* (pp. 1–2). Guilford Press.

Goldman-Rakic, P. S., & Brown, R. M. (1982). Postnatal development of monoamine content and synthesis in the cerebral cortex of rhesus monkeys. *Brain Research, 256*, 339–349.

Gonzalez, R., Thompson, E. L., Sanchez, M., Morris, A., Gonzalez, M. R., Feldstein Ewing, S. W., Mason, M. J., Arroyo, J., Howlett, K., Tapert, S. F., & Zucker, R. A. (2021). An update on the assessment of culture and environment in the ABCD Study: Emerging literature and protocol updates over three measurement waves. *Developmental Cognitive Neuroscience, 52*, 101021.

Goodwin, A., Jones, E. J. H., Salomone, S., Mason, L., Holman, R., Begum-Ali, J., Hunt, A., Ruddock, M., Vamvakas, G., Robinson, E., Holden, C. J., Taylor, C., Smith, T. J., Sonuga-Barke, E., Bolton, P., Charman, T., Pickles, A., Wass, S., Johnson, M. H., & the INTERSTAARS team. (2021). INTERSTAARS: Attention training for infants with elevated likelihood of developing ADHD: A proof-of-concept randomised controlled trial. *Translational Psychiatry, 11*, 644.

Gopnik, M. (1990). Feature-blind grammar and dysphasia. *Nature, 344*, 715.

Goren, C. C., Sarty, M., & Wu, P. Y. K. (1975). Visual following and pattern discrimination of face-like stimuli by newborn infants. *Pediatrics, 56*, 544–549.

Goswami, U. (2019). *Cognitive development and cognitive neuroscience: The learning brain.* Routledge.

Goswami, U. (2020). Reading acquisition and developmental dyslexia: Educational neuroscience and phonological skills. In M. S. C. Thomas, D. Mareschal, & I. Dumontheil (Eds.), *Educational neuroscience: Development across the life span* (1st ed., pp. 144–168). Routledge.

Gottlieb, G. (1992). *Individual development and evolution.* Oxford University Press.

Gottlieb, G. (2007). Probabilistic epigenesis. *Developmental Science, 10*, 1–11.

Green, J. (2017). Basic science and treatment intervention. *Journal of Child Psychology and Psychiatry, 58*, 967–969.

Green, J., Wan, M. W., Guiraud, J., Holsgrove, S., McNally, J., Slonims, V., Elsabbagh, M., Charman, T., Pickles, A., Johnson, M. H., & The BASIS Team. (2013). Intervention for infants at risk of developing autism: A case series. *Journal of Autism and Developmental Disorders, 43*, 2502–2514.

Greenberg, F. (1990). Introduction to special issue on Williams syndrome. *American Journal of Medical Genetics. Supplement, 6*, 85–88.

Greenough, W. T., Black, J. E., & Wallace, C. S. (2002). Experience and brain development. In M. H. Johnson, Y. Munakata, & R. Gilmore (Eds.), *Brain development and cognition: A reader* (2nd ed., pp. 186–216). Blackwell.

Grill-Spector, K., Kushnir, T., Edelman, S., Avidan, G., Itzchak, Y., & Malach, R. (1998). Differential processing of objects under various viewing conditions in the human lateral occipital complex. *Neuron, 24*(1), 187–203.

Grossmann, T., Lloyd-Fox, S., & Johnson, M. H. (2013). Brain responses reveal young infants' sensitivity to when a social partner follows their gaze. *Developmental Cognitive Neuroscience, 6*, 155–161.

Grüter, T., Grüter, M., & Carbon, C. C. (2008). Neural and genetic foundations of face recognition and prosopagnosia. *Journal of neuropsychology, 2*(1), 79–97. https://doi.org/10.1348/174866407x231001

Guitton, H. A., Buchtel, H. A., & Douglas, R. M. (1985). Frontal lobe lesions in man cause difficulties in suppressing reflexive glances and in generating goal-directed saccades. *Experimental Brain Research, 58*, 455–472.

Gunnar, M. (2001). Effects of early deprivation: Findings from orphanage-reared infants and children. In C. A. Nelson & M. Luciana (Eds.), *The handbook of developmental cognitive neuroscience* (pp. 617–630). MIT Press.

Guzzetta, A., Pecini, C., Biagi, L., Tosetti, M., Brizzolara, D., Chilosi, A., Cipriani, P., Petacchi, E., & Cioni, G. (2008). Language organization in left perinatal stroke. *Neuropediatrics, 39*, 157–163.

Gweon, H., Dodell-Feder, D., Bedny, M., & Saxe, R. (2012). Theory of mind performance in children correlates with functional specialization of a brain region for thinking about thoughts. *Child Development, 83*, 1853–1868.

Ha, S., Sohn, I.-J., Kim, N., Sim, H. J., & Cheon, K.-A. (2015). Characteristics of brains in autism spectrum disorder: Structure, function and connectivity across the lifespan. *Experimental Neurbiology, 24*, 273–284.

Hackman, D. A., & Farah, M. J. (2009). Socioeconomic status and the developing brain. *Trends in Cognitive Neuroscience, 13*, 65–73.

Hackman, D. A., & Kraemer, D. J. M. (2020). Socioeconomic disparities in achievement: Insights on neurocognitive development and educational interventions. In M. S. C. Thomas, D. Mareschal, & I. Dumontheil (Eds.), *Educational neuroscience: Development across the life span* (1st ed., pp. 88–119). Routledge.

Haist, F., Adamo, M., Wazny, J. H., Lee, K., & Stiles, J. (2013). The functional architecture for face-processing expertise: FMRI evidence of the developmental trajectory of the core and the extended face systems. *Neuropsychologia, 51*(13), 2893–2908. https://doi.org/10.1016/j.neuropsychologia.2013.08.005

Haith, M. M., & Benson, J. B. (1998). Infant cognition. In D. Kuhn & R. Siegler (Eds.), *Handbook of child psychology: Cognition, perception, and language* (5th ed., Vol. 2, pp. 199–254). Wiley.

Haith, M. M., Hazan, C., & Goodman, G. S. (1988). Expectation and anticipation of dynamic visual events by 3.5-month-old babies. *Child Development, 59*, 467–479.

Halberda, J., Mazzocco, M. M., & Feigenson, L. (2008). Individual differences in non-verbal number acuity correlate with maths achievement. *Nature, 455*, 665–668.

Halit, H., Csibra, G., Volein, A., & Johnson, M. H. (2004). Face-sensitive cortical processing in early infancy. *Journal of Child Psychology and Psychiatry, 45*, 1228–1234.

Halit, H., de Haan, M., & Johnson, M. H. (2003). Cortical specialisation for face processing: Face-sensitive event-related potential components in 3- and 12-month-old infants. *NeuroImage, 19*, 1180–1193.

Hallett, P. E. (1978). Primary and secondary saccades to goals defined by instructions. *Vision Research, 18*, 1270–1296.

Hamasaki, T., Leingartner, A., Ringstedt, T., & O'Leary, D. D. M. (2004). EMX2 regulates sizes and positioning of the primary sensory and motor areas in neocortex by direct specification of cortical progenitors. *Neuron, 43*, 359–372.

Happé, F. (1994). *Autism: An introduction to psychological theory.* UCL Press.

Happé, F., Ronald, A., & Plomin, R. (2006). Time to give up on a single explanation of autism. *Nature Neuroscience, 9*, 1218–1220.

Hare, T. A., Tottenham, N., Galvan, A., Voss, H. U., Glover, G. H., & Casey, B. J. (2008). Biological substrates of emotional reactivity and regulation in adolescence during an emotional go-nogo task. *Biological Psychiatry, 63*(10), 927–934.

Hari, R., Forss, N., Avikainen, S., Kirveskari, E., Salenius, S., & Rizzolatti, G. (1998). Activation of human primary motor cortex during action observation: A neuromagnetic study. *Proceedings of National Academy of Sciences of the United States of America, 95*, 15061–15065.

Hari, R., & Salmelin, R. (1997). Human cortical oscillations: A neuromagnetic view through the skull. *Trends in Neurosciences, 20*, 44–49.

Harlow, E. G., Till, S. M., Russell, T. A., Wijetunge, L. S., Kind, P., & Contractor, A. (2010). Critical period plasticity is disrupted in the barrel cortex of FMR1 knockout mice. *Neuron, 65*(3), 385–398.

Hart, H., Radua, J., Nakao, T., Mataix-Cols, D., & Rubia, K. (2013). Meta-analysis of functional magnetic resonance imaging studies of inhibition and attention in attention-deficit/ hyperactivity disorder: Exploring task-specific, stimulant medication, and age effects. *JAMA Psychiatry, 70*(2), 185–198. https://doi.org/10.1001/jamapsychiatry.2013.277

Hartshorn, K., Rovee-Collier, C., Gerhardstein, P., Bhatt, R. S., Wondoloski, T. L., Klein, P., Glich, J., Wurtzel, M., & Campos-de-Carvalho, M. (1998). The ontogeny of long-term memory over the first year and a half of life. *Developmental Psychobiology, 32*, 69–89.

Hauser, M. D., MacNeilage, P., & Ware, M. (1996). Numerical representations in primates. *Proceedings of the National Academy of Sciences of the United States of America, 93*, 1514–1517.

Haxby, J. V., Gobbini, M. I., Furey, M. L., Ishai, A., Schouten, J. L., & Pietrini, P. (2001). Distributed and overlapping representations of faces and objects in ventral temporal cortex. *Science, 293*, 2425–2430.

Heckman, J. J. (2007). The economics, technology and neuroscience of human capability formation. *Proceedings of the National Academy of Sciences, 104*, 13250–13255.

Held, R. (1985). Binocular vision: Behavioral and neuronal development. In J. Mehler & R. Fox (Eds.), *Neonate cognition: Beyond the blooming, buzzing confusion* (pp. 37–44). Lawrence Erlbaum.

Held, R. (1993). Development of binocular vision revisited. In M. H. Johnson (Ed.), *Brain development and cognition: A reader* (pp. 159–166). Blackwell.

Hendry, A., Greenhalgh, I., Bailey, R., Fiske, A., Dvergsdal, H., & Holmboe, K. (2021). Development of directed global inhibition, competitive inhibition and behavioural inhibition during the transition between infancy and toddlerhood. *Developmental Science, 23*, e13193.

Henrich, J., Heine, S. J., & Norenzayan, A. (2010). The weirdest people in the world? *The Behavioral and Brain Sciences, 33*(2–3), 61–83; discussion 83–135. https://doi.org/10.1017/S0140525X0999152X

Henry, L. A., & MacLean, M. (2003). Relationships between working memory, expressive vocabulary and arithmetical reasoning in children with and without intellectual disabilities. *Educational and Child Psychology, 20*, 51–64.

Hensch, T. K. (2005). Critical period plasticity in local cortical circuits. *Nature Reviews Neuroscience, 6*(11), 877–888.

Hensch, T. K., Fagiolini, M., Mataga, N., Stryker, M. P., Baekkeskov, S., & Kash, S. F. (1998). Local GABA circuit control of experience-dependent plasticity in developing visual cortex. *Science, 282*(5393), 1504–1508.

Henson, R. N., Rylands, A., Ross, E., Vuilleumier, P., & Rugg, M. D. (2004). The effect of repetition lag on electrophysiological and hameodynamic correlates of visual object priming. *NeuroImage, 21*, 1674–1689.

Herbert, M. R., Ziegler, D. A., Makris, N. B., Dakardjiev, A., Hodgson, J., Adrien, K. T., Kennedy, D. N., Filipek, P. A., & Caviness, V. S., Jr. (2003). Larger brain and white matter volumes in children with developmental language disorder. *Developmental Science, 6*(4), F11–F22.

Hernik, M., & Broesch, T. (2019). Infant gaze following depends on communicative signals: An eye-tracking study of 5- to 7-month-olds in Vanuatu. *Developmental Science, 22*(4), e12779.

Heron-Delaney, M., Anzures, G., Herbet, J. S., Quinn, P. C., Slater, A. M., Tanaka, J. W., Lee, K., & Pascalis, O. (2011). Perceptual training prevents the emergence of the other race effect during infancy. *PLoS One, 6*(5), e19858.

Hillyard, S. A., Mangun, G. R., Woldorff, M. G., & Luck, S. J. (1995). Neural systems mediating selective attention. In M. S. Gazzaniga (Ed.), *The cognitive neurosciences* (pp. 665–681). MIT Press.

Hinde, R. A. (1974). *Biological bases of human social behaviour*. McGraw-Hill.

Hinshelwood, J. (1907). Four cases of cogenital word-blindness occurring in the same family. *British Medical Journal, 2*, 1229–1232.

Hobson, R. P. (1993). Understanding persons: The role of affect. In S. Baron-Cohen, H. Tager-Flusberg, & D. J. Cohen (Eds.), *Understanding other minds* (pp. 204–227). Oxford University Press.

Holloway, I. D., & Ansari, D. (2010). Developmental specialization in the right intraparietal sulcus for the abstact representation of numerical magnitude. *Journal of Cognitive Neuroscience, 22*, 2627–2637.

Holmboe, K., Fearon, P. R. M., Csibra, G., Tucker, L. A., & Johnson, M. H. (2008). Freeze-frame: A new infant inhibition task and its relation to frontal cortex tasks during infancy and early childhood. *Journal of Experimental Child Psychology, 100*, 89–114.

Holmboe, K., Nemoda, Z., Fearon, R. M. P., Csibra, G., Sasvari-Szeleky, M., & Johnson, M. H. (2010). Polymorphisms in dopamine system genes are associated with individual differences in attention in infancy. *Developmental Psychology, 46*, 404–416.

Holmes, J., & Adams, J. W. (2006). Working memory and children's mathematical skills: Implications for mathematical development and mathematics curricula. *Educational Psychology, 26*, 339–366.

Holmes, J., Gathercole, S. E., & Dunning, D. L. (2010). Poor working memory: Impact and interventions. *Advances in Child Development and Behavior, 39*, 1–43.

Hood, B. (1993). Inhibition of return produced by covert shifts of visual attention in 6-month-old infants. *Infant Behavior and Development, 16*(2), 245–254. https://doi.org/10.1016/0163-6383(93)80020-9

Hood, B. (1995). Shifts of visual attention in the human infant: A neuroscientific approach. In C. Rovee-Collier & L. Lipsitt (Eds.), *Advances in infancy research* (Vol. 9, pp. 163–216). Ablex.

Hood, B. & Atkinson, J. (1991). *Shifting covert attention in infants*. Paper presented at the Absracts of the Society for Research in Child Development, Seattle, WA.

Hood, B. M., Willen, J. D., & Driver, J. (1998). Adult's eyes trigger shifts of visual attention in human infants. *Psychological Science, 9*, 131–134.

Houdé, O., Rossi, S., Lubin, A., & Joliot, M. (2010). Mapping numerical processing, reading, and executive functions in the developing brain: An fMRI meta-analysis of 52 studies including 842 children. *Developmental Science, 13*(6), 876–885.

Howard, M. A., Rubenstein, J. L. R., & Baraban, S. C. (2013). Bidirectional homeostatic plasticity induced by interneuron cell death and transplantation in vivo. *Proceedings of the National Academy of Sciences, 111*(1), 492–497.

Huttenlocher, P. R. (1990). Morphometric study of human cerebral cortex development. *Neuropsychologia, 28*, 517–527.

Huttenlocher, P. R. (1994). Synaptogenesis, synapse elimination, and neural plasticity in human cerebral cortex: Threats to optimal development. In C. A. Nelson (Ed.), *The Minnesota symposia on child psychology* (Vol. 27, pp. 35–54). Lawrence Erlbaum.

Huttenlocher, P. R. (2002). Morphometric study of human cerebral cortex development. In M. H. Johnson, Y. Munakata, & R. Gilmore (Eds.), *Brain development and cognition: A reader* (2nd ed., pp. 117–128). Blackwell.

Huttenlocher, P. R., & Dabholkar, A. S. (1997a). Regional differences in synaptogenesis in human cerebral cortex. *Journal of Comparative Neurology, 387*, 167–178.

Huttenlocher, P. R., & Dabholkar, J. C. (1997b). Developmental anatomy of prefrontal cortex. In N. A. Krasnegor, G. R. Lyon, & P. S. Goldman-Rakic (Eds.), *Development of the prefrontal cortex: Evolution, neurobiology, and behavior* (pp. 69–84). Paul. H. Brookes.

Huttenlocher, P. R., de Courten, C., Garey, L. G., & Van der Loos, H. (1982). Synaptogenesis in human visual cortex – evidence for synapse elimination during normal development. *Neuroscience Letters, 33*, 247–252.

Hwang, K., Velanova, K., & Luna, B. (2010). Strengthening of top-down frontal cognitive control networks underlying the development of inhibitory control: A functional magnetic resonance imaging effective connectivity study. *The Journal of Neuroscience, 30*, 15535–15545.

Hyde, D. C., & Spelke, E. S. (2011). Neural signatures of number processing in infants: Evidence for two core systems underlying numerical cognition. *Developmental Science, 14*, 360–371.

Hykin, J., Moore, R., Duncan, K., Clare, S., Baker, P., Johnson, I., Bowtell, R., Mansfield, P., & Gowland, P. (1999). Fetal brain activity demonstrated by functional magnetic resonance imaging. *The Lancet, 354*, 645–646.

Iliescu, B. F., & Dannemiller, J. L. (2008). Brain-behavior relationships in early visual development. In C. A. Nelson & M. Luciana (Eds.), *Handbook of developmental cognitive neuroscience* (2nd ed., pp. 127–146). MIT Press.

Insel, T. R. (2010). The challenge of translation in social neuroscience: A review of oxytocin, vasopressin, and affiliative behavior. *Neuron, 65,* 768–779.

Isa, T., & Yoshida, M. (2021). Neural mechanism of blindsight in a macaque model. *Neuroscience, 469,* 138–161. https://doi.org/10.1016/j.neuroscience.2021.06.022

Ishai, A., Ungerleider, L. G., Martin, A., Schouten, J. L., & Haxby, J. V. (1999). Distributed representation of objects in the human ventral visual pathway. *Proceedings of the National Academy of Sciences of the United States of America, 96*(16), 9379–9384.

Iuculano, T. (2016). Neurocognitive accounts of developmental dyscalculia and its remediation. *Progress in Brain Research, 227,* 305–333.

Jacobs, R. A. (2002). What determines visual cue reliability? *Trends in Cognitive Sciences, 6,* 345–350.

Jacobs, R. A., Jordan, M. I., & Barto, A. G. (1991). Task decomposition through competition in a modular connectionist architecture: The what and where vision tasks. *Cognitive Science, 15,* 219–250.

Jaffe, J., Stern, D. N., & Peery, J. C. (1973). "Conversational" coupling of gaze behavior in prelinguistic human development. *Journal of Psycholinguistic Research, 2,* 321–329.

Jandó, G., Mikó-Barátha, E., Markóa, K., Hollódyb, K., Törökc, B., & Kovacs, I. (2012). Early-onset binocularity in preterm infants reveals experience-dependent visual development in humans. *Proceedings of the National Academy of Sciences of the United States of America, 109,* 11049–11052.

Jarvis, H. L., & Gathercole, S. J. (2003). Verbal and non-verbal working memory achievements on National Curriculum tests at 11 and 14 years of age. *Educational and Child Psychology, 20,* 123–140.

Jensen, S. K. G., Obradović, J., & Nelson, C. A., III. (2019). Introduction to special issue on global child development studies. *Developmental Science, 22*(5), e12888.

Jernigan, T. L., & Bellugi, U. (1994). Neuroanatomical distinctions between Williams and Down syndromes. In S. Broman & J. Grafman (Eds.), *Atypical cognitive deficits in developmental disorder: Implications for brain function* (pp. 57–66). Lawrence Erlbaum.

Johannsen, W. (1911). The genotype conception of heredity. *The American Naturalist, 45,* 129–159.

Johnson, E. G., Prabhakar, J., Mooney, L. N., & Ghetti, S. (2020). Neuroimaging the sleeping brain: Insight on memory functioning in infants and toddlers. *Infant Behavior and Development, 58,* 101427.

Johnson, K. A., Kelly, S. P., Bellgrove, M. A., Barry, E., Cox, E., & Gill, M. (2007). Response variability in attention deficit hyperactivity disorder: Evidence for neuropsychological heterogeneity. *Neuropsychologia, 45*(4), 630–638.

Johnson, M. B., Imamura Kawasawa, Y., Mason, C. E., Krsnik, Z., Coppola, G., Bogdanovic, D., Geschwind, D. H., Mane, S. M., State, M. W., & Sestan, N. (2009). Functional and evolutionary insights into human brain development through global transcriptome analysis. *Neuron, 62,* 494–509.

Johnson, M. H. (1990). Cortical maturation and the development of visual attention in early infancy. *Journal of Cognitive Neuroscience, 2,* 81–95.

Johnson, M. H. (Ed.) (1993). *Brain development and cognition: A reader.* Blackwell.

Johnson, M. H. (1994). Visual attention and the control of eye movements in early infancy. In C. Umilta & M. Moscovitch (Eds.), *Attention and perfomance XV: Conscious and nonconscious information processing* (pp. 291–310). MIT Press.

Johnson, M. H. (1995). The inhibition of automatic saccades in early infancy. *Developmental Psychobiology, 28*, 281–291.

Johnson, M. H. (2001). Functional brain development in humans. *Nature Reviews Neuroscience, 2*, 475–483.

Johnson, M. H. (2002). The development of visual attention: A cognitive neuroscience perspective. In M. H. Johnson, Y. Munakata, & R. Gilmore (Eds.), *Brain development and cognition: A reader* (pp. 134–150). Blackwell.

Johnson, M. H. (2005). Sub-cortical face processing. *Nature Reviews Neuroscience, 6*, 766–774.

Johnson, M. H. (2011). Interactive specialization: A domain-general framework for human functional brain development? *Developmental Cognitive Neuroscience, 1*(1), 7–21.

Johnson, M. H., & Bolhuis, J. J. (1991). Imprinting, predispositions and filial preference in the chick. In R. J. Andrew (Ed.), *Neural and behavioral plasticity* (pp. 133–156). Oxford University Press.

Johnson, M. H., Bolhuis, J. J., & Horn, G. (1985). Interaction between acquired preferences and developing predispositions during imprinting. *Animal Behaviour, 33*, 1000–1006.

Johnson, M. H., de Haan, M., Oliver, A., Smith, W., Hatzakis, H., Tucker, L. A., & Csibra, G. (2001). Recording and analyzing high-density event-related potentials with infants using the Geodesic Sensor Net. *Developmental Neuropsychology, 19*(3), 295–323.

Johnson, M. H., Dziurawiec, S., Bartrip, J., & Morton, J. (1992). The effects of movement of internal features on infants' preferences for face-like stimuli. *Infant Behavior and Development, 15*, 129–136.

Johnson, M. H., Dziurawiec, S., Ellis, H. D., & Morton, J. (1991). Newborns' preferential tracking of face-like stimuli and its subsequent decline. *Cognition, 40*, 1–19.

Johnson, M. H., & Fearon, P. (2011). Disengaging the infant mind: Genetic dissociation of attention and cognitive skills in infants – reflections on Leppanen et al. (2011). *Journal of Child Psychology & Psychiatry, 52*, 1153–1154.

Johnson, M. H., Gilmore, R. O., Tucker, L. A., & Minister, S. L. (1995). *Cortical development and saccadic control: Vector summation in young infants* (Developmental Cognitive Neuroscience Tech. Report 95-3). MRC Cognitive Development Unit.

Johnson, M. H., Gliga, T., Jones, E. J. H., & Charman, T. (2015). Infant development, autism and ADHD: Early pathways to emerging disorders. *Journal of Child Psychology and Psychiatry, 56*, 228–247.

Johnson, M. H., Halit, H., Grice, S., & Karmiloff-Smith, A. (2002). Neuroimaging of typical and atypical development: A perspective from multiple levels of analysis. *Development and Psychopathology, 14*, 521–536.

Johnson, M. H., Jones, E., & Gliga, T. (2015). Brain adaptation and alternative developmental trajectories. *Development and Psychopathology, 27*(2), 425–442. https://doi.org/10.1017/S0954579415000073

Johnson, M. H., Mareschal, D., & Csibra, G. (2008). The development and integration of the dorsal and ventral visual pathways in object processing. In M. H. Johnson, Y. Munakata, & R. Gilmore (Eds.), *Brain development and cognition: A reader* (pp. 467–478). Blackwell.

Johnson, M. H., & Morton, J. (1991a). CONSPEC and CONLERN: A two-process theory of infant face recognition. *Psychological Review, 98*(2), 164–181. https://doi.org/10.1037/0033-295X.98.2.164

Johnson, M. H., & Morton, J. (1991b). *Biology and cognitive development: The case of face recognition*. Blackwell.

Johnson, M. H., & Munakata, Y. (2005). Processes of change in brain and cognitive development. *Trends in Cognitive Science, 9*, 152–158.

Johnson, M. H., Munakata, Y., & Gilmore, R. (Eds.) (2002). *Brain development and cognition: A reader* (2nd ed.). Blackwell.

Johnson, M. H., Posner, M. I., & Rothbart, M. K. (1991). Components of visual orienting in early infancy: Contingency learning, anticipatory looking, and disengaging. *Journal of Cognitive Neuroscience, 3*(4), 335–344.

Johnson, M. H., Senju, A., & Tomalski, P. (2015). The two-process theory of face processing: Modifications based on two decades of data from infants and adults. *Neuroscience and Biobehavioral Reviews, 50*, 169–179.

Johnson, M. H., & Tucker, L. A. (1996). The development and temporal dynamics of spatial orienting in infants. *Journal of Experimental Child Psychology, 63*, 171–188.

Johnson, M. H., Tucker, L. A., Stiles, J., & Trauner, D. (1998). Visual attention in infants with perinatal brain damage: Evidence of the importance of anterior lesions. *Developmental Science, 1*, 53–58.

Johnson, S. P., & Aslin, R. N. (1995). Perception of object unity in 2-month-old infants. *Developmental Psychology, 31*, 739–745.

Johnson, S. P., & Aslin, R. N. (1996). Perception of object unity in young infants: The roles of motion, depth and orientation. *Cognitive Development, 11*, 161–180.

Johnson, S. P., & Nanez, J. (1995). Young infants' perception of object unity in two-dimensional displays. *Infant Behavior & Development, 18*, 133–143.

Johnston, M. V., McKinney, M., & Coyle, J. T. (1979). Evidence for a cholinergic projection to neocortex from neurons in basal forebrain. *Proceedings of the National Academy of Sciences of the United States of America, 76*, 5392–5396.

Johnston, T. D. (1988). Developmental explanation and the ontogeny of birdsong: Nature/nurture redux. *Behavioral and Brain Sciences, 11*, 617–663.

Jones, C. R. G., Simonoff, E., Baird, G., Pickles, A., Marsden, A. J. S., Tregay, J., Happé, F., & Charman, T. (2018). The association between theory of mind, executive function, and the symptoms of autism spectrum disorder. *Autism Research, 11*(1), 95–109.

Jonides, J., Smith, E. E., Koeppe, R. A., Awh, E., Minoshima, S., & Mintun, M. A. (1993). Spatial working memory in humans as revealed by PET. *Nature, 363*, 623–625.

Jonkman, L. M., Sniedt, F. L. F., & Kemner, C. (2007). Source localization of the nogo-N2: A developmental study. *Clinical Neurophysiology: Official Journal of the International Federation of Clinical Neurophysiology, 118*(5), 1069–1077.

Just, M. A., Cherkassky, V. L., Keller, T. A., Kana, R. K., & Minshew, N. J. (2007). Functional and anatomical cortical underconnectivity in autism: Evidence from an fMRI study of an executive function task and corpus callosum morphometry. *Cerebral Cortex, 17*, 951–961.

Kaldy, Z., & Sigala, N. (2004). The neural mechanisms of object working memory: What is where in the infant brain? *Neuroscience and Biobehavioral Reviews, 28*, 113–121.

Kaller, M. S., Lazari, A., Blanco-Duque, C., Sampaio-Baptista, C., & Johansen-Berg, H. (2017). Myelin plasticity and behaviour – connecting the dots. *Current Opinion in Neurobiology, 47*, 86–92.

Kalsbeek, A., Voorn, P., Buijs, R. M., Pool, C. W., & Uylings, H. B. (1988). Development of the dopaminergic innervation in the prefrontal cortex of rat. *Journal of Comparative Neurology*, *269*, 58–72.

Kamps, F. S., Richardson, H., Murty, N. A. R., Kanwisher, N., & Saxe, R. (2022). Using child-friendly movie stimuli to study the development of face, place, and object regions from age 3 to 12 years. *Human Brain Mapping*, *43*(9), 2782–2800. https://doi.org/10.1002/hbm.25815

Kang, H. J., Kawasawa, Y. I., Cheng, F., Zhu, Y., Xu, X., Li, M., Sousa, A. M. M., Pletikos, M., Meyer, K. A., Sedmak, G., Guennel, T., Shin, Y., Johnson, M. B., Krsnik, Ž., Mayer, S., Fertuzinhos, S., Umlauf, S., Lisgo, S. N., Vortmeyer, A., . . . Šestan, N. (2011). Spatio-temporal transcriptome of the human brain. *Nature*, *478*, 483–489.

Kanwisher, N., McDermott, J., & Chun, M. M. (1997). The fusiform face area: A module in human extrastriate cortex specialized for face perception. *The Journal of Neuroscience*, *17*(11), 4302–4311.

Kanwisher, N., & Yovel, G. (2006). The fusiform face area: A cortical region specialized for the perception of faces. *Philosophical Transactions of the Royal Society of London. Series B, Biological Sciences*, *361*(1476), 2109–2128.

Karatekin, C. (2001). Developmental disorders of attention. In C. A. Nelson & M. Luciana (Eds.), *The handbook of developmental cognitive neuroscience* (pp. 561–576). MIT Press.

Karatekin, C. (2008). Eye tracking studies of normative and atypical development. In C. A. Nelson & M. Luciana (Eds.), *Handbook of developmental cognitive neuroscience* (2nd ed., pp. 263–300). MIT Press.

Karlsgodt, K. H., Sun, D., Jimenez, A. M., Lutkenhoff, E. S., Willhite, R., van Erp, T. G. M., & Cannon, T. D. (2008). Developmental disruptions in neural connectivity in the pathophysiology of schizophrenia. *Development and Psychopathology*, *20*(4), 1297–1327.

Karmiloff-Smith, A. (1992). *Beyond modularity: A developmental perspective on cognitive science*. MIT Press/Bradford Books.

Karmiloff-Smith, A. (1998). Development itself is the key to understanding developmental disorders. *Trends in Cognitive Sciences*, *2*, 389–398.

Karmiloff-Smith, A. (2002). Development itself is the key to understanding developmental disorders. In M. H. Johnson, Y. Munakata, & R. Gilmore (Eds.), *Brain development and cognition: A reader* (pp. 336–374). Blackwell.

Karmiloff-Smith, A. (2008). Research into Williams syndrome: The state of the art. In C. A. Nelson & M. Luciana (Eds.), *Handbook of developmental cognitive neuroscience* (2nd ed., pp. 691–700). MIT Press.

Karmiloff-Smith, A., Grant, J., Ewing, S., Carette, M. J., Metcalfe, K., Donnai, D., Read, A. P., & Tassabehji, M. (2003). Using case study comparisons to explore genotype/phenotype correlations in Williams syndrome. *Journal of Medical Genetics*, *40*, 136–140.

Karmiloff-Smith, A., & Johnson, M. H. (1994). Thinking on one's feet (review of *A dynamic systems approach to the development of cognition and action*, by Esther Thelan and Linda Smith). *Nature*, *372*, 53–54.

Karmiloff-Smith, A., Klima, E., Bellugi, U., Grant, J., & Baron-Cohen, S. (1995). Is there a social module? Language, face processing and theory of mind in individuals with Williams syndrome. *Journal of Cognitive Neuroscience*, *7*, 196–208.

Kasamatsu, T., & Pettigrew, J. W. (1976). Depletion of brain catecholamines: Failure of monocular dominance shift after monocular conclusion in kittens. *Science*, *194*, 206–209.

Katus, L., Hayes, N. J., Mason, L., Blasi, A., McCann, S., Darboe, M. K., de Haan, M., Moore, S. E., Lloyd-Fox, S., & Elwell, C. E. (2019). Implementing neuroimaging and eye tracking methods to assess neurocognitive development of young infants in low- and middle-income countries. *Gates Open Research, 3*, 1113. https://doi.org/10.12688/gatesopenres.12951.2

Katus, L., Mason, L., Milosavljevic, B., McCann, S., Rozhko, M., Moore, S. E., Elwell, C. E., Lloyd-Fox, S., de Haan, M., & BRIGHT project team. (2020). ERP markers are associated with neurodevelopmental outcomes in 1-5 month old infants in rural Africa and the UK. *Neuroimage, 210*, 116591. https://doi.org/10.1016/j.neuroimage.2020.116591

Katus, L., Milosavljevic, B., Rozhko, M., McCann, S., Mason, L., Mbye, E., Touray, E., Moore, S. E., Elwell, C. E., Lloyd-Fox, S., de Haan, M., & The Bright Study Team. (2022). Neural marker of habituation at 5 months of age associated with deferred imitation performance at 12 months: A longitudinal study in the UK and The Gambia. *Children (Basel), 9*(7), 988. https://doi.org/10.3390/children9070988

Kaufman, J., Csibra, G., & Johnson, M. H. (2003). Representing occluded objects in the human infant brain. *Proceedings of the Royal Society B: Biological Sciences, 270*(Suppl 2), S140–S143.

Kaufman, J., Csibra, G., & Johnson, M. H. (2005). Oscillatory activity in the infant brain reflects object maintenance. *Proceedings of National Academy of Sciences USA, 102*, 15271–15274.

Kaufman, J., Mareschal, D., & Johnson, M. H. (2003). Graspability and object processing in infants. *Infant Behavior and Development, 26*(4), 516–528.

Kavsek, M. J. (2002). The perception of static subjective contours in infancy. *Child Development, 73*(2), 331–344.

Kawabata, H., Gyoba, J., Inoue, H., & Ohtsubo, H. (1999). Visual completion of partly occluded grating in infants under one month of age. *Vision Research, 39*, 3586–3591.

Keehn, B., Lincoln, A. J., Müller, R. A., & Townsend, J. (2010). Attentional networks in children and adolescents with autism spectrum disorder. *Journal of Child Psychology and Psychiatry, 51*(11), 1251–1259.

Kellman, P. J., & Spelke, E. S. (1983). Perception of partly occluded objects in infancy. *Cognitive Psychology, 15*(4), 483–524.

Kelly, D. J., Jack, R. R., Miellet, S., De Luca, E., Foreman, K., & Caldara, R. (2011). Social experience does not abolish cultural diversity in eye movements. *Frontiers in Psychology, 2*(May), 95.

Kelly, D. J., Quinn, P. C., Slater, A. M., Lee, K., Ge, L., & Pascalis, O. (2007). The other-race effect develops during infancy: Evidence of perceptual narrowing. *Psychological Science, 18*, 1084–1089.

Kersey, A. J., Clark, T. S., Lussier, C. A., Mahon, B. Z., & Cantlon, J. F. (2016). Development of tool representations in the dorsal and ventral visual object processing pathways. *NeuroImage, 26*, 3135–3145.

Kerszberg, M., Dehaene, S., & Changeux, J.-P. (1992). Stabilization of complex input–output functions in neural clusters formed by synapse selection. *Neural Networks, 5*, 403–413.

Keverne, E. B., Martel, F. L., & Nevison, C. M. (1996). Primate brain evolution: Genetic and functional considerations. *Proceedings of the Royal Society B: Biological Sciences, 262*, 689–696.

Kihara, M., de Haan, M., Garrashi, H. H., Neville, B. G. R., & Newton, C. R. J. C. (2010). Atypical brain response to novelty in rural African children with a history of severe falciparum malaria. *Journal of the Neurological Sciences, 296*, 88–95.

Kihara, M., Hogan, A. M., Newton, C. R., Garrashi, H. H., Neville, B. R., & de Haan, M. (2010). Auditory and visual novelty processing in normally-developing Kenyan children. *Clinical Neurophysiology, 12*, 564–576.

Killackey, H. P. (1990). Neocortical expansion: An attempt toward relating phylogeny and ontogeny. *Journal of Cognitive Neuroscience, 2*, 1–17.

Killgore, W. D. S., Oki, M., & Yergelin-Todd, A. (2001). Sex-specific developmental changes in amygdala response to affective faces. *Neuroreport, 12*, 427–433.

Kilner, J. M., & Blakemore, S. J. (2007). How does the mirror neuron system change during development? *Developmental Science, 10*, 524–526.

Kilner, J. M., Vargas, C., Duval, S., Blakemore, S. J., & Sirigu, A. (2004). Motor activation prior to observation of a predicted movement. *Nature Neuroscience, 7*, 1299–12301.

Kingsbury, M. A., & Finlay, B. L. (2001). The cortex in multidimensional space: Where do cortical areas come from? *Developmental Science, 4*, 125–142.

Klaver, P., Lichtensteiger, J., Burcher, K., Dietrich, T., Loenneker, T., & Martin, E. (2008). Dorsal stream development in motion and structure-from-motion perception. *NeuroImage, 39*, 1815–1823.

Kleinke, C. L. (1986). Gaze and eye contact; A research review. *Psychological Bulletin, 100*, 78–100.

Klima, E., & Bellugi, U. (1979). *The signs of language.* Harvard University Press.

Klingberg, T. (2006). Development of a superior frontal-intraparietal network for visuo-spatial working memory. *Neuropsychologia, 44*, 2171–2177.

Klingberg, T. (2008). White matter maturation and cognitive development during childhood. In C. A. Nelson & M. Luciana (Eds.), *Handbook of developmental cognitive neuroscience* (2nd ed., pp. 237–244). MIT Press.

Klingberg, T., Forssberg, H., & Westerberg, H. (2002). Increased brain activity in frontal and parietal cortex underlies the development of visuospatial working memory capacity during childhood. *Journal of Cognitive Neuroscience, 14*, 1–10.

Knecht, S., Drager, B., Deppe, M., Bobe, L., Lohmann, H., Floel, A., Ringelstein, E.-B., & Henningsen, H. (2000). Handedness and hemispheric language dominance in health humans. *Brain, 123*, 2515–2518.

Knopik, V. S., Neiderhiser, J. N., DeFries, J. C., & Plomin, R. (2016). *Behavioral genetics* (7th ed.). MacMillan Learning.

Kobayashi, C., Glover, G., & Temple, E. (2007). Cultural and linguistic effects on neural bases of 'Theory of Mind' in American and Japanese children. *Brain Research, 1164*, 95–107.

Koch, F.-S., Sundqvist, A., Thornberg, U. B., Nyberg, S., Lum, J. A. G., Ullmart, M. T., Barr, R., Rudner, M., & Heimann, M. (2020). Procedural memory in infancy: Evidence from implicit sequence learning in an eye-tracking paradigm. *Journal of Experimental Child Psychology, 191*, 104733.

Konrad, K., Neufang, S., Thiel, C. M., Specht, K., Hanisch, C., Fan, J., Herpertz-Dahlmann, B., & Fink, G. R. (2005). Development of attentional networks: An fMRI study with children and adults. *NeuroImage, 28*(2), 429–439.

Kosakowski, H. L., Cohen, M. A., Takahashi, A., Keil, B., Kanwisher, N., & Saxe, R. (2022). Selective responses to faces, scenes and bodies in the ventral visual pathway of infants. *Current Biology, 32*(2), 265–274.e5.

Kostovic, I., Judaš, M., & Petanjek, Z. (2008). Structural development of the human prefrontal cortex. In C. A. Nelson & M. Luciana (Eds.), *Handbook of developmental cognitive neuroscience* (2nd ed., pp. 213–235). MIT Press.

Kostovic, I., Petanjek, Z., & Judas, M. (1993). Early areal differentiation of the human cerebral cortex: Entorhinal area. *Hippocampus, 3*(4), 447–445.

Kovács, Á. M., & Mehler, J. (2009). Cognitive gains in 7-month-old bilingual infants. *Proceedings of the National Academy of Sciences, 106*(16), 6556–6560. https://doi.org/10.1073/pnas.0811323106

Kozorovitskiy, Y., & Gould, E. (2008). Adult neurogenesis in the hippocampus. In C. A. Nelson & M. Luciana (Eds.), *Handbook of developmental cognitive neuroscience* (2nd ed., pp. 51–62). MIT Press.

Krol, K. M., Monakhov, V., Lai, P. S., Ebstein, R., & Grossmann, T. (2015). Genetic variation in CD38 and breastfeeding experience interact to impact infants' attention to social eye cues. *Proceedings of the National Academy of Sciences of the United States of America, 112*, E5434–E4442.

Krol, K. M., Namaky, N., Monakhov, M. V., Lai, P. S., Ebstein, R., & Grossmann, T. (2021). Genetic variation in the oxytocin system and its link to social motivation in human infants. *Psychoneuroendocrinology, 131*, 105290.

Krubitzer, L. A. (1998). What can monotremes tell us about brain evolution? *Philosophical Transactions of the Royal Society of London. Series B, Biological Sciences, 353*, 1127–1146.

Kucian, K., Loenneker, T., Dietrich, T., Martin, E., & von Aster, M. G. (2005). Development of neural networks for number processing: An fMRI study in children and adults. *NeuroImage, 26*(Suppl. 1), 46.

Kuefner, D., de Heering, A., Jacques, C., Palermo-Soler, E., & Rossion, B. (2010). Early visually evoked electrophysiological responses over the human brain (P1, N170) show stable patterns of face-sensitivity from 4 years to adulthood. *Frontiers in Human Neuroscience, 3*, 67.

Kuhl, P. K. (2000). A new view of language acquisition. *Proceedings of the National Academy of Science, 97*, 11850–11857.

Kuhl, P. K., Conboy, B. T., Coffey-Corina, S., Padden, D., Fivera-Gaxiola, M., & Nelson, T. (2008). Phonetic learning as a pathway to language: New data and native language magnet theory expanded (NLM-e). *Philosophical Transactions of the Royal Society B, 363*, 979–1000.

Kvavilashvili, L., Messer, D. J., & Ebdon, P. (2001). Prospective memory in children: The effects of age and task interruption. *Developmental Psychology, 37*(3), 418–430.

Kyttälä, M., Aunio, P., Lehto, J. E., & Luit, J. V. (2003). Visuospatial working memory and early numeracy. *Educational and Child Psychology, 20*, 65–76.

Lai, C. S. L., Fisher, S. E., Hurst, J. A., Vargha-Khadem, F., & Monaco, A. P. (2001). A forkhead-domain gene is mutated in a severe speech and language disorder. *Nature, 413*, 519–523.

Landry, O., & Parker, A. (2013). A meta-analysis of visual orienting in autism. *Frontiers in Human Neuroscience, 7*, 833.

Landry, R., & Bryson, S. E. (2004). Impaired disengagement of attention in young children with autism. *Journal of Child Psychology and Psychiatry, 45*, 1115–1122.

Langton, S. R. H., & Bruce, V. (1999). Reflexive visual orienting in response to the social attention of others. *Visual Cognition, 6*, 541–567.

Larsen, B., & Luna, B. (2018). Adolescence as a neurobiological critical period for the development of higher-order cognition. *Neuroscience & Biobehavioral Reviews, 94*, 179–195.

Lavenex, P., Banta Lavenex, P., & Amaral, D. G. (2007). Postnatal development of the primate hippocampal formation. *Developmental Neuroscience, 29,* 179–192.

Le Grand, R., Mondloch, C. J., Maurer, D., & Brent, H. P. (2001). Early visual experience and face processing. *Nature, 410,* 890.

Le, T. M., Huang, A. S., O'Rawe, J., & Leung, H. C. (2020). Functional neural network configuration in late childhood varies by age and cognitive state. *Developmental Cognitive Neuroscience, 45,* 100862.

Leamey, C. A., Glendining, K. A., Kreiman, G., Kang, N.-D., Wang, K. H., Fassler, R., Sawatari, A., Tonegawa, S., & Sur, M. (2008). Differential gene expression between sensory neocortical areas: Potential roles for Ten_m3 and Bc16 in patterning visual and somatosensory pathways. *Cerebral Cortex, 18,* 53–66.

Lebel, C., & Beaulieu, C. (2011). Longitudinal development of human brain wiring continues from childhood into adulthood. *Journal of Neuroscience, 31*(30), 10937–10947.

Lebel, C., & Deoni, S. (2018). The development of brain white matter microstructure. *NeuroImage, 182,* 207–218.

LeBlanc, J. J., & Fagiolini, N. (2011). Autism: A 'critical period' disorder? *Neural Plasticity,* 921680. https://doi.org/10.1155/2011/921680

Lee, J. J., Wedow, R., Okbay, A., Kong, E., Maghzian, O., Zacher, M., Nguyen-Viet, T. A., Bowers, P., Sidorenko, J., Karlsson Linnér, R., Fontana, M. A., Kundu, T., Lee, C., Li, H., Li, R., Royer, R., Timshel, P. N., Walters, R. K., Willoughby, E. A., . . . Cesarini, D. (2018). Gene discovery and polygenic prediction from a genome-wide association study of educational attainment in 1.1 million individuals. *Nature Genetics, 50*(8), 1112–1121.

Lee, K., Bull, R., & Ho, R. M. H. (2013). Developmental changes in executive functioning. *Child Development, 84*(6), 1933–1953.

Lenneberg, E. (1967). *Biological foundations of language.* Wiley.

Leonard, C. M., & Eckert, M. A. (2008). Asymmetry and dyslexia. *Developmental Neuropsychology, 33,* 663–681.

Leppanen, J., Peltola, M., Mantymaa, M., Puura, K., Mononen, N., & Lehtimaki, T. (2011). Serotonin and early cognitive development: Variation in the tryptophan hydroxylase 2 gene is associated with visual attention in 7-month-old infants. *Journal of Child Psychology and Psychiatry, 52*(11), 1144–1152.

Lewis, D. A. (2012). Cortical dysfunction and cognitive deficits in schizophrenia-implications for preemptive interventions. *European Journal of Neuroscience, 35,* 1871–1878.

Lewis, T. L., Maurer, D., & Milewski, A. (1979). The development of nasal detection in young infants. *Investigative Opthalmology and Visual Science Supplement, 18,* 271.

Liégeois, F., Baldeweg, T., Connelly, A., Gadian, D. G., Mishkin, M., & Vargha-Khadem, F. (2003). Language fMRI abnormalities associated with FOXP2 gene mutation. *Nature Neuroscience, 6*(11), 1230–1237.

Liégeois, F., Connelly, A., Cross, J. H., Boyd, S. G., Gadian, D. G., Vargha-Khadem, F., & Baldeweg, T. (2004). Language reorganization in children with early onset lesions of the left hemisphere: An fMRI study. *Brain, 127,* 1229–1236.

Liegeois, F., Mayes, A., & Morgan, A. (2014). Neural correlates of developmental speech and language disorders: Evidence from neuroimaging. *Current Developmental Disorders Reports, 1,* 215–227.

Liegeois, F. J., Hildebrand, M. S., Bonthrone, A., Turner, S. J., Scheffer, I. E., Bahlo, M., Connelly, A., & Morgan, T. (2016). Early neuroimaging markers of FOXP2 intragenic deletion. *Scientific Reports, 6*, 35192.

Lipsitt, L. P. (1990). Learning processes in the human newborn: Sensitization, habituation, and classical conditioning. *Annals of the New York Academy of Sciences, 608*, 113–127.

Lipton, J. S., & Spelke, E. (2003). Origins of number sense: Large-number discrimination in human infants. *Psychological Science, 14*(5), 396–401.

Liston, C., Watts, R., Tottenham, N., Davidson, M. C., Niogi, S., Ulug, A. M., & Casey, B. J. (2006). Frontostriatal microstructure modulates efficient recruitment of cognitive control. *Cerebral Cortex, 16*(4), 553–560.

Livingstone, M. S., Rosen, G. D., Drislane, F. W., & Galaburda, A. M. (1991). Physiological and anatomical evidence for a magnocellular defect in developmental dyslexia. *Proceedings of the National Academy of Sciences of the United States of America, 88*, 7943–7947.

Lloyd-Fox, S., Begus, K., Halliday, D., Pirazolli, L., Blasi, A., Papademtriou, M., Darboe, M. K., Prentice, A., Johnson, M. H., Moore, S. E., & Elwell, C. (2017). Cortical specialisation to social stimuli from the first days to the second year of life: A rural Gambian cohort. *Developmental Cognitive Neuroscience, 25*, 92–104.

Lloyd-Fox, S., Blasi, A., & Elwell, C. E. (2010). Illuminating the developing brain: The past, present and future of functional near infrared spectroscopy. *Neuroscience and Biobehavioural Reviews, 34*(3), 269–284.

Lloyd-Fox, S., Blasi, A., Elwell, C. E., Charman, T., Murphy, D., & Johnson, M. H. (2013). Reduced neural sensitivity to social stimuli in infants at risk for autism. *Proceedings of the Royal Society B, 280*, 20123026.

Lloyd-Fox, S., Blasi, A., Mercure, E., Elwell, C. E., & Johnson, M. H. (2012). The emergence of cerebral specialization for the human voice over the first months of life. *Social Neuroscience, 7*(3), 317–330. https://doi.org/10.1080/17470919.2011.614696

Lloyd-Fox, S., Blasi, A., Pasco, G., Gliga, T., Jones, E. J. H., Murphy, D. G. M., Elwell, C. E., Charman, T., Johnson, M. H., & The BASIS Team. (2017). Cortical responses before six months of life associate with later autism. *European Journal of Neuroscience, 47*, 736–749.

Lloyd-Fox, S., Blasi, A., Volein, A., Everdell, N., Elwell, C., & Johnson, M. H. (2009). Social perception in infancy: A near infrared spectroscopy study. *Child Development, 80*, 986–999.

Lloyd-Fox, S., Papademetriou, M., Darboe, M. K., Everdell, N. L., Wegmuller, R., Prentice, A. M., Moore, S. E., & Elwell, C. E. (2014). Functional near infrared spectroscopy (fNIRS) to assess cognitive function in infants in rural Africa. *Scientific Reports, 4*, 4740. https://doi.org/10.1038/srep04740

Logue, S., Chein, J., Gould, T., Holliday, E., & Steinberg, L. (2014). Adolescent mice, unlike adults, consume more alcohol in the presence of peers than alone. *Developmental Science, 17*(1), 79–85. https://doi.org/10.1111/desc.12101

Loo, S. K., Hale, T. S., Macion, J., Hanada, G., McGough, J. J., McCracken, J. T., & Smalley, S. L. (2009). Cortical activity patterns in ADHD during arousal, activation and sustained attention. *Neuropsychologia, 47*, 2114–2119.

Lorenz, K. (1965). *Evolution and the modification of behavior.* University of Chicago Press.

Luciana, M. (2003). The neural and functional development of human prefrontal cortex. In M. de Haan & M. H. Johnson (Eds.), *The cognitive neuroscience of development* (pp. 157–174). Psychology Press.

Luciana, M., & Nelson, C. A. (1998). The functional emergence of prefrontally-guided working memory systems in four- to eight-year-old children. *Neuropsychologia, 36*(3), 273–293.

Luciana, M., & Nelson, C. A. (2000). Neurodevelopmental assessment of cognitive function using the Cambridge Neuropsychological Testing Automated Battery (CANTAB): Validation and future goals. In M. Ernst & J. M. Rumsey (Eds.), *Functional neuroimaging in child psychiatry* (pp. 379–397). Cambridge University Press.

Luna, B., Garver, K. E., Urban, T. A., Lazar, N. A., & Sweeney, J. A. (2004). Maturation of cognitive processes from late childhood to adulthood. *Child Development, 75*, 1357–1372.

Luna, B., Marek, S., Larsen, B., Tervo-Clemmens, B., & Chahal, R. (2015). An integrative model of the maturation of cognitive control. *Annual Review of Neuroscience, 38*, 151–170.

Luna, B., Thulborn, K. R., Munoz, D. P., Merriam, E. P., Garver, K. E., Minshew, N. J., Keshavan, M. S., Genovese, C. R., Eddy, W. F., & Sweeney, J. A. (2001). Maturation of widely distributed brain function subserves cognitive development. *NeuroImage, 13*(5), 786–793.

Lynch, K. M., Shi, Y., Toga, A. W., Clark, K. A., & Pediatric Imaging, Neurocognition and Genetics Study. (2019). Hippocampal shape maturation in childhood and adolescence. *Cerebral Cortex, 229*(9), 3651–3665.

Mackey, A. P., Finn, A. S., Leonard, J. A., Jacoby-Senghor, D. S., West, M. R., Gabrieli, C. F. O., & Gabrieli, J. D. E. (2015). Neuroanatomical correlates of the income-achievement gap. *Psychological Science, 26*(6), 925–933. https://doi.org/10.1177/0956797615572233

MacNeill, L. A., Ram, N., Bell, M. A., Fox, N. A., & Pérez-Edgar, K. (2018). Trajectories of infants' biobehavioral development: Timing and rate of A-not-B performance gains and EEG maturation. *Child Development, 89*(3), 711–724. https://doi.org/10.1111/cdev.13022

MacSweeney, M., Capek, C. M., Campbell, R., & Woll, B. (2008). The signing brain: The neurobiology of sign language. *Trends in Cognitive Sciences, 12*, 438–440.

Maggioni, E., Squarcina, L., Dusi, N., Diwadkar, V. A., & Brambilla, P. (2020). Twin MRI studies on genetic and environmental determinants of brain morphology and function in the early lifespan. *Neuroscience and Biobehavioral Reviews, 109*, 139–149. https://doi.org/10.1016/j.neubiorev.2020.01.003

Magis-Weinberg, L., Custers, R., & Dumontheil, I. (2020). Sustained and transient processes in event-based prospective memory in adolescence and adulthood. *Journal of Cognitive Neuroscience, 32*(10), 1924–1945. https://doi.org/10.1162/jocn_a_01604

Maguire, E. A., Vargha-Khadem, F., & Mishkin, M. (2001). The effects of bilateral hippocampal damage on fMRI regional activations and interactions during memory retrieval. *Brain, 124*, 1156–1170.

Mahon, B. Z., Milleville, S., Negri, G. A. L., Rumiati, R. I., Martin, A., & Caramazza, A. (2007). Action-related properties of objects shape object representations in the ventral stream. *Neuron, 55*(3), 507–520.

Malanchini, M., Rimfeld, K., Allegrini, A. G., Ritchie, S. J., & Plomin, R. (2020). Cognitive ability and education: How behavioural genetic research has advanced our knowledge and understanding of their association. *Neuroscience and biobehavioral reviews, 111*, 229–245.

Mancini, J., Casse-Perrot, C., Giusiano, B., Girard, N., Camps, R., Deruelle, C., & de Schonen, S. (1998). Face processing development after a perinatal unilateral brain lesion. *Human Frontiers Science Foundation Developmental Cognitive Neuroscience Technical Report Series*, No. 98.6.

Marcus, G. F., & Fisher, S. E. (2003). FOXP2 in focus: What can genes tell us about speech and language? *Trends in Cognitive Sciences, 7*(6), 257–262.

Marcusson, J. O., Morgan, D. G., Winblad, B., & Finch, C. E. (1984). Serotonin-2 binding sites in human frontal cortex and hippocampus: Selective loss of S-2 A sites with age. *Brain Research, 311*, 51–56.

Mareschal, D. (2010). Computational perspectives on cognitive development. *Wiley Interdisciplinary Reviews: Cognitive Science, 1*(5), 696–708.

Mareschal, D., Butterworth, B., & Tolmie, A. (2013). *Educational neuroscience.* Wiley-Blackwell.

Mareschal, D., & Johnson, M. H. (2003). The "what" and "where" of infant object representations. *Cognition, 88*, 259–276.

Mareschal, D., Johnson, M. H., Sirois, S., Spratling, M., Thomas, M., & Westermann, G. (2007). *Neuroconstuctivism: How the brain constructs cognition.* Oxford University Press.

Mareschal, D., Plunkett, K., & Harris, P. (1999). A computational and neuropsychological account of object-oriented behaviours in infancy. *Developmental Science, 2*, 306–317.

Mareschal, D., & Thomas, M. S. C. (2007). Computational modelling in developmental psychology. *IEEE Transactions on Evolutionary Computation (Special Issue on Autonomous Mental Development), 11*, 137–150.

Marin-Padilla, M. (1990). The pyramidal cell and its local-circuit interneurons: A hypothetical unit of the mammalian cerebral cortex. *Journal of Cognitive Neuroscience, 2*, 180–194.

Markant, J., Cicchetti, D., Hetzel, S., & Thomas, K. M. (2014). Relating dopaminergic and cholinergic polymorphisms to spatial attention in infancy. *Developmental Psychology, 50*, 360–369.

Markham, J., & Greenough, W. T. (2004). Experience-driven brain plasticity: Beyond the synapse. *Neuron Glia Biology, 1*, 351–364.

Marr, D. (1982). *Vision.* W.H. Freeman.

Martin, K., Ketchabaw, W. T., & Turkeltaub, P. E. (2022). Plasticity of the language system in children and adults. *Handbook of Clinical Neurology, 184*, 397–414.

Martinos, M., Yoong, M., Patil, S., Chin, R. F., Neville, B. G., Scott, R. C., & de Haan, M. (2012). Recognition memory is impaired in children after prolonged febrile seizures. *Brain, 135*, 3153–3164.

Matejko, A. A., Hutchison, J. E., & Ansari, D. (2019). Developmental specialization of the left intraparietal sulcus for symbolic ordinal processing. *Cortex, 114*, 41–53.

Matsuzawa, J. (1985). Color naming and classification in a chimpanzee (*Pan troglodytes*). *Journal of Human Evolution, 14*, 283–291.

Matsuzawa, T. (1991). Nesting cups and metatools in chimpanzees. *Behavioral and Brain Sciences, 14*, 570–571.

Matsuzawa, T. (2007). Comparative cognitive development. *Developmental Science, 10*, 97–103.

Maurer, D. (1985). Infants' perception of facedness. In T. N. Field & N. Fox (Eds.), *Social perception in infants* (pp. 73–100). Ablex.

Maurer, D., & Barrera, M. (1981). Infants' perception of natural and distorted arrangements of a schematic face. *Child Development, 47*, 523–527.

Maurer, D., Lewis, T. L., & Mondloch, C. J. (2008). Plasticity of the visual system. In C. A. Nelson & M. Luciana (Eds.), *Handbook of developmental cognitive neuroscience* (2nd ed., pp. 415–438). MIT Press.

Maurer, U., Brem, S., Bucher, K., & Brandeis, D. (2005). Emerging neurophysiological specialization for letter strings. *Journal of Cognitive Neuroscience, 17*, 1532–1552.

Maurer, U., Brem, S., Bucher, K., Kranz, F., Benz, R., Halder, P., Steinhausen, H.-C., & Brandeis, D. (2007). Impaired tuning of a fast occipito-temporal response to print in dyslexic children learning to read. *Brain, 130*, 3200–3210.

Maurer, U., Brem, S., Kranz, F., Bucher, K., Benz, R., Halder, P., Steinhausen, H.-C., & Brandels, D. (2006). Coarse neural tuning for print peaks when children learn to read. *NeuroImage, 33*, 749–758.

Maylor, E. A. (1985). Facilitory and inhibitory components of orienting in visual space. In M. I. Posner & O. M. Marin (Eds.), *Attention and performance XI* (pp. 189–204). Erlbaum.

McCarthy, E. K., Murray, D. M., & Kiely, M. E. (2022). Iron deficiency during the first 1000 days of life: Are we doing enough to protect the developing brain? *The Proceedings of the Nutrition Society, 81*(1), 108–118.

McCarthy, G., Puce, A., Gore, J. C., & Alison, T. (1997). Face-specific processing in the human fusiform gyrus. *Journal of Cognitive Neuroscience, 9*, 605–610.

McDonough, L., Mandler, J. M., McKee, R. D., & Squire, L. R. (1995). The deferred imitation task as a nonverbal measure of declarative memory. *Proceedings of the National Academy of Sciences of the United States of America, 92*, 7580–7584.

Meaburn, E., Dale, P. S., Craig, I. W., & Plomin, R. (2002). Language-impaired children: No sign of the FOXP2 mutation. *Neuroreport, 13*, 1075–1077.

Méary, D., Mermillod, M., & Pascalis, O. (2014). Binocular correlation model of face preference: How good, how simple? *Developmental Science, 17*(6), 828–830. https://doi.org/10.1111/desc.12201

Meek, J. H. (2002). Basic principles of optical imaging and application to the study of infant development. *Developmental Science, 5*(3), 371–380.

Meek, J. H., Firbank, M., Elwell, C. E., Atkinson, J., Braddick, O., & Wyatt, J. S. (1998). Regional hemodynamic responses to visual stimulation in awake infants. *Paediatric Research, 43*, 840–843.

Mehler, J., Nespor, M., Gervain, J., Endress, A., & Shukla, M. (2008). Mechanisms of language acquisition: Imaging and behavioural evidence. In C. A. Nelson & M. Luciana (Eds.), *The handbook of developmental cognitive neuroscience* (2nd ed., pp. 325–336). MIT Press.

Melby-Lervåg, M., Redick, T. S., & Hulme, C. (2016). Working memory training does not improve performance on measures of intelligence or other measures of "far transfer": Evidence from a meta-analytic review. *Perspectives on Psychological Science, 11*(4), 512–534.

Meltzoff, A. N., & Moore, M. K. (1977). Imitation of facial and manual gestures by human neonates. *Science, 198*, 74–78.

Menon, V., Boyett-Anderson, J. M., & Reiss, A. L. (2005). Maturation of medial temporal lobe response and connectivity during memory encoding. *Brain Research. Cognitive Brain Research, 25*, 379–385.

Mercure, E., Evans, S., Perazzoli, L., Goldberg, L., Bowden-Howl, H., Coulson-Thaker, K., Beedi, I., Lloyd-Fox, S., Johnson, M. H., & MacSweeney, M. (2020). Language experience impacts brain activation for spoken and signed language in infancy: Insights from unimodal and bimodal bilinguals. *Neurobiology of Language (Cambridge), 1*, 9–32.

Merigan, W., & Maunsell, J. (1993). How parallel are the primate visual pathways? *Annual Review of Neuroscience, 16*, 369–402.

Merzenich, M. M., Wright, B. A., Jaenkins, W., Xerri, C., Byl, N., & Miller, S. L. (2002). Cortical plasticity underlying perceptual, motor, and cognitive skill development: Implications for neurorehabilitation. In M. H. Johnson, Y. Munakata, & R. Gilmore (Eds.), *Brain development and cognition: A reader* (2nd ed., pp. 292–304). Blackwell.

Meulemans, T., Van der Linden, M., & Perruchet, P. (1998). Implicit sequence learning in children. *Journal of Experimental Child Psychology, 69*, 199–221.

Meyer-Lindenberg, A., & Weinberger, D. R. (2006). Intermediate phenotypes and genetic mechanisms of psychiatric disorders. *Nature Reviews, 7*(10), 818–827. https://doi.org/10.1038/nrn1993

Micheletti, S., Corbett, F., Atkinson, J., Barddick, O., Mattei, P., Galli, J., Calza, S., & Fazzi, E. (2021). Dorsal and ventral stream function in children with developmental coordination disorder. *Frontiers in Human Neuroscience, 15*, 703217.

Miller, J. G., & Kinsbourne, M. (2012). Culture and neuroscience in developmental psychology: Contributions and challenges. *Child Development Perspectives, 6*(1), 35–41.

Mills, K. L., Dumontheil, I., Speekenbrink, M., & Blakemore, S.-J. (2015). Multitasking during social interactions in adolescence and early adulthood. *Royal Society Open Science, 2*(11). https://doi.org/10.1098/rsos.150117

Mills, K. L., Goddings, A.-L., Clasen, L. S., Giedd, J. N., & Blakemore, S.-J. (2014). The developmental mismatch in structural brain maturation during adolescence. *Developmental Neuroscience, 36*(3–4), 147–160.

Mills, K. L., Goddings, A.-L., Herting, M. M., Meuwese, R., Blakemore, S.-J., Crone, E. A., Dahl, R. E., Gurlglu, B., Raznahan, A., Sowell, E. R., & Tamnes, C. K. (2016). Structural brain development between childhood and adulthood: Convergence across four longitudinal samples. *NeuroImage, 141*, 273–281.

Milner, A. D., & Goodale, M. A. (1995). *The visual brain in action.* Oxford University Press.

Milner, B. (1982). Some cognitive effects of frontal-lobe lesions in man. *Philosophical Transactions of the Royal Society of London. Series B, Biological Sciences, 298*, 211–226.

Milosavljevic, B., Vellekoop, P., Maris, H., Halliday, D., Drammeh, S., Sanyang, L., Darboe, M. K., Elwell, C., Moore, S. E., & Lloyd-Fox, S. (2019). Adaptation of the Mullen scales of early learning for using among infants aged 5- to 24-months in rural Gambia. *Developmental Science, 22*(5), e12808.

Minagawa-Kawai, Y., van der Lely, H., Ramus, F., Sato, Y., Mazuka, R., & Dupoux, E. (2011). Optical brain imaging reveals general auditory and language-specific processing in early infant development. *Cerebral Cortex, 21*(2), 254–261. https://doi.org/10.1093/cercor/bhq082

Minshew, N. J., & Williams, D. L. (2007). The new neurobiology of autism: Cortex, connectivity, and neuronal organization. *Archives of Neurology, 64*, 945–950.

Mishkin, M., Suzuki, W. A., Gadian, D. G., & Vargha-Khadem, F. (1997). Hierarchical organization of cognitive memory. *Philosophical Transactions of the Royal Society, London B Biological Sciences, 352*, 1461–1467.

Miyashita-Lin, E. M., Hevner, R., Wassarman, K. M., Martinez, S., & Rubenstein, J. L. (1999). Early neocortial regionalization in the absence of thalamic innervation. *Science, 285*, 906–909.

Molapour, T., Hagan, C. C., Silston, B., Wu, H., Ramstead, M., Friston, K., & Mobbs, D. (2021). Seven computations of the social brain. *Social Cognitive and Affective Neuroscience, 16*(8), 745–760. https://doi.org/10.1093/scan/nsab024

Molnar, Z., & Blakemore, C. (1991). Lack of regional specificity for connections formed between thalamus and cortex in coculture. *Nature, 351,* 475–477.

Monette, S., Bigras, M., & Guay, M.-C. (2011). The role of executive functions in school achievement at the end of Grade 1. *Journal of Experimental Child Psychology, 109,* 158–173.

Moore, R. J., Vadeyar, S. H., Fulford, J., Tyler, D. J., Gribben, C., Baker, P. N., James, D., & Gowland, P. A. (2001). Antenatal determination of fetal brain activity in response to an acoustic stimulus using functional magnetic resonance imaging. *Human Brain Mapping, 12,* 94–99.

Morton, J., & Johnson, M. H. (1991). CONSPEC and CONLERN: A two-process theory of infant face recognition. *Psychological Review, 98*(2), 164–181. https://doi.org/10.1037/0033-295X.98.2.164

Morton, J., Mehler, J., & Jusczyk, P. W. (1984). On reducing language to biology. *Cognitive Neuropsychology, 1,* 83–116.

Mosconi, M., McCarthy, G., & Pelphrey, K. A. (2005). Taking an intentional stance on eye gaze shifts: A functional neuroimaging study of social perception in children. *NeuroImage, 27,* 247–252.

Mounoud, P., Duscherer, K., Moy, G., & Perraudin, S. (2007). The influence of action perception on object recognition: A developmental study. *Developmental Science, 10,* 836–852.

Mulder, H., Van Houdt, C. A., Van der Ham, I. J. M., Van der Stigchel, S., & Oudegenoeg-Paz, O. (2020). Attentional flexibility predicts A-not-B task performance in 14-month-old infants: A head-mounted eye tracking study. *Brain Sciences, 10*(5), 279.

Munakata, Y., McClelland, J.L., Johnson, M.H. & Siegler, R.S. (1994). *Now you see it, now you don't: A gradualistic framework for understanding infants' successes and failures in object permanence tasks* (Technical Report No. PDP.CNS.94.2). Carnegie Mellon University.

Munakata, Y., Stedron, J. M., Chatham, C. H., & Kharitonova, M. (2008). Neural network models of cognitive development. In C. A. Nelson & M. Luciana (Eds.), *The handbook of developmental cognitive neuroscience* (2nd ed., pp. 367–382). MIT Press.

Munawar, K., Kuhn, S. K., & Haque, S. (2018). Understanding the reminiscence bump: A systematic review. *PLoS One, 13*(12), e0208595. https://doi.org/10.1371/JOURNAL.PONE.0208595

Mundy, P., & Jarrold, W. (2010). Infant joint attention, neural networks and social cognition. *Neural Networks, 23*(8–9), 985–997. https://doi.org/10.1016/j.neunet.2010.08.009

Munoz, M., Chadwick, M., Perez-Hernandez, E., Vargha-Khadem, F., & Mishkin, M. (2011). Novelty preference in patients with developmental amnesia. *Hippocampus, 21,* 12668–11276.

Murty, V. P., Calabro, F., & Luna, B. (2016). The role of experience in adolescent cognitive development: Integration of executive, memory, and mesolimbic systems. *Neuroscience and Biobehavioral Reviews, 70,* 46–58.

Nadig, A. S., Ozonoff, S., Young, G. S., Rozga, A., Sigman, M., & Rogers, S. J. (2007). A prospective study of response to name in infants at risk for autism. *The Journal of Pediatrics, 161*(4), 378–383.

Nagy, Z., Westergerg, H., & Klingberg, T. (2004). Maturation of white matter is associated with the development of cognitive functions during childhood. *Journal of Cognitive Neuroscience, 16,* 1227–1233.

Naito, M. (2003). The relationship between theory of mind and episodic memory: Evidence for the development of autonoetic consciousness. *Journal of Experimental Child Psychology, 85*(4), 312–336. https://doi.org/10.1016/s0022-0965(03)00075-4

Naito, M., & Koyama, K. (2006). The development of false-belief understanding in Japanese children: Delay and difference? *International Journal of Behavioral Development, 30*(4), 290–304. https://doi.org/10.1177/0165025406063622

Nakao, T., Radua, J., Rubia, K., & Mataix-Cols, D. (2011). Gray matter volume abnormalities in ADHD: Voxel-based meta-analysis exploring the effects of age and stimulant medication. *The American Journal of Psychiatry, 168*(11), 1154–1163.

Nelson, C. A. (1995). The ontogeny of human memory: A cognitive neuroscience perspective. *Developmental Psychology, 31*, 723–738.

Nelson, C. A. (2003). The development of face recognition reflects an experience-expectant and activity-dependent process. In O. Pascalis & A. Slater (Eds.), *The development of face processing in infancy and early childhood: Current perspectives* (pp. 79–98). Nova Science Publishers.

Nelson, C. A., de Haan, M., & Thomas, K. M. (2006). *Neuroscience and cognitive development: The role of experience and the developing brain*. Wiley.

Nelson, C. A., & Luciana, M. (Eds.) (2008). *The handbook of developmental cognitive neuroscience* (2nd ed.). MIT Press.

Nelson, C. A., & Ludemann, P. M. (1989). Past, current and future trends in infant face perception research. *Canadian Journal of Psychology, 43*, 183–198.

Nelson, C. A., & Webb, S. J. (2003). A cognitive neuroscience perspective on early memory development. In M. de Haan & M. H. Johnson (Eds.), *The cognitive neuroscience of development* (pp. 99–126). Psychology Press.

Nelson, E. E., Leibenluft, E., McClure, E. B., & Pine, D. S. (2005). The social re-orientation of adolescence: A neuroscience perspective on the process and its relation to psychopathology. *Psychological Medicine, 35*, 163–174.

Neville, H. J., & Bavelier, D. (2002). Specificity and plasticity in neurocognitive development in humans. In M. H. Johnson, Y. Munakata, & R. Gilmore (Eds.), *Brain development and cognition: A reader* (2nd ed., pp. 251–270). Blackwell.

Neville, H. J., Bavelier, D., Corina, D., Rauschecker, J. P., Karni, A., Lalwani, A., Braun, A., Clark, V., Jezzard, P., & Turner, R. (1998). Cerebral organization for language in deaf and hearing subjects: Biological constraints and effects of experience. *Proceedings of the National Academy of Sciences of the United States of America, 95*, 922–929. https://doi.org/10.1073/pnas.95.3.92

Neville, H. J., Stevens, C., Pakulak, E., Bell, T. A., Fanning, J., Klein, S., & Isbell, E. (2013). Family-based training program improves brain function, cognition, and behavior in lower socioeconomic status preschoolers. *Proceedings of the National Academy of Sciences of the United States of America, 110*(29), 12138–12143.

Newsome, W. T., Wurtz, R. H., & Komatsu, H. (1988). Relation of cortical areas MT and MST to pursuit eye movements. II. Differentiation of retinal from extraretinal inputs. *Journal of Neurophysiology, 60*(2), 604–620.

Niego, A., & Benítez-Burraco, A. (2022). Autism and Williams syndrome: Truly mirror conditions in the socio-cognitive domain? *International Journal of Developmental Disabilities, 68*(4), 399–415. https://doi.org/10.1080/20473869.2020.1817717

Nielsen, M., Haun, D., Kärtner, J., & Legare, C. H. (2017). The persistent sampling bias in developmental psychology: A call to action. *Journal of Experimental Child Psychology, 162*, 31–38.

Northam, B. G., Liegeois, F., Tournier, J. D., Croft, L. J., Johns, P. N., Chong, W. K., Wyatt, J. S., & Baldeweg, T. (2012). Interhemispheric temporal lobe connectivity predicts language impairment in adolescents born preterm. *Brain, 135*, 3781–3798.

Nowakowski, R. S. (1987). Basic concepts of CNS development. *Child Development, 58*, 568–595.

Nydén, A., Niklasson, L., Stahlberg, O., Anckarsäter, H., Wentz, E., Rastam, M., & Gillberg, C. (2010). Adults with autism spectrum disorder and ADHD neuropsychological aspects. *Research in Developmental Disabilities, 31*(6), 1659–1668. https://doi.org/10.1016/j.ridd.2010.04.010

Nystrom, P., Ljunghammar, T., Rosander, K., & von Hofsten, C. (2011). Using mu rhythm desynchronization to measure mirror neuron activity in infants. *Developmental Science, 14*, 327–335.

O'Hare, E. D., & Sowell, E. R. (2008). Imaging human developmental changes in the grey and white matter of the human brain. In C. A. Nelson & M. Luciana (Eds.), *The handbook of developmental cognitive neuroscience* (pp. 23–38). MIT Press.

O'Hearn, K., Roth, J. K., Courtney, S. M., Luna, B., Street, W., Terwillinger, R., & Landau, B. (2011). Object recognition in Williams syndrome: Uneven ventral stream activation. *Developmental Science, 14*(3), 549–565.

Ojeda, J., & Avlia, A. (2019). Early actions of neurotransmitters during cortex development and maturation of reprogrammed neurons. *Frontiers in Synaptic Neuroscience, 11*, 33. https://doi.org/10.3389/fnsyn.2019.00033

O'Leary, D. D. M. (2002). Do cortical areas emerge from a protocortex? In M. H. Johnson, Y. Munakata, & R. O. Gilmore (Eds.), *Brain development and cognition: A reader* (pp. 217–230). Blackwell.

O'Leary, D. D. M., & Nakagawa, Y. (2002). Patterning centers, regulatory genes and extrinsic mechanisms controlling arealization of the neocortex. *Current Opinion in Neurobiology, 12*, 14–25.

Olesen, P. J., Nagy, Z., Westerberg, H., & Klingberg, T. (2003). Combined analysis of DTI and fMRI data reveals a joint maturation of white and grey matter in a fronto-parietal network. *Cognitive Brain Research, 18*, 48–57.

Olesen, P. J., Westerberg, H., & Klingberg, T. (2004). Increased prefrontal and parietal activity after training of working memory. *Nature Neuroscience, 7*, 75–79.

Oliver, A., Johnson, M. H., Karmiloff-Smith, A., & Pennington, B. (2000). Deviations in the emergence of representations: A neuroconstructivist framework for analyzing developmental disorders. *Developmental Science, 3*, 1–23.

Oliver, A., Johnson, M. H., & Shrager, J. (1996). The emergence of hierarchical clustered representations in a Hebbian neural network model that simulates aspects of development in the neocortex. *Network: Computation in Neural Systems, 7*, 291–299.

Olson, E. A., & Luciana, M. (2008). The development of prefrontal cortex functions in adolescence: Theoretical models and a possible dissociation of dorsal versus ventral subregions. In C. A. Nelson & M. Luciana (Eds.), *The handbook of developmental cognitive neuroscience* (2nd ed., pp. 575–590). MIT Press.

O'Reilly, M. A., Bathelt, J., Sakkalou, E., Sakki, H., Salt, A., Dale, N. J., & de Haan, M. (2017). Frontal EEG asymmetry and later behavior vulnerability in infants with congenital visual impairment. *Clinical Neurophysiology, 128*(11), 2191–2199.

O'Reilly, R. C. (1998). Six principles for biologically based computational models of cortical cognition. *Trends in Cognitive Sciences, 2*, 455–462.

O'Reilly, R. C. (2006). Biologically based computational models of high-level cognition. *Science, 314*, 91–94.

Overman, W., Bachevalier, J., Turner, M., & Peuster, A. (1992). Object recognition versus object discrimination: Comparison between human infants and infant monkeys. *Behavioral Neuroscience, 106*, 15–29. https://doi.org/10.1037/0735-7044.106.1.15

Owen, A. M. (1997). Tuning into the temporal dynamics of brain activation using functional magnetic resonance imaging (fMRI). *Trends in Cognitive Sciences, 1*(4), 123–125.

Oyama, S. (2000). *The ontogeny of information: Developmental systems and evolution*(2 rev. ed.). Duke University Press.

Pahs, G., Rankin, P., Cross, H. J., Croft, L., Northam, G. B., Liegeois, F., Greenway, S., Harrison, S., Vargha-Khadem, F., & Baldeweg, T. (2013). Asymmetry of planum temporale constrains interhemispheric language plasticity in children with focal epilepsy. *Brain, 136*, 3163–3175.

Pallas, S. L. (2001). Intrinsic and extrinsic factors shaping cortical identity. *Neurosciences, 24*, 417–423.

Papageorgiou, K. A., Smith, T. J., Wu, R., Johnson, M. H., Kirkham, N. Z., & Ronald, A. (2014). Individual differences in infant fixation duration relate to attention and behavioral control in childhood. *Psychological Science*. Advance online publication, *25*(7), 1371–1379. https://doi.org/10.1177/0956797614531295

Pascalis, O., de Haan, M., & Nelson, C. A. (2002). Is face processing species-specific during the first year of life? *Science, 14*, 199–209.

Pascalis, O., de Haan, M., Nelson, C. A., & de Schonen, S. (1998). Long-term recognition memory for faces assessed by visual paired comparison in 3- and 6-month-old infants. *Journal of Experimental Psychology: Learning, Memory, and Cognition, 24*, 249–260.

Pascalis, O., & de Schonen, S. (1994). Recognition memory in 3- to 4-day-old human neonates. *Neuroreport, 5*, 1721–1724.

Passarotti, A. M., Paul, B. M., Bussiere, J. R., Buxton, R. B., Wong, E. C., & Stiles, J. (2003). The development of face and location processing: An fMRI study. *Developmental Science, 6*(1), 100–117.

Passarotti, A. M., Smith, J., DeLano, M., & Huang, J. (2007). Developmental differences in the neural bases of the face inversion effect show progressive tuning of face-selective regions to the upright orientation. *NeuroImage, 34*, 1708–1722.

Passolunghi, M. C., Cornoldi, C., & De Liberto, S. (1999). Working memory and intrusions of irrelevant information in a group of specific poor problem solvers. *Memory and Cognition, 27*, 779–790.

Pastorelli, C., Zuffiano, A., Lansford, J. E., Thartori, E., Bornstein, M. H., Chang, L., Deater-Deckard, K., Giunta, L., Dodge, K., Gurdal, S., Liu, Q., Long, Q., Oburu, P., Skinner, A., Sorbring, E., Steinberg, L., Tapanya, S., Tirado, L., Yotanyamaneewong, S., & Bacchini, D. (2021). Positive youth development: Parental warmth, values and prosocial development in 11 cultural groups. *Journal of Youth Development, 16*, 379–401. https://doi.org/10.5195/jyd.2021.1026

Patai, E. Z., Gadian, D. G., Cooper, J. M., Dzieciol, A. M., Mishkin, M., & Vagha-Khadem, F. (2015). Extent of hippocampal atrophy predicts degree of deficit in recall. *Proceedings of the National Academy of Sciences of the United States of America, 112*, 12830–12833. https://doi.org/10.1073/pnas.1511904112

Patel, Y., Shin, J., Abé, C., Agartz, I., Alloza, C., Alnæs, D., Ambrogi, S., Antonucci, L. A., Arango, C., Arolt, V., Auzias, G., Ayesa-Arriola, R., Banaj, N., Banaschewski, T., Bandeira, C., Başgöze, Z., Basso Cupertino, R., Bau, C. H.d., Bauer, J., . . . Paus, T. (2022). Virtual ontogeny of cortical growth preceding mental illness. *Biological Psychiatry, 92*(4), 299–313. https://doi.org/10.1016/j.biopsych.2022.02.959

Paulesu, E., Danelli, L., & Berlingeri, M. (2014). Reading the dyslexic brain: Multiple dysfunctional routes revealed by a new meta-analysis of PET and fMRI activation studies. *Frontiers in Human Neuroscience, 8*(830). https://doi.org/10.3389/fnhum.2014.00830

Paulesu, E., Frith, U., Snowling, M., Gallagher, A., Morton, J., Frackowiak, R. S. J., & Frith, C. D. (1996). Is developmental dyslexia a disconnection syndrome? Evidence from PET scanning. *Brain, 119*, 143–157. https://doi.org/10.1093/brain/119.1.143

Pearson, D. A., & Lane, D. M. (1990). Visual attention movements: A developmental study. *Child Development, 61*, 1779–1795.

Peelen, M. V., Glaser, B., Vuilleumier, P., & Eliez, S. (2009). Differentia development of selectivity for faces and bodies in the fusiform gyrus. *Developmental Science, 12*, 16–25.

Peña, M., Maki, A., Kovacic, D., Dehaene-Lambertz, G., Koizumi, H., Bouquet, F., & Mehler, J. (2003). Sounds and silence: An optical topography study of language recognition at birth. *Proceedings of the National Academy of Sciences of the United States of America, 100*, 11702–11705.

Peña, M., Werker, J. F., & Dehaene-Lambertz, G. (2012). Earlier speech exposure does not accelerate speech acquisition. *Journal of Neuroscience, 32*, 11159–11163.

Pennington, B. (2001). Genetic methods. In C. A. Nelson & M. Luciana (Eds.), *The handbook of developmental cognitive neuroscience* (pp. 149–158). MIT Press.

Pennington, B. (2002). Genes and brain: Individual differences and human universals. In M. H. Johnson, Y. Munakata, & R. O. Gilmore (Eds.), *Brain development and cognition: A reader* (2nd ed., pp. 494–508). Blackwell.

Pennington, B., & Welsh, M. (1995). Neuropsychology and developmental psychopathology. In D. Cicchetti & D. J. Cohen (Eds.), *Developmental psychopathology, Vol. 1: Theory and methods* (pp. 254–290). Wiley.

Penzes, P., Cahill, M. E., Jones, K. A., VanLeeuwen, J.-E., & Woolfrey, K. M. (2011). Dendritic spine pathology in neuropsychiatric disorders. *Nature Neuroscience, 14*(3), 285–293.

Pepperberg, I. M. (1987). Acquisition of the same–different concept by an African Gray Parrot (*Psittacus erthacus*) – learning with respect to categories of color, shape and material. *Animal Learning and Behavior, 15*, 423–432.

Perdue, K. L., Jensen, S. K. G., Kumar, S., Richards, J. E., Kakon, S. H., Haque, R., Petri, W. A., Jr., Lloyd-Fox, S., Elwell, C., & Nelson, C. A., III (2019). Using functional near-infrared spectroscopy to assess social information processing in poor urban Bangladeshi infants and todllers. *Developmental Science, 22*, e12839.

Perisco, A. M., & Bourgeron, T. (2006). Searching for ways out of the autism maze: Genetic, epigenetic and environmental clues. *Trends in Neurosciences, 29*, 349–358. https://doi.org/10.1016/j.tins.2006.05.010

Petanjek, Z., Judas, M., Simic, G., Rasin, M. R., Uylings, H. B. M., Rakic, P., & Kostovic, I. (2011). Extraordinary neoteny of synaptic spines in the human prefrontal cortex. *Proceedings of the National Academy of Sciences, 108*(32), 13281–13286.

Peters, L., & Ansari, D. (2019). Are specific learning disorders truly specific, and are they disorders? *Trends in Neuroscience and Education, 17*, 100115.

Peters, L., Bulthé, J., Daniels, N., Op de Beeck, H., & De Smedt, B. (2018). Dyscalculia and dyslexia: Different behavioral, yet similar brain activity profiles during arithmetic. *Neuroimage Clinical, 18*, 663–674.

Peters, L., & De Smedt, B. (2018). Arithmetic in the developing brain: A review of brain imaging studies. *Developmental Cognitive Neuroscience, 30*, 265–279.

Petersen, S. E., & Posner, M. I. (2012). The attention system of the human brain: 20 years after. *Annual Review of Neuroscience, 35*, 73–89.

Pfeifer, J. H., Lieberman, M. D., & Dapretto, M. (2007). "I know you are but what am I?!": Neural bases of self- and social knowledge retrieval in children and adults. *Journal of Cognitive Neuroscience, 19*, 1323–1337.

Phillips, M. L., Drevets, W. C., Rauch, S. L., & Lane, R. (2003). Neurobiology of emotion perception. The neural basis of emotion perception. *Biological Psychiatry, 54*, 504–514.

Piaget, J. (1954). *The construction of reality in the child* (M. Cook, Trans.). Basic Books.

Piaget, J. (2002). The epigenetic system and the development of cognitive functions. In M. H. Johnson, Y. Munakata, & R. Gilmore (Eds.), *Brain development and cognition: A reader* (2nd ed., pp. 29–35). Blackwell.

Piccardi, E. S., Begum Ali, J., Jones, E. J. H., Mason, L., Charman, T., Johnson, M. H., Gliga, T., & BASIS/STAARS Team. (2021). Behavioural and neural markers of tactile sensory processing in infants at elevated likelihood of autism spectrum disorder and/or attention deficit hyperactivity disorder. *Journal of Neurodevelopmental Disorders, 13*(1), 1.

Piven, J., Berthier, M. L., Starkstein, S. E., Nehme, E., Pearlson, G., & Folstein, S. (1990). Magnetic resonance imaging evidence for a deficit of cerebral cortical development in autism. *American Journal of Psychiatry, 147*, 734–739.

Plomin, R., DeFries, J. C., Knopik, V. S., & Neiderhiser, J. M. (2016). Top 10 replicated findings from behavioral genetics. *Perspectives on Psychological Science, 11*(1), 3–23.

Polleux, F., Whitford, K. L., Dijkhuizen, P. A., Vitalis, T., & Ghosh, A. (2002). Control of cortical interneuron migration by neurotrophins and PI3-kinase signalling. *Development, 129*, 3147–3160.

Popejoy, A. B., & Fullerton, S. M. (2016). Genomics is failing on diversity. *Nature, 538*(7624), 161–164.

Posner, M. I. (1988). Structures and functions of selective attention. In T. Boll & B. Bryant (Eds.), *Clinical neuropsychology and brain function: Research, measurement, and practice* (pp. 171–202). American Psychological Association.

Posner, M. I., & Cohen, Y. (1980). Attention and the control of movements. In G. E. Stelmach & J. Roguiro (Eds.), *Tutorials in motor behavior* (pp. 243–258). North Holland.

Posner, M. I., & Cohen, Y. (1984). Components of visual orienting. In H. Bouma & D. G. Bouwhis (Eds.), *Attention and performance* (pp. 531–556). Lawrence Erlbaum.

Posner, M. I., & Petersen, S. E. (1990). The attention system of the human brain. *Annual Review of Neuroscience, 13*, 25–42.

Posner, M. I., Rafal, R. D., Choate, L., & Vaughan, J. (1985). Inhibition of return: Neural basis and function. *Cognitive Neuropsychology, 2*, 211–228.

Posner, M. I., & Rothbart, M. K. (1981). The development of attentional mechanisms. In J. H. Flower (Ed.), *Nebraska symposium on motivation* (pp. 1–51). University of Nebraska Press.

Powell, J., Lewis, P. A., Roberts, N., Garcia-Finana, M., & Dunbar, R. I. M. (2012). Orbital prefrontal cortex volume predicts social network size: An imaging study of individual differences in humans. *Proceedings of the Royal Society B, 279*(1736), 2157–2162. https://doi.org/10.1098/rspb.2011.2574

Powell, L. J., Kosakowski, H. L., & Saxe, R. (2018). Social origins of cortical face areas. *Trends in Cognitive Sciences, 22*(9), 752–763.

Power, A. J., Colling, L. C., Mead, N., Barnes, L., & Goswami, U. (2016). Neural encoding of the speech envelope by children with developmental dyslexia. *Brain and Language, 160*, 1–10.

Prabhakar, J., Johnson, E. G., Nordahl, C. W., & Ghetti, S. (2018). Memory-related hippocampal activation in the sleeping toddler. *Proceedings of the National Academy of Sciences of the United States of America, 115*, 6500–6505.

Puce, A., Allison, T., Bentin, S., Gore, J. C., & McCarthy, G. (1998). Temporal cortex activation in human viewing eye and mouth movements. *Journal of Neuroscience, 18*, 2188–2199.

Purpura, D. P. (1975). Normal and aberrant neuronal development in the cerebral cortex of human fetus and young infant. In N. A. Buchwald & M. A. B. Brazier (Eds.), *Brain mechanisms of mental retardation* (pp. 141–169). Academic Press.

Pylyshyn, Z. W., & Storm, R. W. (1988). Tracking multiple independent targets: Evidence for a parallel tracking mechanism. *Spatial Vision, 3*, 179–197.

Quinn, P. C., Lee, K., & Pascalis, O. (2019). Face processing in infancy and beyond: The case of social categories. *Annual Review of Psychology, 70*, 165–189. https://doi.org/10.1146/annurev-psych-010418-102753

Rabinowicz, T. (1979). The differential maturation of the human cerebral cortex. In F. Falkner & J. M. Tanner (Eds.), *Human growth, Vol. 3: Neurobiology and nutrition* (pp. 141–169). Plenum Press.

Rafal, R., Smith, J., Krantz, J., Cohen, A., & Brennan, C. (1990). Extrageniculate vision in hemianopic humans: Saccade inhibition by signals in the blind field. *Science, 250*, 1507–1518.

Ragsdale, C. W., & Grove, E. A. (2001). Patterning in the mammalian cerebral cortex. *Current Opinion in Neurobiology, 11*, 50–58.

Rakic, P. (1987). Intrinsic and extrinsic determinants of neocortical parcellation: A radial unit model. In P. Rakic & W. Singer (Eds.), *Neurobiology of neocortex. Report of the Dahelm workshop on Neurobiology of Neocortex, Berlin, 17–22 May 1987* (pp. 5–27). John Wiley & Sons.

Rakic, P. (1988). Specification of cerebral cortical areas. *Science, 241*, 170–176.

Rakic, P. (1995). Corticogenesis in human and nonhuman primates. In M. S. Gazzaniga (Ed.), *The cognitive neurosciences* (pp. 127–145). MIT Press.

Rakic, P. (2002). Intrinsic and extrinsic determinants of neocortical parcellation: A radial unit model. In M. H. Johnson, Y. Munakata, & R. Gilmore (Eds.), *Brain development and cognition: A reader* (2nd ed., pp. 57–82). Blackwell.

Rakic, P., Bourgeois, J.-P., Eckenhoff, M. F., Zecevic, N., & Goldman-Rakic, P. S. (1986). Concurrent overproduction of synapses in diverse regions of primate cerebral cortex. *Science, 232*, 153–157.

Rapee, R. M., Oar, E. L., Johnco, C. J., Forbes, M. K., Fardouly, J., Magson, N. R., & Richardson, C. E. (2019). Adolescent development and risk for the onset of social-emotional disorders: A review and conceptual model. *Behaviour Research and Therapy, 123*, 103501.

Rauschecker, J. P., & Singer, W. (1981). The effects of early visual experience on the cat's visual cortex and their possible explanation by Hebb synapses. *Journal of Psychology (London), 310*, 215–239.

Ravikumar, B. V., & Sastry, P. I. (1985). Muscarinic cholinergic receptors in human foetal brain: Characterization and ontogeny of [3H] quinuclidinyl benzilate bind sites in frontal cortex. *Journal of Neurochemistry, 44*, 240–246.

Rayner, K. (1998). Eye movements in reading and information processing: 20 years of research. *Psychological Bulletin, 124*(3), 372–422.

Reilly, J., Bates, E., & Marchman, V. (1998). Narrative discourse in children with early focal brain injury. *Brain and Language, 61*, 335–375.

Reiss, A. L., Eliez, J., Schmitt, E., Straus, E., Lai, Z., Jones, W., & Bellugi, U. (2000). Neuratomy of Williams syndrome: A high resolution MRI study. *Journal of Cognitive Neuroscience, 12*(Supplement), 65–73. https://doi.org/10.1162/089892900561986

Reynolds, G. D., & Richards, J. E. (2005). Familiarization, attention and recognition memory in infancy: An event-related potential and cortical source localization study. *Developmental Psychology, 41*, 598–615.

Reynolds, G. D., & Richards, J. E. (2009). Cortical source localization of infant cognition. *Developmental Neuropsychology, 34*(3), 312–329.

Riccio, C. A., & Reynolds, C. R. (2001). Continuous performance tests are sensitive to ADHD in adults but lack specificity. A review and critique for differential diagnosis. *Annals of the New York Academy of Sciences, 931*, 113–139.

Richards, J. E. (1991). Infant eye movements during peripheral visual stimulus localization as a function of central stimulus attention status. *Psychophysiology, 28*, S4.

Richards, J. E. (2001). Attention in young infants: A developmental psychophysiological perspective. In C. A. Nelson & M. Luciana (Eds.), *The handbook of developmental cognitive neuroscience* (pp. 321–338). MIT Press.

Richards, J. E. (2003). The development of visual attention and the brain. In M. de Haan & M. H. Johnson (Eds.), *The cognitive neuroscience of development* (pp. 73–93). Psychology Press.

Richards, J. E. (2008). Attention in young infants: A developmental psychophysiological perspective. In C. A. Nelson & M. Luciana (Eds.), *The handbook of developmental cognitive neuroscience* (2nd ed., pp. 479–498). MIT Press.

Richardson, H., Lisandrelli, G., Riobueno-Naylor, A., & Saxe, R. (2018). Development of the social brain from age three to twelve years. *Nature Communications, 9*(1), 1–12.

Richardson, H., Taylor, J., Kane-Grade, F., Powell, L., Bosquet Enlow, M., & Nelson, C. (2021). Preferential responses to faces in superior temporal and medial prefrontal cortex in three-year-old children. *Developmental Cognitive Neuroscience, 50*, 100984. https://doi.org/10.1016/j.dcn.2021.100984

Richmond, J., & Nelson, C. A. (2007). Accounting for change in declarative memory: A cognitive neuroscience perspective. *Developmental Review, 27*, 349–373.

Risch, N., Hoffmann, T. J., Anderson, M., Croen, L. A., Grether, J. K., & Windham, G. H. (2014). Familial recurrence of autism spectrum disorder: Evaluating genetic and environmental contributions. *The American Journal of Psychiatry, 171*(11), 1206–1213. https://doi.org/10.1176/appi.ajp.2014.13101359

Rivera, S. M., Reiss, A. L., Eckert, M. A., & Menon, V. (2005). Developmental changes in mental arithmetic: Evidence for increased functional specialization in the left inferior parietal cortex. *Cerebral Cortex, 15*(11), 1779–1790.

Rivera-Gaxiola, M., Silva-Pereyra, J., & Kuhl, P. K. (2005). Brain potentials to native and non-native speech contrasts in 7- and 11-month-old American infants. *Developmental Science, 8,* 162–172.

Rizzolatti, G., & Craighero, J. (2004). The mirror-neuron system. *Annual Review of Neuroscience, 27,* 169–192.

Rodman, H. R., Skelly, J. P., & Bross, C. G. (1991). Stimulus selectivity and state dependence of activity in inferior temporal cortex in infant monkeys. *Proceedings of the National Academy of Sciences of the United States of America, 88,* 7572–7575.

Rogers, S. J., & Pennington, B. F. (1991). A theoretical approach to deficits in infantile autism. *Development and Psychopathology, 3,* 137–162.

Rollins, L., & Riggins, T. (2021). Adapting event-related potential research paradigms for children: Considerations from research on the development of recognition memory. *Developmental Psychobiology, 63*(6), e22159.

Rossion, B. (2014). Understanding face perception by means of prosopagnosia and neuroimaging. *Frontiers in Bioscience, 6*(2), 258–307.

Rovee-Collier, C., & Cuevas, K. (2009). Multiple memory systems are unnecessary to account for infant memory development: An ecological model. *Developmental Psychology, 45*(1), 160–174. https://doi.org/10.1037/a0014538

Rovee-Collier, C., & Giles, A. (2010). Why a neuromaturational model of memory fails: Exuberant learning in early infancy. *Behavioural Processes, 83,* 197–206.

Rumsey, J. M., & Ernst, M. (2000). Functional neuroimaging of autistic disorders. *Mental Retardation and Developmental Disabilities Research Reviews, 6,* 171–179.

Rutter, M. (1998). Developmental catch-up, and deficit, following adoption after severe global early privation. *The Journal of Child Psychology and Psychiatry and Allied Disciplines, 39,* 465–476.

Rutter, M., Andersen-Wood, L., Beckett, C., Bredenkamp, D., Castle, J., Groothues, C., Kreppner, J., Keaveney, L., Lord, C., & O'Connor, T. G. (1999). Quasi-autistic patterns following severe early global privation. *Journal of Child Psychology and Psychiatry, and Allied Disciplines, 40*(4), 537–549.

Sacrey, L. A., Bryson, S. E., & Zwaigenbaum, L. (2013). Prospective examination of visual attention during play in infants at high-risk for autism spectrum disorder: A longitudinal study from 6 to 36 months of age. *Behavioural Brain Research, 256,* 441–450.

Saez de Urabain, I. R., Nuthamann, A., Johnson, M. H., & Smith, T. J. (2017). Disentangling the mechanisms underlying infant fixation durations in scene perception: A computational account. *Vision Research, 134,* 43–59.

Sampaio, R. C., & Truwit, C. L. (2001). Myelination in the developing human brain. In C. A. Nelson & M. Luciana (Eds.), *The handbook of developmental cognitive neuroscience* (2nd ed., pp. 35–45). MIT Press.

Sanai, N., Nguyen, T., Ihrie, R. A., Mirzadeh, Z., Tsai, H. H., Wong, M., Gupta, N., Berger, M. S., Huang, E., Garcia-Verdugo, J. M., Rowitch, D. H., & Alvarez-Buylla, A. (2011). Corridors of migrating neurons in the human brain and their decline during infancy. *Nature, 478*(7369), 382–386. https://doi.org/10.1038/nature10487

Sanes, D. H., Reh, T. A., & Harris, W. A. (2006). *The development of the nervous system* (2nd ed.). Elsevier Academic Press.

Saxe, R. R., Whitfield-Gabrieli, S., Scholz, J., & Pelphrey, K. A. (2009). Brain regions for perceiving and reasoning about other people in school-aged children. *Child Development, 80*(4), 1197–1209. https://doi.org/10.1111/j.1467-8624.2009.01325.x

Scarr, S. (1992). Developmental theories for the 1990s: Development and individual differences. *Child Development, 63*, 1–19.

Scerif, G. (2010). Attention trajectories, mechanisms and outcomes: At the interface between developing cognition and environment. *Developmental Science, 13*(1), 805–812.

Schacter, D., & Moscovitch, M. (1984). Infants' amnesia and dissociable memory systems. In M. Moscovitch (Ed.), *Infant memory* (pp. 173–216). Plenum Press.

Schatz, J., Craft, S., Koby, M., & DeBaun, M. (2000). A lesion analysis of visual orienting performance in children with cerebral vascular injury. *Developmental Neuropsychology, 17*, 49–61.

Schatz, J., Craft, S., White, D., Park, T. S., & Figiel, G. (2001). Inhibition of return in children with perinatal brain injury. *Journal of the International Neuropsychological Society, 7*, 275–284.

Scheinost, D., Spann, M. S., McDonough, L., & Peterson, B. S. (2020). Associations between different dimensions of prenatal stress, neonatal hippocampal connectivity and infant memory. *Neuropsychopharmacology, 45*, 1272–1279.

Schellenberg, E. G. (2020). Music training, individual differences, and plasticity. In M. S. C. Thomas, D. Mareschal, & I. Dumontheil (Eds.), *Educational neuroscience* (pp. 415–441). Routledge.

Scherf, K. S., Behrmann, M., Humphreys, K., & Luna, B. (2007). Visual category-selectivity for faces, places and objects emerges along different developmental trajectories. *Developmental Science, 10*, F15–F30.

Scherf, K. S., Smyth, J. M., & Delgado, M. R. (2013). The amygdala: An agent of change in adolescent neural networks. *Hormones & Behavior, 64*, 293–313.

Schiller, P. H. (1985). A model for the generation of visually guided saccadic eye movements. In D. Rose & V. G. Dobson (Eds.), *Models of the visual cortex* (pp. 62–70). Wiley.

Schlaggar, B. L., Brown, T. T., Lugar, H. M., Visscher, K. M., Miezin, F. M., & Petersen, S. E. (2002). Functional neuroanatomical differences between adults and school-age children in the processing of single words. *Science, 296*, 1476–1479.

Schlaggar, B. L., & McCandliss, B. D. (2007). Development of neural systems for reading. *Annual Review of Neuroscience, 30*, 475–503.

Schlaggar, B. L., & O'Leary, D. D. M. (1993). Patterning of the barrel field in somatosensory cortex with implications for the specification of neocortical areas. *Perspectives on Developmental Neurobiology, 1*(2), 81–91.

Schliebs, R., Kullman, E., & Bigl, V. (1986). Development of glutamate binding sites in the visual structures of the rat brain: Effect of visual pattern deprivation. *Biomedica Biochemica Acta, 45*, 495–506.

Schneider, W., Knopf, M., & Stefanek, J. (2002). The development of verbal memory in childhood and adolescence: Findings from the Munich Longitudinal Study. *Journal of Educational Psychology, 94*(4), 751–761.

Schoechlin, C., & Engel, R. R. (2005). Neuropsychological performance in adult attention-deficit hyperactivity disorder: Meta-analysis of empirical data. *Archives of Clinical Neuropsychology, 20,* 727–744.

Scott, G. D., Karns, C. M., Dow, M. W., Stevens, C., & Neville, H. J. (2014). Enhanced peripheral visual processing in congenitally deaf humans is supported by multiple brain regions, including primary auditory cortex. *Frontiers in Human Neuroscience, 8,* 177.

Senju, A., & Johnson, M. H. (2009). The eye contact effect: Mechanisms and development. *Trends in Cognitive Sciences, 13,* 127–134.

Senju, A., Vernetti, A., Kikuchi, Y., Akechi, H., Hasegawa, T., & Johnson, M. H. (2013). Cultural background modulates how we look at other persons' gaze. *International Journal of Behavioral Development, 37*(2), 131.

Seress, L. (2001). Morphological changes of the human hippocampal formation from mid-gestation to early childhood. In C. A. Nelson & M. Luciana (Eds.), *The handbook of developmental cognitive neuroscience* (2nd ed., pp. 45–58). MIT Press.

Seress, L., & Abraham, H. (2008). Pre- and postnatal morphological development of the human hippocampal formation. In C. A. Nelson & M. Luciana (Eds.), *Handbook of developmental cognitive neuroscience* (2nd ed., pp. 187–212). MIT Press.

Seybold, B. A., Stanco, A., Cho, K. K., Potter, G. B., Kim, C., Sohal, V. S., Rubenstein, J. L., & Schreiner, C. E. (2012). Chronic reduction in inhibition reduces receptive field size in mouse auditory cortex. *Proceedings of the National Academy of Sciences of the United States of America, 109*(34), 13829–13834.

Shackman, J. E., Wismer Fries, A. B., & Pollak, S. D. (2008). Environmental influences on brain-behavioral development: Evidence from child abuse and neglect. In C. A. Nelson & M. Luciana (Eds.), *The handbook of developmental cognitive neuroscience* (2nd ed., pp. 869–882). MIT Press.

Shafer, V. L., Yu, Y. H., & Datta, H. (2011). The development of English vowel perception in monolingual and bilingual infants: Neurophysiological correlates. *Journal of Phonetics, 39*(4), 527–545. https://doi.org/10.1016/j.wocn.2010.11.010

Shankle, W. R., Kimball, R. A., Landing, B. H., & Hara, J. (1998). Developmental patterns in the cytoarchitecture of the human cerebral cortex from birth to 6 years examined by correspondence analysis. *Proceedings of the National Academy of Sciences of the United States of America, 95,* 4023–4028.

Shatz, C. J. (2002). Emergence of order in visual system development. In M. H. Johnson, Y. Munakata, & R. Gilmore (Eds.), *Brain development and cognition: A reader* (2nd ed., pp. 231–244). Blackwell.

Shaw, P., Eckstrand, K., Sharp, W., Blumenthal, J., Lerch, J. P., Greenstein, D., Clasen, L., Evans, A., Giedd, J., & Rapoport, J. L. (2007). Attention-deficit/hyperactivity disorder is characterized by a delay in cortical maturation. *Proceedings of the National Academy of Sciences of the United States of America, 104*(49), 19649–19654.

Shaw, P., Greenstein, D., Lerch, J., Clasen, L., Lenroot, R., Gogtay, N., Evans, A., Rapoport, J., & Giedd, J. (2006). Intellectual ability and cortical development in children and adolescents. *Nature, 440,* 676–679.

Shaw, P., Kabani, N. J., Lerch, J. P., Eckstrand, K., Lenroot, R., Gogtay, N., Greenstein, D., Clasen, L., Evans, A., Rapoport, J. L., Giedd, J. N., & Wise, S. P. (2008). Neurodevelopmental trajectories of the human cerebral cortex. *The Journal of Neuroscience, 28,* 3586–3594.

Shephard, E., Zuccolo, P. F., Idrees, I., Godoy, P. B. G., Salomone, E., Ferrante, C., Sorgato, P., Catão, L. F. C. C., Goodwin, A., Bolton, P. F., Tye, C., Groom, M. J., & Polanczyk, G. V. (2022). Systematic review and meta-analysis: The science of early-life precursors and interventions for attention-deficit/hyperactivity disorder. *Journal of the American Academy of Child and Adolescent Psychiatry, 61*, 187–226. https://doi.org/10.1016/j.jaac.2021.03.016

Shepherd, G. M. (1972). The neuron doctrine: A revision of functional concepts. *Yale Journal of Biology and Medicine, 45*, 584–599.

Sheridan, M. A., Fox, N. A., Zeanah, C. H., McLaughlin, K. A., & Nelson, C. A. (2012). Variation in neural development as a result of exposure to institutionalization early in childhood. *Proceedings of the National Academy of Sciences of the United States of America, 109*(32), 12927–12932.

Shimojo, S., Birch, E., & Held, R. (1983). Development of vernier acuity assessed by preferential looking. *Supplement: Investigative Ophthalmology & Visual Science, 24*, 93.

Shrager, J., & Johnson, M. H. (1995). Waves of growth in the development of cortical function: A computational model. In I. Kovacs & B. Julesz (Eds.), *Maturational windows and adult cortical plasticity* (pp. 31–44). Addison-Wesley.

Shultz, T. R., & Rivest, F. (2001). Knowledge-based cascade-correlation: Using knowledge to speed learning. *Connection Science, 13*, 43–72.

Shultz, T. R., Rivest, F., Egri, L., Thivierge, J.-P., & Dandurand, F. (2007). Could knowledge-based neural learning be useful in developmental robotics? The case of KBCC. *International Journal of Humanoid Robotics, 4*, 245–279.

Siegel, M., Körding, K. P., & König, P. (2000). Integrating top-down and bottom-up sensory processing by somato-dendritic interactions. *Journal of Computational Neuroscience, 8*, 161–173.

Silva, A. J., Paylor, R., Wehner, J. M., & Tonegawa, S. (1992). Impaired spatial learning in alpha-calcium-calmodulin kinase II mutant mice. *Science, 257*, 206–211.

Silva, A. J., Stevens, C. F., Tonegawa, S., & Wang, Y. (1992). Deficient hippocampal long-term potentiation in a-calcium-calmodulin kinase II mutant mice. *Science, 257*, 201–206.

Simion, F., Macchi Cassia, V., Turati, C., & Valenza, E. (2003). Non-specific perceptual biases at the origins of face processing. In O. Pascalis & A. Slater (Eds.), *The development of face processing in infancy and early childhood: Current perspectives* (pp. 13–26). Nova Science Publishers.

Simion, F., Valenza, E., Umilta, C., & Dalla Barba, B. (1995). Inhibition of return in newborns is temporo-nasal asymmetrical. *Infant Behavior and Development, 18*, 189–194.

Simion, F., Valenza, E., Umilta, C., & Dalla Barba, B. (1998). Preferential orienting to faces in newborns: A temporal-nasal asymmetry. *Journal of Experimental Psychology. Human Perception and Performance, 24*(5), 1399–1405.

Singer, W., & Gray, C. M. (1995). Visual feature integration and the temporal correlation hypothesis. *Annual Review of Neuroscience, 18*, 555–586.

Slater, A. M., Mattock, A., & Brown, E. (1990). Size constancy at birth: Newborn infants' responses to retinal and real size. *Journal of Experimental Child Psychology, 49*, 314–322.

Slater, A. M., Morison, V., & Rose, D. (1982). Perception of shape by the new-born baby. *British Journal of Developmental Psychology, 1*, 135–142.

Slattery, E. J., O'Callaghan, E., Ryan, P., Fortune, D. G., & McAvinue, L. P. (2022). Popular interventions to enhance sustained attention in children and adolescents: A critical systematic review. *Neuroscience & Biobehavioral Reviews, 137*, 104633.

Sluzenski, J., Newcombe, M., & Ottinger, W. (2004). Changes in reality monitoring and episodic memory in early childhood. *Developmental Science, 7*, 225–245.

Smith, A. R., Steinberg, L., Strang, N., & Chein, J. (2015). Age differences in the impact of peers on adolescents' and adults' neural response to reward. *Developmental Cognitive Neuroscience, 11*, 75–82. https://doi.org/10.1016/j.dcn.2014.08.010

Sniekers, S., Stringer, S., Watanabe, K., Jansen, P. R., Coleman, J. R. I., Krapohl, E., Taskesen, E., Hammerschlag, A. R., Okbay, A., Zabaneh, D., Amin, N., Breen, G., Cesarini, D., Chabris, C. F., Iacono, W. G., Ikram, M. A., Johannesson, M., Koellinger, P., Lee, J. J., . . . Posthuma, D. (2017). Genome-wide association meta-analysis of 78,308 individuals identifies new loci and genes influencing human intelligence. *Nature Genetics, 49*(7), 1107–1112.

Söderqvist, S., Matsson, H., Peyrard-Janvid, M., Kere, J., & Klingberg, T. (2014). Polymorphisms in the dopamine receptor 2 gene region influence improvements during working memory training in children and adolescents. *Journal of Cognitive Neuroscience, 26*, 54–62.

Söderqvist, S., Nutley, S. B., Ottersen, J., Grill, K. M., & Klingberg, T. (2012). Computerized training of non-verbal reasoning and working memory in children with intellectual disability. *Frontiers in Human Neuroscience, 6*, 271.

Sokolowski, H. M., Fias, W., Bosah Ononye, C., & Ansari, D. (2017). Are numbers grounded in a general magnitude processing system? A functional neuroimaging meta-analysis. *Neuropsychologia, 105*, 50–69.

South, M., Ozonoff, S., & Schultz, R. T. (2008). Neurocognitive development in autism. In C. A. Nelson & M. Luciana (Eds.), *The handbook of developmental cognitive neuroscience* (pp. 701–716). MIT Press.

Southgate, V., Johnson, M. H., Osborne, T., & Csibra, G. (2009). Predictive motor activation during action observation in human infants. *Biology Letters, 5*(6), 769–772.

Spelke, E. S., Breinlinger, K., Macomber, J., & Jacobsen, K. (1992). Origins of knowledge. *Psychological Review, 99*(4), 605–632.

Spencer, J. P., Thomas, M. S. C., & McClelland, J. L. (2009). *Toward a unified theory of development: Connectionism and dynamic systems.* Oxford University Press.

Spiridon, M., & Kanwisher, N. (2002). How distributed is visual category information in human occipital-temporal cortex? An fMRI study. *Neuron, 35*(6), 1157–1165.

Spratling, M., & Johnson, M. H. (2004). A feedback model of visual attention. *Journal of Cognitive Neuroscience, 16*, 219–237.

Spratling, M., & Johnson, M. H. (2006). A feedback model of perceptual learning and categorization. *Visual Cognition, 13*, 129–165.

Spreen, O., Risser, A. T., & Edgell, D. (1995). *Developmental neuropsychology.* Oxford University Press.

Squire, L. R., Stark, C. E., & Clark, R. F. (2004). The medial temporal lobe. *Annual Review of Neuroscience, 27*, 279–306.

St. Clair-Thompson, H. L., & Gathercole, S. E. (2006). Executive functions and achievements in school: Shifting updating inhibition and working memory. *Quarterly Journal of Experimental Psychology, 59*, 745–759.

Stanwood, G. D., & Levitt, P. (2008). The effects of monoamines on the developing nervous system. In C. A. Nelson & M. Luciana (Eds.), *Handbook of developmental cognitive neuroscience* (2nd ed., pp. 83–94). MIT Press.

Stechler, G., & Latz, E. (1966). Some observations on attention and arousal in the human infant. *Journal of the American Academy of Child and Adolescent Psychiatry, 5*, 517–525.

Steinberg, L. (2008). A social neuroscience perspective on adolescent risk-taking. *Developmental Review, 28*(1), 78–106.

Steinberg, L., Icenogle, G., Shulman, E. P., Breiner, K., Chein, J., Bacchini, D., Chang, L., Chaudhary, N., Di Giunta, L., Dodge, K. A., Fanti, K. A., Lansford, J. E., Malone, P. S., Oburu, P., Pastorelli, C., Skinner, A. T., Sorbring, E., Tapanya, S., Tirado, L. M. U., . . . Takash, H. M. S. (2018). Around the world, adolescence is a time of heightened sensation seeking and immature self-regulation. *Developmental Science, 21*(2), e12532. https://doi.org/10.1111/desc.12532

Stern, J. A. (1977). *The first relationship: Infant and mother.* Harvard University Press.

Stiles, J. (2008). *The fundamentals of brain development: Integrating nature and nurture.* Harvard University Press.

Stiles, J., Bates, E., Thal, D., Trauner, D., & Reilly, J. (2002). Linguistic and spatial cognitive development in children with pre- and perinatal focal brain injury: A ten-year overview from the San Diego Longitudinal Project. In M. H. Johnson, Y. Munakata, & R. Gilmore (Eds.), *Brain development and cognition: A reader* (2nd ed., pp. 272–291). Blackwell.

Stiles, J., Brown, T. T., Haist, F., & Jernigan, T. L. (2015). Brain and cognitive development. In R. M. Lerner (Ed.), *Handbook of child psychology and developmental science* (7th ed., pp. 9–62). Wiley.

Stiles, J., & Thal, D. (1993). Linguistic and spatial cognitive development following early focal brain injury: Patterns of deficit and recovery. In M. H. Johnson (Ed.), *Brain development and cognition: A reader* (pp. 643–664). Blackwell.

Streit, P. (1984). Glutamate and aspartate as transmitter candidates for systems of the cerebral cortex. In E. G. Jones & A. Peters (Eds.), *Cerebral cortex: Functional properties of cortical cells* (Vol. 2, pp. 119–143). Plenum Press.

Stryker, M. P., & Harris, W. (1986). Binocular impulse blockade prevents the formation of ocular dominance columns in cat visual cortex. *Journal of Neuroscience, 6*, 2117–2133.

Stuss, D. T. (1992). Biological and psychological development of executive functions. *Brain and Cognition, 20*, 8–23.

Supekar, K., Musen, M., & Menon, V. (2009). Development of large-scale functional brain networks in children. *PLoS Biology, 7*, e1000157.

Supekar, K., Swigart, A. G., Tenison, C., Jolles, D. D., Rosenberg-Lee, M., Fuchs, L., & Menon, V. (2013). Neural predictors of individual differences in response to math tutoring in primary-grade school children. *Proceedings of the National Academy of Sciences of the United States of America, 110*(20), 8230–8235.

Sur, M., & Leamey, C. A. (2001). Development and plasticity of cortical areas and networks. *Nature Reviews Neuroscience, 2*(4), 251–262.

Sur, M., & Rubenstein, J. L. R. (2005). Patterning and plasticity of cerebral cortex. *Science, 310*, 805–810.

Swaab-Barneveld, H., de Sonneville, L., Cohen-Kettenis, P., Gielen, A., Buitelaar, J., & van Engeland, H. (2000). Visual sustained attention in a child psychiatric population. *Journal of the American Academy of Child and Adolescent Psychiatry, 39*, 651–659.

Swaminathan, S., & Schellenberg, E. G. (2020). Musical ability, music training, and language ability in childhood. *Journal of Experimental Psychology: Learning, Memory, and Cognition, 46*(12), 2340–2348.

Swanson, H. L., Howard, C., & Saez, L. (2006). Do different components of working memory underlie different subgroups of reading disabilities. *Journal of Learning Disabilities, 39,* 252–268.

Sydnor, V. J., Larsen, B., Bassett, D. R., Alexaner-Boch, A., Fair, D. A., Liston, C., Mackey, A. P., Milham, M. P., Pines, A., Roalf, D. R., Seidlitz, J., Xu, T., Raznhan, A., & Satterthwaite, T. D. (2021). Neurodevelopment of the association cortices: Patterns, mechanisms, and implications for psychopathology. *Neuron, 109*(18), 2820–2846.

Symeonidou, I., Dumontheil, I., Chow, W.-Y., & Breheny, R. (2016). Development of online use of theory of mind during adolescence: An eye-tracking study. *Journal of Experimental Child Psychology, 149,* 81–97.

Symons, L. A., Hains, S. M. J., & Muir, D. W. (1998). Look at me: Five-months-old infants' sensitivity to very small deviations in eye-gaze during social interactions. *Infant Behavior and Development, 21,* 531–536.

Tadic, V., Pring, L., & Dale, N. (2009). Attentional processes in young children with congenital visual impairment. *British Journal of Developmental Psychology, 27*(2), 311–330.

Tadic, V., Pring, L., & Dale, N. (2010). Are language and social communication intact in children with congenital visual impairment at school age? *Journal of Child Psychology and Psychiatry, 51*(6), 696–705.

Taga, G., Asakawa, K., Maki, A., Konishi, Y., & Koizumi, H. (2003). Brain imaging in awake infants by near-infrared optical topography. *Proceedings of the National Academy of Sciences of the United States of America, 100,* 10722–10727.

Tager-Flusberg, H. (2003). Developmental disorders of genetic origin. In M. de Haan & M. H. Johnson (Eds.), *The cognitive neuroscience of development* (pp. 237–261). Psychology Press.

Tajik-Parvinchi, D. J., & Sandor, P. (2013). Enhanced antisaccade abilities in children with Tourette syndrome: The Gap-effect Reversal. *Frontiers in Human Neuroscience, 7,* 768.

Tallal, P., Miller, S. L., Bedi, G., Byma, G., Wang, X., Nagarajan, S. J., Schreiner, C., Jenkins, W. M., & Merzenich, M. M. (1996). Language comprehension in language-learning impaired children improved with acoustically modified speech. *Science, 271,* 81–84. https://doi.org/10.1126/science.271.5245.81

Tallal, P., & Stark, R. E. (1980). Speech perception of language-delayed children. In G. H. Yeni-Komshian, J. F. Kavanagh, & C. A. Ferguson (Eds.), *Child phonology: Perception* (Vol. 2, pp. 155–171). Academic Press.

Tallal, P., Stark, R. E., Clayton, K., & Mellits, D. (1980). Developmental dysphasia: Relation between acoustic processing deficits and verbal processing. *Neuropsychologia, 18*(3), 273–284.

Tallon-Baudry, C., Bertrand, O., Peronnet, F., & Pernier, J. (1998). Induced-band activity during the delay of a visual short-term memory task in humans. *Journal of Neuroscience, 18,* 4244–4254.

Tamm, L., Menon, V., & Reiss, A. L. (2002). Maturation of brain function associated with response inhibition. *Journal of the American Academy of Child & Adolescent Psychiatry, 41*(10), 1231–1238.

Tamm, L., Narad, M. E., Antonini, T. N., O'Brien, K. M., Hawk, L. W., Jr., & Epstein, J. N. (2012). Reaction time variability in ADHD: A review. *Neurotherapeutics, 9*(3), 500–508.

Tamnes, C. K., Roalf, D. R., Goddings, A.-L., & Lebeld, C. (2018). Diffusion MRI of white matter microstructure development in childhood and adolescence: Methods, challenges and progress. *Developmental Cognitive Neuroscience, 33*, 161–175. https://doi.org/10.1016/j.dcn.2017.12.002

Tamnes, C. K., Walhovd, K. B., Grydeland, H., Holland, D., Østby, Y., Dale, A. M., & Fjell, A. M. (2013). Longitudinal working memory development is related to structural maturation of frontal and parietal cortices. *Journal of Cognitive Neuroscience, 25*(10), 1611–1623.

Tang, Y.-Y., & Liu, Y. (2009). Numbers in the cultural brain. *Progress in Brain Research, 178*, 151–157.

Thapar, A., & Riglin, L. (2020). The importance of a developmental perspective in psychiatry: What do recent genetic-epidemiological findings show? *Molecular Psychiatry, 25*(8), 1631–1639.

Thatcher, R. W. (1992). Cyclic cortical reorganization during early childhood. Special Issue: The role of frontal lobe maturation in cognitive and social development. *Brain and Cognition, 20*, 24–50.

Thelen, E., & Smith, L. B. (1994). *A dynamic systems approach to the development of cognition and action.* MIT Press.

Thivierge, J.-P., Totine, D., & Shultz, T. R. (2005). Simulating frontotemporal pathways involved in lexical ambiguity resolution. In *Proceedings of the twenty-seventh annual conference of the cognitive science society* (pp. 2178–2183). Erlbaum.

Thomas, K. M., Drevets, W. C., Dahl, R. E., Ryan, N. D., Birmaher, B., Eccard, C. H., Axelson, D., Whalen, P. J., & Casey, B. J. (2001). Amygdala response to fearful faces in anxious and depressed children. *Archives of General Psychiatry, 58*, 1057–1063. https://doi.org/10.1001/archpsyc.58.11.1057

Thomas, K. M., Hunt, R. H., Vizueta, N., Sommer, T., Durston, S., Yang, Y., & Worden, M. S. (2004). Evidence of developmental differences in implicit sequence learning: An fMRI study of children and adults. *Journal of Cognitive Neuroscience, 16*, 1339–1351. https://doi.org/10.1162/0898929042304688

Thomas, K. M., & Nelson, C. A. (2001). Serial reaction time learning in preschool- and school-age children. *Journal of Experimental Child Psychology, 79*, 364–387.

Thomas, K. M., & Tseng, A. (2008). Functional MRI methods in developmental cognitive neuroscience. In C. A. Nelson & M. Luciana (Eds.), *Handbook of developmental cognitive neuroscience* (3rd ed., pp. 311–324). MIT Press.

Thomas, M., & Johnson, M. H. (2008). New advances in understanding sensitive periods in brain development. *Current Directions in Psychological Science, 17*, 1–5.

Thomas, M. S., Knowland, V. C., & Karmiloff-Smith, A. (2011). Mechanisms of developmental regression in autism and the broader phenotype: A neural network modelling approach. *Psychological Review, 118*, 637–654.

Thomas, M. S. C., & Johnson, M. H. (2006). The computational modelling of sensitive periods. *Developmental Psychobiology, 48*, 337–344.

Thomas, M. S. C., & Johnson, M. H. (2008). New advances in understanding sensitive periods in development. *Current Directions in Psychological Science, 17*(1), 1–5.

Thompson, D. W. (1917). *On growth and form.* Cambridge University Press.

Tillema, J. M., Byars, A. W., Jacla, L. M., Schapiro, M. B., Schmithorst, V. J., Szaflarski, J. P., & Holland, S. K. (2008). Cortical reorganization of language functioning following perinatal left MCA stroke. *Brain and Language, 105*, 99–111.

Tinbergen, N. (1951). *The study of instinct.* Oxford University Press.

Tipper, S. P., Bourque, T. A., Anderson, S. H., & Brehaut, J. C. (1989). Mechanisms of attention: A developmental study. *Journal of Experimental Child Psychology, 48*, 353–378.

Toga, A. W., Thompson, P. M., & Sowell, E. R. (2006). Mapping brain maturation. *Trends in Neurosciences, 29*, 148–158.

Tole, S., Goudreau, G., Assimacopoulos, S., & Grove, E. A. (2000). Emx2 is required for growth of the hippocampus but not for hippocampal field specification. *The Journal of Neuroscience, 20*(7), 2618–2625.

Tomalski, P., & Johnson, M. H. (2010). The effects of early adversity on the adult and developing brain. *Current Opinion in Psychiatry, 23*, 233–238.

Tomasello, M. (2018). How children come to understand false beliefs: A shared intentionality account. *Proceedings of the National Academy of Sciences of the United States of America, 115*(34), 8491–8498.

Tong, X., & McBride, C. (2020). Neuroscience in reading and reading difficulties. In M. S. C. Thomas, D. Mareschal, & I. Dumontheil (Eds.), *Educational neuroscience* (pp. 123–143). Routledge.

Towgood, K. J., Meuwese, J. D. I., Gilbert, S. J., Turner, M. S., & Burgess, P. W. (2009). Advantages of the multiple case series approach to the study of cognitive deficits in autism spectrum disorder. *Neuropsychologia, 47*(13), 2981–2988. https://doi.org/10.1016/j.neuropsychologia.2009.06.028

Tranel, D., & Damasio, A. R. (1985). Knowledge without awareness: An autonomic index of facial recognition by prosopagnosics. *Science, 228*, 1453–1454.

Trick, L. M., & Pylyshyn, Z. W. (1994). Why are small and large numbers enumerated differently? A limited-capacity preattentive stage in vision. *Psychological Review, 101*, 80–102.

Turesky, T. K., Vanderauwera, J., & Gaab, N. (2021). Imaging the rapidly developing brain: Current challenges for MRI studies in the first five years of life. *Developmental Cognitive Neuroscience, 47*, 100893. https://doi.org/10.1016/j.dcn.2020.100893

Turkewitz, G., & Kenny, P. A. (1982). Limitations on input as a basis for neural organizaton and perceptual development: A preliminary theoretical statement. *Developmental Psychobiology, 15*, 357–368.

Twomey, T., Price, C. J., Waters, D., & MacSweeney, M. (2020). The impact of early language exposure on the neural systems supporting language in hearing and deaf adults. *NeuroImage, 209*, 116411.

Tzourio-Mazoyer, N., de Schonen, S., Crivello, F., Reutter, B., Aujard, Y., & Mazoyer, B. (2002). Neural correlates of woman face processing by 2-month-old infants. *NeuroImage, 15*, 454–461.

Udwin, O., & Yule, W. (1991). A cognitive and behavioural phenotype in Williams syndrome. *Journal of Clinical and Experimental Neuropsychology, 13*, 232–244.

Ullman, H., Almeida, R., & Klingberg, T. (2014). Structural brain maturation and brain activity predict future working memory capacity during childhood development. *Journal of Neuroscience, 34*, 1592–1598.

Urakawa, S., Takamoto, K., Ishikawa, A., Ono, T., & Nishijo, H. (2015). Selective medial prefrontal cortex responses during live mutual gaze interactions in human infants: An fNIRS study. *Brain Topography, 28*(5), 691–701. https://doi.org/10.1007/s10548-014-0414-2

van der Mark, S., Bucher, K., Maurer, U., Schiulz, E., Brem, S., Buckelmuller, J., Kronbichler, M., Loenneker, T., Klaver, P., Martin, E., & Brandeis, D. (2009). Children with dyslexia lack multiple specializations along the visual word-form (VWF) system. *NeuroImage, 47*, 1940–1949.

van Elk, M., van Schie, H. T., Hunnius, S., Vesper, C., & Bekkering, H. (2008). You'll never crawl alone: Neurophysiological evidence for experience-dependent motor resonance in infancy. *NeuroImage, 43*, 808–814.

van Essen, D. C., Anderson, C. H., & Felleman, D. J. (1992). Information processing in the primate visual system: An integrated systems perspective. *Science, 255*, 419–423.

Vargha-Khadem, F., & Cacucci, F. (2021). A brief history of developmental amnesia. *Neuropsychologia, 150*, 107689.

Vargha-Khadem, F., Gadian, D. G., Watkins, K. E., Connelly, A., van Paesschen, W., & Mishkin, M. (1997). Differential effects of early hippocampal pathology on episodic and semantic memory. *Science, 277*, 376–380.

Vargha-Khadem, F., Issacs, E., & Muter, V. (1994). A review of cognitive outcome after unilateral lesions sustained during childhood. *Child Neurology, 9*(Supplement), 2S67–2S73.

Vargha-Khadem, F., Watkins, K., Alcock, K. J., Fletcher, P., & Passingham, R. E. (1995). Praxic and nonverbal cognitive deficits in a large family with a genetically transmitted speech and language disorder. *Proceedings of the Natonal Academy of Sciences of the United States of America, 92*, 930–933.

Vaughan, H. G., & Kurtzberg, D. (1989). Electrophysiological indices of normal and aberrant cortical maturation. In P. Kellaway & J. Noebels (Eds.), *Problems and concepts of developmental neurophysiology* (pp. 263–287). Johns Hopkins University Press.

Vecera, S. P., & Johnson, M. H. (1995). Eye gaze detection and the cortical processing of faces: Evidence from infants and adults. *Visual Cognition, 2*, 101–129.

Visscher, P. M., Wray, N. R., Zhang, Q., Sklar, P., McCarthy, M. I., Brown, M. A., & Yang, J. (2017). 10 years of GWAS discovery: Biology, function, and translation. *American Journal of Human Genetics, 101*(1), 5–22. https://doi.org/10.1016/j.ajhg.2017.06.005

Waddington, C. H. (1975). *The evolution of an evolutionist.* Cornell University Press.

Wainwright, A., & Bryson, S. E. (2002). The development of exogenous orienting: Mechanisms of control. *Journal of Experimental Child Psychology, 82*, 141–155.

Wallace, R. B., Kaplan, R., & Werboff, J. (1977). Hippocampus and behavioral maturation. *International Journal of Neuroscience, 7*, 185.

Walsh, V. (2003). A theory of magnitude: Common cortical metrics of time, space and quantity. *Trends in Cognitive Sciences, 7*(11), 483–488.

Wang, A. T., Lee, S. S., Sigman, M., & Dapretto, M. (2006). Neural basis of irony comprehension in children with autism: The role of prosody and context. *Brain, 129*, 932–943.

Wass, S. V. (2018). How orchids concentrate? The relationship between physiological stress reactivity and cognitive performance during infancy and early childhood. *Neuroscience & Biobehavioral Reviews*, 90, 34–49. https://doi.org/10.1016/j.neubiorev.2018.03.029

Wass, S. V., Porayska-Pomsta, K., & Johnson, M. H. (2011). Training attentional control in infancy. *Current Biology, 21*(18), 1543–1547.

Watson, J. D., & Crick, F. H. (1953). The structure of DNA. *Cold Spring Harbor Symposia on Quantitative Biology, 18*, 123–131.

Wattam-Bell, J. (1990). The development of maximum velocity limits for direction discrimination in infancy. *Perception, 19*, 369.

Wattam-Bell, J. (1991). Development of motion-specific cortical responses in infancy. *Vision Research, 31*, 287–297.

Way, B. M., & Lieberman, M. D. (2010). Is there a genetic contribution to cultural differences? Collectivism, individualism and genetic markers of social sensitivity. *Social Cognitive and Affective Neuroscience, 5*(2–3), 203–211. https://doi.org/10.1093/scan/nsq059

Weaver, I. C. G., Cervoni, N., Champagne, F. A., Alessio, A. C. D., Sharma, S., Seckl, J. R., Dymov, S., Szyf, M., & Meaney, M. J. (2004). Epigenetic programming by maternal behavior. *Nature Neuroscience, 7*, 847–854. https://doi.org/10.1038/nn1276

Webb, S. J., & Nelson, C. A. (2001). Perceptual priming for upright and inverted faces in infants and adults. *Journal of Experimental Child Psychology, 79*(1), 1–22.

Webster, M. J., Bachevalier, J., & Ungeleider, L. G. (1995). Transient subcortical connections of inferior temporal areas TE and TEO in infant macaque monkeys. *Journal of Comparative Neurology, 352*, 213–226.

Wellman, H. M., Cross, D., & Watson, J. (2001). Meta-analysis of theory-of-mind development: The truth about false belief. *Child Development, 72*(3), 655–684.

Welsh, M., DeRoche, K., & Gilliam, D. (2008). Neurocognitive models of early treated phenylketonuria: Insights from meta-analysis and new molecular genetic findings. In C. A. Nelson & M. Luciana (Eds.), *The handbook of developmental cognitive neuroscience* (3rd ed., pp. 677–690). MIT Press.

Wen, H., Xu, T., Wang, X., Yu, X., & Bi, Y. (2022). Brain intrinsic connection patterns underlying tool processing in human adults are present in neonates and not in macaques. *NeuroImage, 258*, 119339.

Werker, J. F., & Polka, L. (1993). Developmental changes in speech perception: New challenges and new directions. *Journal of Phonetics, 21*, 83–101.

Werker, J. F., & Tees, R. C. (1999). Influences of infant speech processing: Toward a new synthesis. *Annual Review of Psychology, 50*, 509–535.

Werker, J. F., & Vouloumanos, A. (2001). Speech and language processing in infancy: A neurocognitive approach. In C. A. Nelson & M. Luciana (Eds.), *The handbook of developmental cognitive neuroscience* (2nd ed., pp. 269–280). MIT Press.

Whalen, J., Gallistel, C. R., & Gelman, R. (1999). Nonverbal counting in humans: The psychophysics of number representation. *Psychological Science, 10*, 130–137.

White, T., & Hilgetag, C. C. (2008). Gyrification and development of the human brain. In C. A. Nelson & M. Luciana (Eds.), *The handbook of developmental cognitive neuroscience* (pp. 39–50). MIT Press.

Wilkinson, N., Paikan, A., Gredebäck, G., Rea, F., & Metta, G. (2014). Staring us in the face? An embodied theory of infant face preferences. *Developmental Science, 17*, 809–825.

Wilson, A., Ahmed, H., Mead, N., Noble, H., Richardson, U., Wolpert, M. A., & Goswami, U. (2021). Neurocognitive predictors of response to intervention with GraphoGame Rime. *Frontiers in Education, 6*, 125. https://doi.org/10.3389/feduc.2021.639294

Wimmer, H., & Perner, J. (1983). Beliefs about beliefs: Representation and constraining function of wrong beliefs in young children's understanding of deception. *Cognition, 13*, 103–128.

Wood, F., Flowers, L., Buchsbaum, M., & Tallal, P. (1991). Investigation of abnormal left temporal functioning in dyslexia through rCBF, auditory evoked potentials, and positron emission tomography. *Reading and Writing: An Interdisciplinary Journal, 3*, 379–393.

Wynn, K. (1992). Addition and subtraction by human infants. *Nature, 358*, 749–750.

Wynn, K. (1998). Psychological foundations of number: Numerical competence in human infants. *Trends in Cognitive Sciences, 2*, 296–303.

Xie, W., Mallin, B. M., & Richards, J. E. (2019). Development of brain functional connectivity and its relation to infant sustained attention in the first year of life. *Developmental Science, 22*(1), e12793.

Xu, F., & Spelke, S. (2000). Large number discrimination in 6-month-old infants. *Cognition, 74*, B1–B11.

Yakovlev, P. I., & Lecours, A. (1967). The myelogenetic cycles of regional maturation of the brain. In A. Minokowski (Ed.), *Regional development of the brain in early life* (pp. 3–70). Davis.

Yamada, H., Sadato, N., Konishi, M., Muramoto, S., Kimura, K., Tanaka, M., Yonekura, Y., Ishii, Y., & Itoh, H. (2000). A milestone for normal development of the infantile brain detected by functional MRI. *Neurology, 55*, 218–223. https://doi.org/10.1212/wnl.55.2.218

Yamada, H., Sadato, N., Konishi, Y., Kimura, K., Tanaka, M., Yonekura, Y., & Ishii, Y. (1997). A rapid brain metabolic change in infants detected by fMRI. *Neuroreport, 8*, 3775–3778. https://doi.org/10.1097/00001756-199712010-00024

Yates, T. S., Ellis, C. T., & Turk-Browne, N. B. (2021). Emergence and organization of adult brain function throughout child development. *Neuroimage, 226*, 117606. https://doi.org/10.1016/j.neuroimage.2020.117606

Yerys, B. E., Jankowski, K. F., Shook, D., Rosenberger, L. R., Barnes, K. A., Berl, M. M., Ritzl, E. K., VanMeter, J., Vaidya, C. J., & Gaillard, W. D. (2009). The fMRI success rate of children and adolescents: Typical development, epilepsy, attention deficit/hyperactivity disorder, and autism spectrum disorders. *Human Brain Mapping, 30*(10), 3426–3435.

Yovel, G., & Duchaine, B. (2006). Specialized face perception mechanisms extract both part and space information: Evidence from developmental prosopagnosia. *Journal of Cognitive Neuroscience, 18*, 580–593.

Zeamer, A., Heuer, E., & Bachevalier, J. (2010). Developmental trajectory of object recognition memory in infant rhesus macaques with and without neonatal hippocampal lesions. *The Journal of Neuroscience, 30*(27), 9157–9165.

Zeamer, A., Richardson, R. L., Weiss, A. R., & Bachevalier, J. (2015). The development of object recognition memory in rhesus macaques with neonatal lesions of the perirhinal cortex. *Developmental Cognitive Neuroscience, 11*, 31–41. https://doi.org/10.1016/j.dcn.2014.07.002

Zhang, T. Y., & Meaney, M. J. (2010). Epigenetics and the environmental regulation of the genome and its function. *Annual Review of Psychology, 61*, 439–466.

Zhao, T. C., Boorom, C., Kuhl, P. K., & Gordon, R. (2021). Infants' neural speech discrimination predicts individual differences in grammar ability at 6 years. *Developmental Cognitive Neuroscience, 48*, 100949.

Zhou, D., Lebel, C., Treit, S., Evans, A., & Beaulieu, C. (2015). Accelerated longitudinal cortical thinning in adolescence. *NeuroImage, 104*, 138–145.

Zipser, D., & Andersen, R. A. (1988). A back-propagation programmed network that simulates response properties of a subset of posterior parietal neurons. *Nature, 331*, 679–684.

Zwaigenbaum, L., Bryson, S., Roberts, W., Rogers, T., Brian, J., & Szatmari, P. (2005). Behavioral markers of autism in the first year of life. *International Journal of Developmental Neuroscience, 23*, 143–152.

Index

Page numbers in italics indicate figures. Page numbers in bold indicate tables. Figures listed instead of a page number appear in the color plate section.